International Conference on Trends in Mechanical Engineering Sciences (ICTMES 2020)

IOP Conference Series: Materials Science and Engineering
Volume 1189

Hassan, India
6 – 7 August 2020

Editors:

Ezhil Vannan S.
Dushyanth Kumar K. R.
Shashank Lingappa M.

ISBN: 978-1-7138-4649-9
ISSN: 1757-8981

Printed from e-media with permission by:

Curran Associates, Inc.
57 Morehouse Lane
Red Hook, NY 12571

Some format issues inherent in the e-media version may also appear in this print version.

Institute of Physics
Dirac House, Temple Back
Bristol BS1 6BE UK

Phone: 44 1 17 929 7481
Fax: 44 1 17 920 0979

techtracking@iop.org

Additional copies of this publication are available from:

Curran Associates, Inc.
57 Morehouse Lane
Red Hook, NY 12571 USA
Phone: 845-758-0400
Fax: 845-758-2633
Email: curran@proceedings.com
Web: www.proceedings.com

TABLE OF CONTENTS

Preface

Peer Review Declaration

Effect of Extrusion on Tensile Behavior of S Glass and Fly Ash Reinforced Al 7005 Hybrid Composites. 1
M Shantharaja, S Praveen Kumar

Impact of Ball Burnishing Process on Residual Stress Distribution in Aluminium 2024 Alloy Using Experimental and Numerical Simulation 6
Harish, D. Shivalingappa, N Raghavendra, V Ganesh

Prediction of Mechanical Properties of Epoxy Composites Containing Mono and Hybrids Particulate Fillers. 11
K Srinivas, M S Bhagyashekar

Studies on Tensile Strength of Notched Woven Fabric Composites for Different Stacking Sequence 17
B Latha Shankar, G S Shivashankar, D Thulasi Durai, K R Dushyanth Kumar

Influence of Heat Treatment on Stress Intensity Factor of Nano Boron Carbide Hybrid Al2219 Composites with MoS2 24
Sagar B Sansuddi, G S Shivashankar, B Latha Shankar, K R Dushyanth Kumar

Experimental Study of Cooling Characteristics to Analyse Quenching Phenomenon of Non-Heat Treatable Material with Different Quenchants 32
M C Yogesh Gowda, H M Sathisha, L Bharath Raj, T S Prasanna Kumar

Wear Behavior of Aluminum-TiB2 Metal Matrix Composites Using Taguchi Method. 38
H.S. Dr. Manjunatha, V T Satish, G. Dr. Mallesh

Investigation on the Influence of Nano Filler on Tribological Behaviour of Glass-Nylon Fiber Reinforced Epoxy Based Hybrid Composites 45
K Dilip Kumar, M Shantharaja, G Santhosh

Characterisation of Aluminium Metal Matrix Composites Reinforced with SiC (Carborundum) & Al2O3 (Corundum) 50
Gangaraj, K. Manjunath

An Overview of Polymer Nanocomposites: Understanding of Mechanical and Tribological Behavior 59
Qurat Ul Ain, Rakesh Sehgal, M F Wani, M K Singh

Preparation and Characterisation of Temperature Regulating Silk Cocoon Extract (Sericin)/PVA Based Smart Fabric 65
Vinutha P, Divakar H N, Subbaiah K M, Savitha M

Microstructure and Wear Behavior of Microwave Treated WC-10Co-4Cr Composite Coating on AISI 4140 Alloy Steel 71
D G Pradeep, H S Nithin, B N Sharath, K S Madhu, C V Venkatesh

Computational Study of Effect of Ceramic Coatings and Piston Bowl Depth on Heat Fluxes of Re-Entrant Type Piston 79
B Suresha, Anantha Padmanabha, R Amith, S Ishwara Prasanna, I M Janamdar

Study on Performance and Emission Characteristics of Four Stroke Gasoline Engine Under Formulated Neem Oil as Base Lubricant 95
A R Girish, Y M Shashidhara

A Review on Ceramic Coatings for Low Carbon Steel Methods Materials and Applications 101
L Umashanker, T P Bharathesh, C Roopashree, R Saravanan

Influence of Ageing Time on Microstructure, Hardness and Corrosion of UNS Zeron 100 Super Duplex Stainless Steel 108
D Disha, M B Davanageri, Ritesh Kumar

Experimental and Numerical Modal Analysis of Short Carbon Fiber Reinforced Polyethersulfone Composites 113
B Harshavardhan, R Ravishankar, B Suresha, S Srinivas

Tribological Suitability of Aluminium Hybrid Composite Above Atmospheric Temperature 122
B.N. Sharath, K.S. Madhu, D.G. Pradeep, C.V. Venkatesh

Evaluation of Mechanical Properties of Ceramic Reinforced Aluminium-7029 Hybrid Composite 128
K S Madhu, B N Sharath, C V Venkatesh, D G Pradeep

Elastic Modulus Measurement of Thermoplastic Composites Through Modal Analysis 134
B Harshavardhan, R Ravishankar, B Suresha, U Arun C Dixit

A Review on the Effect of Surface Roughness and Liquid Slip on Fluid Flow in PDMS Microchannels 141
Rakesh T G, Girisha C, Satish V T

Experimental Investigation on the Machinability Characteristics in Drilling of Syntactic Foams 147
H S Ashrith, T P Jeevan, Mrityunjay Doddamani

A Taguchi Approach in Investigating the Effects of Process Parameters in Drilling of Glass Fibre Reinforced Plastic Composite 153
S. N. Ravi Shankar, Pal Surjya K., ARUN K. Samantaray

Synthesis of Eggshell Based Hydroxyapatite Using Hydrothermal Method 164
Shahid Hussain, Kazi Sabiruddin

Experimental Investigation of Effect of Hybridization of UHMWPE Fibers on the Mechanical and Physical Properties of Basalt Fiber Reinforced Polymer Matrix Composites 169
R Raghavendra Rao, S Pradeep, S Karthik, K R Gopi, B S Shivashankara

Study on Abrasion Wear Behaviour of Hybrid Composites Using Dimensional Analysis and Wear Models 177
B M Rudresh, B N Ravikumar, D Madhu

Investigation of Mechanical Properties of ECAP Processed AL7068 Aluminium Alloy 187
B S Shivashankara, K R Gopi, S Pradeep, R Raghavendra Rao

Effect of Shape of Cut-Out on Natural Frequency of Square Plate 193
H K Jayavardhan, Prasanta Kumar Samal

Experimental Determination of Mode Shapes of a Plate Using Speaker as Excitation Device 199
K Gurukiran, Prasanta Kumar Samal

Theoretical Study of Influence of MHD in an Infinitely Long Rough Porous Journal Bearing 205
G K Kalavathi, B K Yuvaraja, P A Dinesh, M G Vasundhara

Application of Data Science in Automotive Design Optimization for Faster Convergence 214
Rajeshwar S Kadadevaramath, Mohan Sangli

Design and Development of Interactive and Low Cost Autonomous Ergonomic Weed Eliminator 221
Saurabh Suman, Vijay Kamate, S Mohan Kumar, Umakant P Kulkarni

Optimization of Profile Shift Co-Efficient for Highest Contact Ratio in Non-Standard Gearing 227
A R Dr. Rajesh, Gonsalvis Dr. Joseph, K A Dr. Venugopal

Acoustic Emissions from a High-Speed Propeller 233
R Soundararajan, T J S Jothi

Numerical Investigation of Single-Phase Heat Transfer in Converging and Diverging Microchannel 238
R Prajwal Athreya, J Tejas, Mascarenhas Venson, N S Srikanth

Experimental Investigation of Forced Convection Heat Transfer from Vertical Grooved Plates 245
Abdul Razak Kaladgi, K C Vishwanath, Sreenivasalu Reddy, A Chandrashekar

Effect of Copper Oxide Nano Fluids as Coolant on Thermal Performance of Spiral Heat Exchanger 251
Abdul Razak Kaladgi, K. C. Vishwanath, P. Madhu, A. Chandrashekar, B. V. Chaluvaraju

Modelling and Simulation of Virtual NOx Sensor for Diesel Engine Using Thermodynamic Model 258
Shoaib Ahmed, M R Halesh, Narasimha kollaparti, P V Aman, V T Satish

Effect of Flap Deflection Angle on Flow Characteristics of Aerofoil 267
G Abhishek Vinod, T J S Jothi

Vision ASSIT for Autonomous Surgery Robot 271
R Madhu Mohan, C Dr Grisha, M S Kunal, V Lokanatha Reddy, M Mahendra, N Pawan

Author Index

IOP Conf. Series: Materials Science and Engineering 1189 (2021) 011001 doi:10.1088/1757-899X/1189/1/011001

Preface

In the present era, Mechanical Engineering is the back-bone of the economy of a nation. The theme of the conference is to explore the new horizon of innovations by academics, researchers and scientists from worldwide to discuss, exchange knowledge about the advancements in mechanical engineering science. The conference is organized by the academics and researchers belong to various scientific areas of Mechanical Engineering and its allied fields.

ICTMES-2020 focuses on the advancements in the filed of Mechanical Engineering Science applied to the areas of Materials & Manufacturing, Engineering Design, Thermal Engineering and Industrial Engineering & Management. The conference is aimed for creating a cross disciplinary summit that transcends departmental, institutional, industrial, public and private research organizations and global barriers and lends itself to the integration of research and education in the vital field of advancements in Mechanical Engineering Science. In addition to the contributory papers, the conference aims at inviting renowned experts for delivering keynote addresses/ special plenary lectures in the field of Mechanical Engineering Sciences. All the accepted papers will be published in the book of abstracts of the conference.

List of Editors:

1. Dr. Ezhil Vannan S.
2. Dr. Dushyanth Kumar K. R.
3. Dr. Shashank Lingappa M.

IOP Conf. Series: Materials Science and Engineering 1189 (2021) 011002 doi:10.1088/1757-899X/1189/1/011002

Peer review declaration

- **Type of peer review:** Single-blind
- **Describe criteria used by Reviewers when accepting/declining papers. Was there the opportunity to resubmit articles after revisions?**
  Quality of the article along with its relevance to the field of conference theme and the scope of the work are the major criteria considered for accepting/ declining the articles. YES, an opportunity was provided to resubmit the article after revision.
- **Conference submission management system:**
- **Number of submissions received:** 100
- **Number of submissions sent for review:** 100
- **Number of submissions accepted:** 60
- **Acceptance Rate (Number of Submissions Accepted / Number of Submissions Received X 100):** 60 %
- **Average number of reviews per paper:** 03
- **Total number of reviewers involved:** 20
- **Any additional info on review process (ie plagiarism check system):** NA
- **Contact person for queries:**

Dr. Ezhil Vannan S. (evs@mcehassan.ac.in)
Dr. Dushyanth Kumar K. (krd@mcehassan.ac.in)
Dr. Shashank Lingappa M. (msl@mcehassan.ac.in)

Effect of extrusion on tensile behavior of S Glass and Fly ash reinforced Al 7005 hybrid composites.

Shantharaja M[1], Praveen Kumar S[2]
[1,2]Dept.of Mechanical Engineering, UVCE,, Bengaluru, Karnataka- 560056, India.

[1]shantharajam@uvce.co.in

[2]Corresponding Author: praveen9485@gmail.com

Abstract: Metal Matrix Composites (MMCs) are most capable in accomplishing improved mechanical properties and wear properties due to the presence of reinforced particles. Aluminium is used in various applications because of its light weight and corrosion resistant. In the present work, Al7005 matrix alloy is reinforced with different Weight % of S-Glass and Fly-ash were prepared through stir casting technique and hot extruded at a ratio of 5.32:1. The extruded Al 7005 hybrid MMCs were later subjected to tensile test and compared with casted composites. The properties of extruded hybrid MMCs were found to be better than the hybrid MMCs. It is observed that Yield strength (YS) of extruded Al 7005 +5% S-Glass+6% Fly ash hybrid MMCs improved by 7.43% and Ultimate tensile strength (UTS)of extruded Al 7005 +5% S-Glass+6% Fly-ash hybrid MMCs improved by 7.43%, While percentage of elongation is reduced to 14% than casted composites.

Keywords: Al7005 alloy; S-Glass; Fly-ash; hybrid MMCs, Hot Extrusion

1. INTRODUCTION

Composite materials in general are the materials which are engineered combinations of two or more materials tailored to get the desired properties. The matrix phase and reinforcement phase with significantly different physical or chemical properties, which are constituents of any composite material exhibiting properties which are different from the individual components. Metal matrix composites (MMCs) are made of metallic matrix with one or more discontinuous reinforcing phases [1]. Many researchers have carried out research using varieties of reinforcement's ceramics such as carbides, oxides and nitrides [2-6]. These reinforcements are quite expensive. Most of the researchers conduct tests on primary processed methods i.e. liquid metallurgy such as stir casting [7-8]. Very less work has been done on the secondary process such as rolling, extrusion, wire drawing etc. [9]. Several researchers have worked such reinforcements.C.S. Ramesh et al.[10] studied the effect of extrusion ratio on wear behaviour of hot extruded Al6061–SiCp (Ni–P Coated) composites. Vipin K. Sharma et. al [11]. Similarly, several researchers have worked on developing the noble composites with high strength and low cost [12-15]. The present work compares the effect of reinforcement on tensile behaviour of Al 7005 matrix alloy reinforced with S-glass and Fly ash hybrid composites and comparing the results as cast and extruded composites.

2. EXPERIMENTAL DETAILS

In the present study S-glass and fly-ash (5-10μm and 25-30μm respectively) were used as reinforcements with Al 7005 as matrix material. The different compositions of Al 7005 hybrid MMCs were developed and later it was subjected to extrusion process extruded at a ratio of 5.32:1maintaining a ram speed of .5mm/sec with a billet temperature of 500^{0}C as per literature [11].

2.1 Tensile Test Details

The tensile test was conducted as per ASTM E8M-15a standards. The extruded samples were cut and machined and the test was conducted in as per ASTM standards. Three specimens of each compositions are tested and the average of three values are recorded.

3. RESULTS AND DISCUSSIONS

3.1 Comparison of Yield strength of as cast and extruded Al 7005 MMCs

Figure 1 shows the variation of yield strength (YS) of as cast and extruded Al 7005 hybrid MMCs.

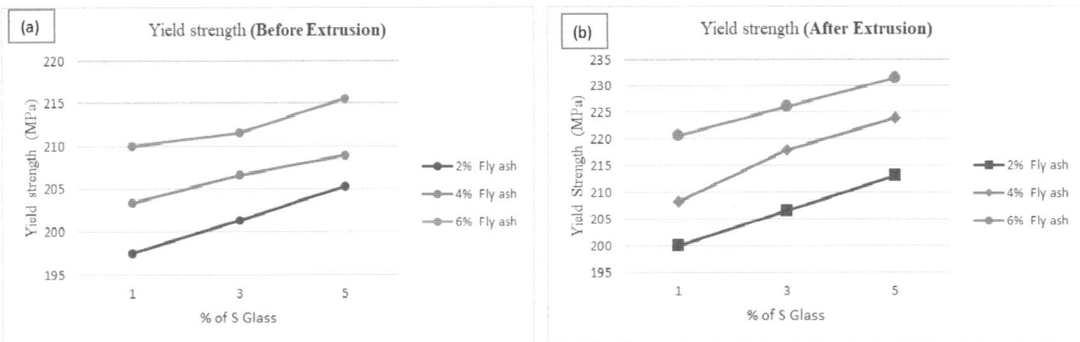

Figure 1: Variation of Yield strength values of as cast and extruded Al 7005 MMCs

It is observed from the graphs that as the Wt % of S-Glass & Fly ash increases the yield strength of all the hybrid composites increases in both the cases. It can be found that Al7005 alloy and its hybrid MMCs has a better yield strength when compared with as cast Al 7005 and its hybrid MMCs after extrusion. The composite having 5% S-Glass with 6% Fly ash gives highest yield strength i.e. 231.51 MPa and this may be due to the good bonding of reinforcements with the matrix material. Similar trends have been observed by several researchers [16-18].The enhanced strength of the extruded MMCs is due to the synergetic effects of both the reinforcements (S-Glass and Fly-ash).

3.2 Comparison of Ultimate Tensile strengthas cast and extruded composites

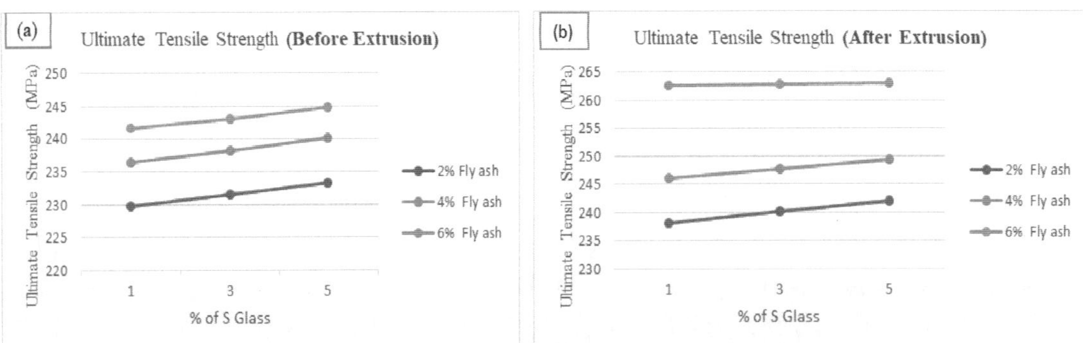

Figure 2 : Variation of Ultimate tensile strengthof as cast and extruded composites

Figure 2 shows the variation of Ultimate tensile strength (UTS) of different Al 7005 hybrid MMCs before and after extrusion. It is observed from the graphs that, as the Wt % of S-Glass & Fly ash increases the ultimate tensile strength of all the hybrid composites has been increased in both as cast and extruded composites). From the results is observed that the extruded composite having 5 % S-Glass with 6% Fly ash gives highest tensile strength of 263.08 MPa, This increase may be due to the dispersion of hard phases of reinforcements in soft ductile matrix and due to the excellent adhesion of S-glass and Fly ash particles to Al 7005 matrix, which resists estrangement during tensile loading [19-20]. This may be ascribed to heavy residual stress induced thru solidification and generation of density dislocation.The enhanced strength of the extruded MMCs is due to the synergetic effects of both the reinforcements (S-Glass and Fly-ash).

3.3 Comparison of Ductility of as cast and extruded composites.

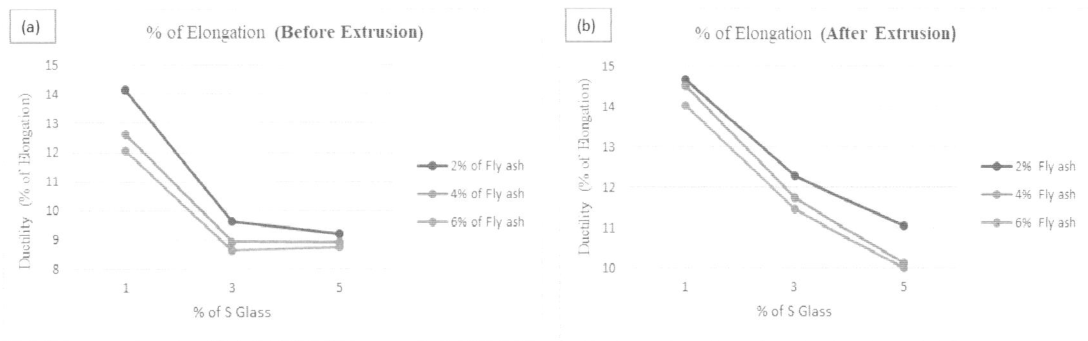

Figure 3: Variation of Ductility of as cast and extruded composites.

From figure 3, it is observed that, as the Wt % of S-Glass & Fly ash increases the ductility of composites decreases in both the cases. It is noticed that, there is a slight improvement in ductility of hybrid MMCs. The composite having 5 % S-Glass with 6% Fly ash gives least ductility and this may be attributed to fact that with enhanced reinforcement i.e. S-glass and Flyash leads to inherently brittle phases and presence of intermetallic phases aids as possible locations for crack nucleation resulting to reduction in ductility under quasi-static loading [19-20].

4. CONCLUSIONS

A through study on tensile properties have been done on the Al 7005 alloy reinforced with S-Glass with 6% Fly ash. Comparison studies have been done on as cast and extruded hybrid MMCs. It can be concluded that the tensile properties i.e. Yield strength (YS) and Ultimate tensile strength (UTS) of extruded Al 7005 hybrid MMCs shows better improvement compared to as-cast composites. It is noticed that the Al 7005 reinforced with 5% S-Glass and 6% Fly-ash shows highest yield strength of 231.51MPa against 215.49MPa and Ultimate tensile strength of 263.08 MPa against 244.88 MPa for extruded and as-cast composites respectively. While the ductility of hybrid composites reduces to 10 % from 16.02 %.

5. REFERENCES

[1] SiddabathulaMadhusudan, Mohammed MoulanaMohiuddinSarcar, NarsipalliBhargava Rama Mohan Rao, “Mechanical properties of Aluminum-Copper(p) composite metallic materials”, Journal of Applied Research and Technology, 293–299, 2016.

[2] L Avinash, T Ram Prabhu, S Bontha "The Effect on the dry sliding wear behavior of gravity cast A357 reinforced with dual size silicon-carbide-particles"-Applied-Mechanics-and-Materials,829,(2016),83-89.

[3] B Sajjan, L Avinash, S Varun, KN Varun Kumar, A Parthasarathy "Investigation of Mechanical Properties and Dry Sliding Wear Behaviour of Graphite Reinforced Al7068 Alloy-"Applied-Mechanics-and-Materials-,867,(2017),-10-18 .

[4] AvinashLakshmikanthan, SrikanthBontha, M.Krishna, Praveennath G. Koppad, T.Ramprabhu "Microstructure, mechanical and wear properties of the A357 composites reinforced with dual sized SiC particles "Journal of Alloys and Compounds,786 (2019) 570-580

[5] L Avinash, T Ram Prabhu, A Parthasarathy, KN Varun Kumar, B Sajjan " Wear and mechanical behaviour of Hypo-eutectic Al-7% Si-0.5% Mg alloy (A357) reinforced with Al_2O_3 particles", Applied Mechanics and Materials ,829,(2016), 66-72

[6] A Parthasarathy, L Avinash, KN Varun Kumar, B Sajjan, S Varun "Fabrication and Characterization of Al-0.4% Si-0.5% Mg-SiCp Using Permanent Mould Casting Technique-"Applied-Mechanics-and-Materials,-867,(2017),34-40.

[7] L Avinash, H Kumar, A Parthasarathy, KN Varun Kumar, B Sajjan, "The Effect of Ceramic Reinforcement on the Microstructure, Mechanical Properties and Tribological Behavior of Al-7.5% Si-0.5% Mg Alloy" Applied Mechanics and Materials, 867,(2017),3-9,

[8] AvinashLakshmikanthan, T. Ram Prabhu, UdayagiriSaiBabu, Praveennath G. Koppad, Manoj Gupta, MunishamaiahKrishna ,SrikanthBontha (2020) "The effect of heat treatment on the mechanical and tribological properties of dual size SiC reinforced A357 matrix composites" Journal of Materials Research and Technology, Volume 9, Issue 3, May–June 2020, Pages 6434-6452

[9] S.C. Tjong, Z.Y. Ma, Microstructural and mechanical characteristics of in situ metal matrix composites, Mat. Sci. Eng. 29, 49-113 (2000).

[10] CS Ramesh, R Keshavamurthy, BH Channabasappa, Abrar Ahmed" Microstructure and mechanical properties of Ni–P coated Si3N4 reinforced Al6061 composites"Materials Science and Engineering: A,Volume 502, Issues 1–2, 25 February 2009, Pages 99-106

[11] Vipin. K Sharma, R. C. Singh, Rajiv Chaudhary, "Effect of fly ash particles with aluminium melt on the wear of aluminium metal matrix composites", Engineering Science and Technology, an International Journal, 2017.

[12] CS Ramesh, Mir Safiulla "Wear behavior of hot extruded Al6061 based composites "Wear Volume 263, Issues 1–6, 10 September 2007, Pages 629-635

[13] A. Evans, C.S. Marchi and A. Mortenson, Metal matrix Composites in Industry, An Introduction and a Survey, Kluwer, Boston (2003)

[14] Qiang Liu, Xiao Zhou ,Haitao Zhou , Xiaohui Fan , Keming Liu "The effect of extrusion conditions on the properties and textures of AZ31B alloy" Journal of Magnesium and Alloys 5 (2017) 202–209

[15] S. J. Niteesh Kumar, R. Keshavamurthy ,M. R. Haseebuddin ,Praveennath G. Koppad "Mechanical Properties of Aluminium-Graphene Composite Synthesized by Powder Metallurgy and Hot Extrusion" Trans Indian Inst Met (2017) 70(3):605–613

[16] Mohammad Sharifi, E., and F. Karimzadeh. 2011. Wear behavior of aluminum matrix hybrid nanocomposites fabricated by powder metallurgy. *Wear* 271 (7–8):1072–79. doi:10.1016/j.wear.2011.05.015

[17] Umanath, K., K. Palanikumar and S. T. Selvamani. 2013. Analysis of dry sliding wear behaviour of Al6061/SiC/Al2O3 hybrid metal matrix composites. Composites: Part B 53:159–168.doi:10.1016/j.compositesb.2013.04.051

[18] H.C. Anilkumar , H.S. Hebbar and K.S. Ravishankar "Mechanical Properties of Fly Ash Reinforced Aluminium Alloy (Al6061) Composites", International Journal of Mechanical and Materials Engineering (IJMME), Vol.6 (2011), No.1, 41-45

[19] Vipin K. Sharma , R.C. Singh , Rajiv Chaudhary "Effect of flyash particles with aluminium melt on the wear of aluminium metal matrix composites" Engineering Science and Technology, an International Journal 20 (2017) 1318–1323

[20] Zuoyong Dou, GaohuiWu, Xiaoli Huang, Dongli Sun and Longtao Jiang "Electromagnetic shielding effectiveness of aluminum alloy–fly ash composites" Composites: Part A 38 (2007) 186. DOI: 10.1016/composites a.2006.01.015

Impact of Ball Burnishing Process on Residual stress distribution in Aluminium 2024 Alloy using Experimental and Numerical Simulation

Harish[1], D. Shivalingappa[1], Raghavendra.N[1], Ganesh V[2]

[1]Department of Mechanical Engineering, B. N. M. Institute of Technology, Bengaluru 560070, India.

[2] Research Engineer , Fracktal Works Pvt.Ltd

Abstract: Deep ball burnishing is surface enhancement process is specifically used to induce compressive residual stress in material sub surface. In this investigation deep ball burnishing surface treatment process was carried out on the periphery of Al 2024 alloy and induced beneficial compressive residual stress was determined by X-Ray Diffraction method. In this work explicit module of the FE-Code ABAQUS-6.17 was used to model the ball burnishing process and analysis is carried out to determine the residual stresses in Aluminium 2024 Alloy by changing instantaneously process variables like burnishing passes and pressure. Comparison was made among experimental and results obtained from FEA, it was observed that there is well agreement between the FE Results and experimental results. Also the results revealed the impact of deep ball burnishing process parameters on distribution of beneficial compressive residual stress.

Keywords : Ball burnishing, Residual stress, Finite element analysis.

1. Introduction

The cyclic loads are the prime reason for failure of aero engine components. The aeroengine parts manufactured mainly fails due to the cyclic loads as they are instantaneously subjected to HCF and LCF. Lack of mirror surface finish of the part would increase the rate of crack initiation, propagation and gradually leads to the failure of component. In commercial practice many surface coating and surface treatment processes are available to enhance surface integrity properties of engineering components like laser shock peening, deep rolling, Burnishing which helps to avoid the fatigue failure of engine components. Aviles R et al.[1] investigated the effect of deep ball burnishing on fatigue life of mechanical components with and without surface treatment. In this work deep ball burnishing surface treatment is employed to improve the surface integrity parameters that are needed to enhance its fatigue life. Rolling process makes the component surface smoother with out producing chip and with less amount of cold work. The integral process rolling variables like pressure, force, speed, feed, types of lubricant also affects the treatment effectiveness. Loh N[2] conducted exhaustive literature work related to the process variable parameters affecting the surface integrity of components in turn increase the fatigue life. Harish et al.[3] conducted the comparative study of influence of deep ball and roller burnishing treatment on surface integral properties mainly its impact on induced residual stresses of Aluminium 2024 Alloy. It was found that ball burnishing process was effective in improving the hardness while roller burnishing process was effective in surface roughness. Beres et al.[4] numerically investigated the residual stress distribution results of low plasticity burnishing process. Harish et al [5]analysed the impact of ball burnishing on surface properties of Aluminium 2024 Alloy. It was found that ball burnishing process has a great impact on the surface enhancement of material.

IOP Conf. Series: Materials Science and Engineering 1189 (2021) 012002 doi:10.1088/1757-899X/1189/1/012002

2. Selection of Test Specimen

In this investigation the deep burnishing process is carried out on Aluminium alloy 2024. Al 2024 alloy is an most widely used element in aero engine component manufacturing industry due to its resistance to fatigue failure and its valuable strength to weight ratio. The main constituent of Al 2024 alloy is Cu (4.95%).Ultimate tensile strength and yield strength was found to be 469 Mpa , 441 Mpa respectively. Details of Material properties of Al 2024 alloy was given in below Table 1.

Elements	Al	Cu	Cr	Mg	Mn	Si
(Wt %)	94.7	4.9	0.1	1.8	0.9	0.5

Table1 Material properties used for simulating Aluminium 2024 Alloy

3. Experimentation `

The deep rolling burnished tool consists of tungsten carbide steel ball, helical spring, hydrostatic bearing seat. The steel ball is pressed against the specimen with hydraulic pressure supplied by hydraulic unit of pressure of maximum capacity 20MPa.The deep rolling tool is fixed in the tool post of conventional lathe machine. Seven equal spaces were machined on the specimen. Later Specimen was placed in between supporting chucks of lathe. Then with the help of burnishing tool, ball is pressed over specimen. Deep rolling process was carried out at different section by varying overturns and pressure to understand the effectiveness of burnishing treatment. Burnishing Process is carried out under the different pressure at different sections. In this investigation treatment is carried out using the variable pressure 10MPa, 15MPa and 20MPa respectively. The maximum overturns carried out at specimen surface is 3. After the completion of surface treatment the beneficial compressive stress is measured by X-Ray diffraction method.

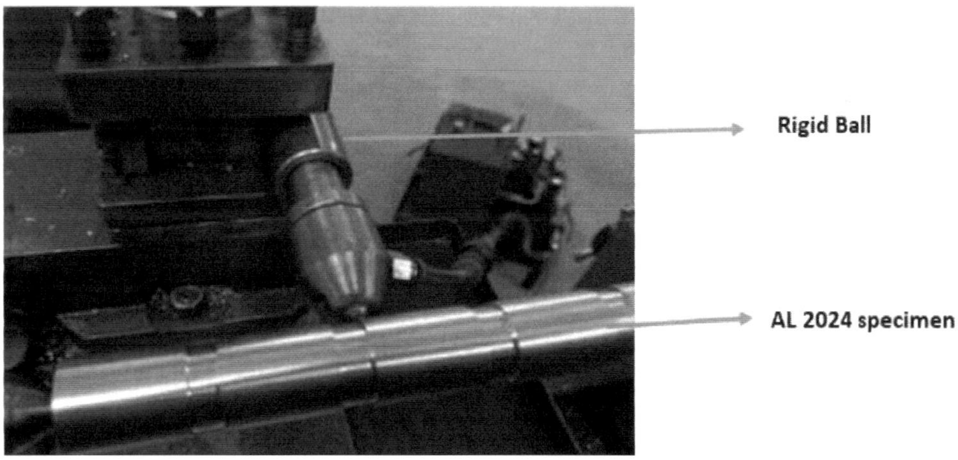

Figure 1 Deep Ball Burnishing process

4. Finite Element Modelling

The ABAQUS 6.17 is used to simulate burnishing process to investigate residual stresses in various depth. The numerical simulation was carried out using two variables i.e burnishing pass and burnishing pressure. The 3D finite element mesh for modelling of the deep ball burnishing process as shown in figure below. The simulation has been carried out by assuming tungsten carbide ball as a rigid and rectangular Al 2024 alloy as deformable. In this analysis the used boundary conditions are rotation and linear displacement for rigid ball and plate is completely constrained in all the directions. In this analysis work piece was meshed using C3D8R elements. Figure 2 shows the meshing strategy of the rectangular specimen. The burnishing process is followed as according to the burnishing pattern.

IOP Conf. Series: Materials Science and Engineering 1189 (2021) 012002 doi:10.1088/1757-899X/1189/1/012002

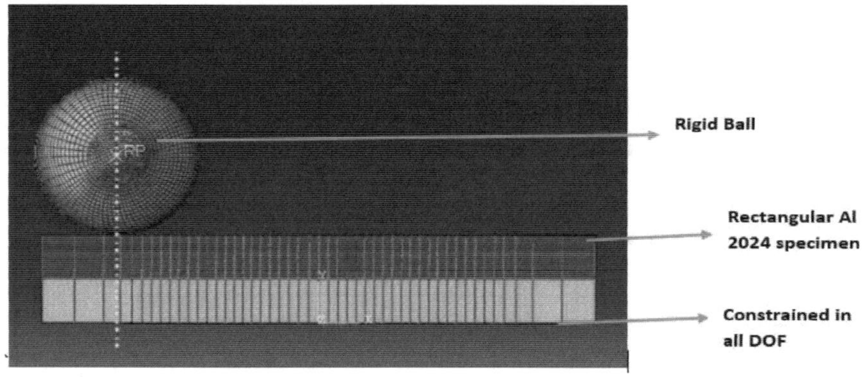

Figure 2 3D finite element mesh for modelling of the Deep Ball Burnishing process.

5. Numerical Results and its Validation.

After the numerical simulation of deep rolling process, induced residual stresses in longitudinal and transverse direction as shown in below figure 3 and 4. Longitudinal and transverse residual stresses are computed by varying process parameters such as overturns and deep rolling pressure are shown in the figure 6,7 and 8.The experimental results obtained in [3] is validated with numerical simulation results. It was observed that experimental results are deviating by numerical results only by 9%

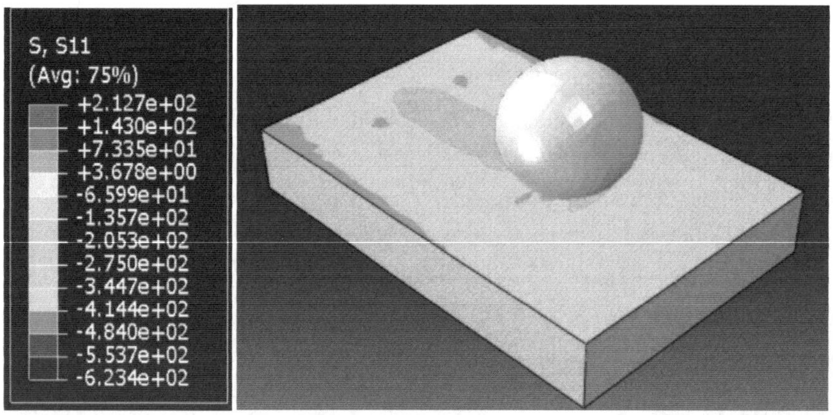

Figure 3 Longitudinal residual stress distribution

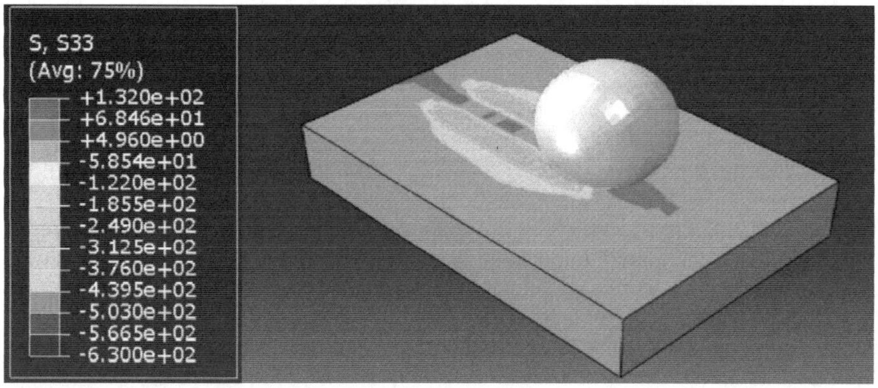

Figure 4 Transverse residual stress distribution

IOP Conf. Series: Materials Science and Engineering 1189 (2021) 012002 doi:10.1088/1757-899X/1189/1/012002

Validation of results:

The experimental results obtained in [3] is validated with numerical simulation results. It was observed that experimental results are deviating by numerical results only by 9%. To verify the accuracy of the numerical simulation of ball burnishing process, the residual stress profile comparison made and it is shown in figure 5.The stress distribution for the rolling variables in longitudinal direction was shown in figures 6, 7, 8 and 9.

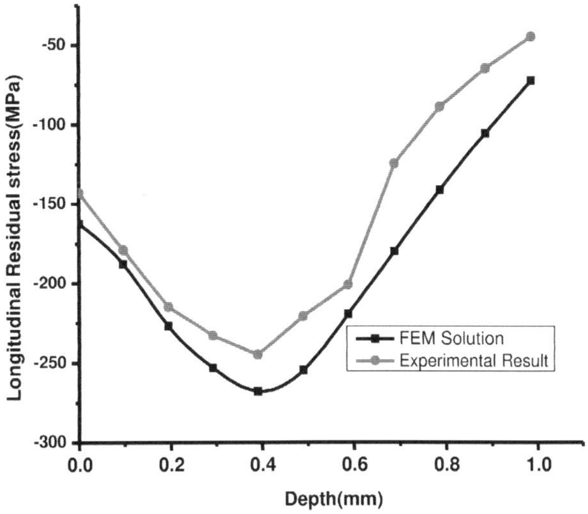

Figure 5 validation of results

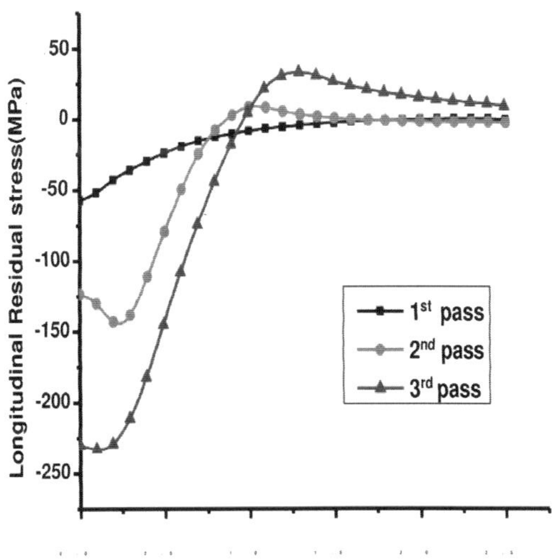

Fig.6 stress distribution for pressure 10 MPa

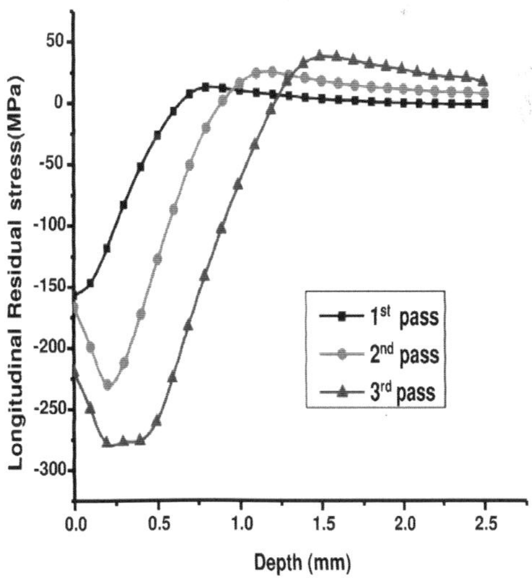

Fig. 7 stress distribution for pressure 15 MPa

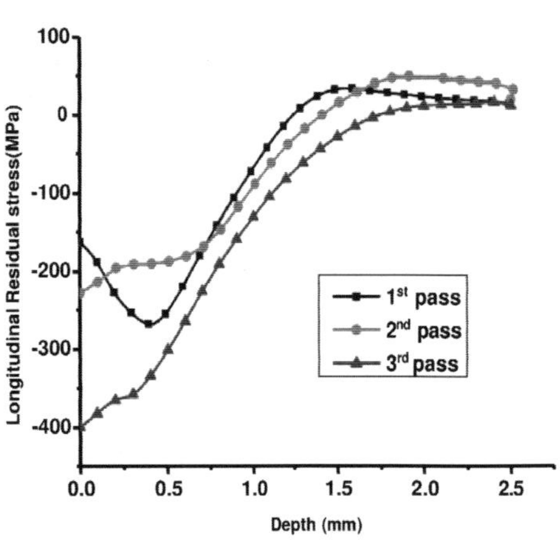

Figure 8 Residual stress distribution for 3rd pass

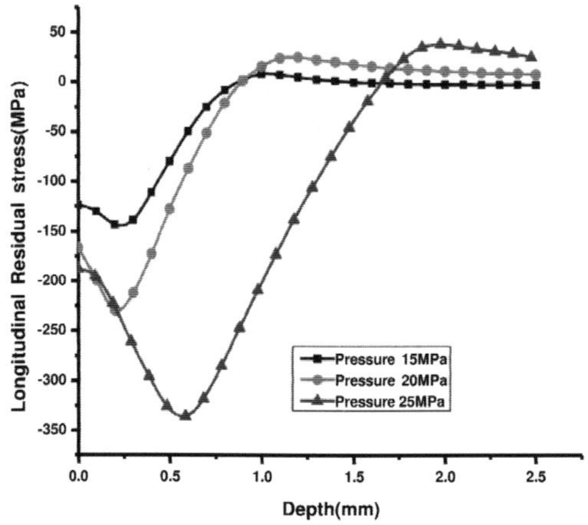

Figure 9 Residual stress distribution for varied pressure.

6. Conclusion :

A comprehensive 3D finite explicit dynamic analysis was conducted to simulate the ball burnishing process .Based on obtained residual stress distribution from FE method and experimental method the following conclusions are drawn.

- It was found that experimental results shows well agreement with the numerical results. The results revealed that increase in overlap of the rolling tracks largely increases the magnitude of the residual stress created in target plate.
- Numerical results revealed that the overlapping effect has a greater impact on residual stress distribution in the given material.
- It is also found that burnishing passes also major impact on the residual stress distribution in Aluminium 2024 Alloy.
- It may also be observed that the depth of residual stress was improved by increasing the burnishing pressure.

References

[1] R. Avilés, J. Albizuri, A. Rodríguez, and L. N. López de Lacalle, "On the fatigue strength of ball burnished mechanical elements," *Mechanisms and Machine Science*, vol. 17, pp. 365–373, 2014, doi: 10.1007/978-94-007-7485-8_45.

[2] N. H. Loh and S. C. Tam, "Effects of ball burnishing parameters on surface finish A literature survey and discussion," 1988.

[3] Harish and D. Shivalingappa, "The influence of ball and roller burnishing process parameters on surface integrity of Al 2024 alloy," *Materials Today: Proceedings*, vol. 27, pp. 1337–1340, 2020, doi: 10.1016/j.matpr.2020.02.614.

[4] W. Beres, J. Li, and P. Patnaik, "GT2004-53925 NUMERICAL SIMULATION OF THE LOW PLASTICITY BURNISHING PROCESS FOR FATIGUE PROPERTY ENHANCEMENT," 2004. [Online]. Available: http://www.asme.org/about-asme/terms-of-use.

[5] Harish, D. Shivalingappa, P. Vishnu, and Sampath Kumaran, "Impact of Ball burnishing process parameters on surface Integrity of an Aluminium 2024 Alloy," in *IOP Conference Series: Materials Science and Engineering*, Jun. 2018, vol. 376, no. 1, doi: 10.1088/1757-899X/376/1/012099.

Prediction of Mechanical properties of Epoxy composites containing mono and hybrids particulate fillers.

K Srinivas[1], M S Bhagyashekar[2]
[1]Independent Researcher Bengaluru India
[2]HKBK College of Engineering Bengaluru India
Email: sripadsanj@gmail.com

Key words: Epoxy, hybrid particulate fillers, composites, Al-Aluminium and Cu-Copper, Strength prediction

Abstract: This paper reports the mechanical properties such as tensile strength and tensile modulus of epoxy composite systems containing micro particulates of Cu, Al and hybrid (Al-Cu) which are randomly distributed. The experimental results are compared with strength prediction equations of Biggs, Ross Murphy &Todd, Langley& Green and Wong. Experimentally determined tensile modulus was compared with Reuss, and Counto equations. The values obtained from above equations was not in line with experiment. Hence in view of the above, a new strength predictive equation is developed which is empirical in nature and exhibits good correlation with experimental data.

1. Introduction

Epoxy is one of the structural polymers which are widely used. Epoxy resins have excellent resistance to heat, moisture, chemicals and good compatibility with many materials [1]. However, cross link structure of epoxies makes them brittle and poor crack resistance [2]. A combination of epoxy and particulate fillers can improve physical, mechanical, wear resistance, thermal and electrical conductivities [3]. The epoxy filled particulate polymer composites have wide acceptance in industrial applications due to its ability to meet specific requirement and ease of manufacturing [4]. Generally, all fillers improve modulus as they are stiffer than matrix whereas similar improvement is not seen with respect to strength [5]. The strength of micron size filler reinforced polymer composites depends on the strength of interface between matrix and reinforcement. The effective load transfer from matrix to reinforcement can happen with good interface. Further filler volume fractions, intrinsic strength of filler and processing method adapted have significant impact on polymer composite strength. The size, shape and distribution of particle in a matrix and interface have a role in strength of a polymer [6, 7]. Prediction of strength and modulus of a polymer composite to an extent reduce experimentation. Prediction models which predict composite strength accurately with one type of filler may not be accurate with other filler. Hence prediction of strength of a particulate filled polymer composites is not at all simple. There are various analytical equations developed by researchers from time to time to predict tensile strength and tensile modulus of particulate filled epoxy composites. Further, the exactness of a predictive model is measured with experimental data to validate its usefulness. The present study compares the effect of reinforcing micron sized particulate metallic fillers of Al and Cu as mono and hybrid (Al-Cu) filler. Further the strength of the epoxy composites found by experiment was compared with strength predictive models. The variation was found to be more. In view of this a new empirical equation is developed which exhibits good correlation with experiment values.

2. Materials and Methods

Epoxy resin of type LY556 and an amine hardener HY951 supplied by Huntsman India Ltd was the host matrix and the fillers aluminium and copper were procured from Loba chemie India ltd. The size of Al fillers is 10-80μm and Cu fillers are of 5-15μm shown in figure 1.

2.1 Composite preparation

The predetermined amounts of reinforcements were preheated to remove any moisture present in it and were dispersed in epoxy. The mixtures were stirred continuously until the reinforcements were properly distributed. The resin and hardener were mixed in the ratio 10:1 by weight. Hardener was added to the mixture with continuous stirring, followed by pouring the mixture in to a prepared mould. The cast specimens were allowed to cure in the mould for 24 hours at room temperature followed by post curing at 50^0 C for 30 minutes, 70^0 C for 60 minutes and 85^0 C for 120 minutes. The particle volume fraction is calculated using the equation. $V_f = \frac{W_f}{W_f+(1-W_f)\frac{\rho_{par}}{\rho_{mat}}}$ (1)

a) Al b) Cu

Figure1: SEM of particulates of aluminium and copper

3 Mechanical Tests

The tensile tests were conducted on a flat dog bone type specimen having dimensions 165 X 20 X 3mm on LYOD'S universal testing machine. ASTM D 638 test procedure was followed to evaluate the tensile properties [8]. The specimens were pulled at a constant loading rate of 1mm/min till the specimen breaks. For each type, 5 specimens were tested. The strength of the composites was determined using the equation

$\sigma_c = P/b\ d$ (2)

Where, σ_c= ultimate tensile strength (MPa), P = load (N), b, d = width and thickness (mm)

The modulus of the composites was determined using the equation

$$E = \frac{(\Delta P)}{(\Delta L)}\frac{L}{bd} \quad (3)$$

P = load (N), b, d = width and thickness (mm), E = young's modulus of elasticity (GPa).

The Scanning electron microscopy of particulate fillers and fractured surfaces were captured by Joel make 5600 LV scanning electron microscope. The predictive equations used are shown in table 1.

4. Results and Discussion:

4.1. Effect of fillers on tensile strength and tensile modulus

Figure 2(a) presents the tensile strength of unfilled epoxy and epoxy filled with particles of Al, Cu and hybrid (Al-Cu). Neat epoxy exhibits a tensile strength of 39MPa. Al-epoxy composites show a fluctuating behavior in strength and strength decreases with increased fillers. The decreased strength of Al-epoxy composites may be attributed to low dense Al fillers (2.4gm/cc), bigger particles (10 to 80μm) and lower inherent strength (100MPa).Al fillers occupy more volume for a given weight fraction leading to agglomeration, making less coverage of particles by resin, creating weak interface which cannot sustain the applied load and fails at loads lower than neat epoxy [9]. Cu-epoxy composites show an improvement in tensile strength up to 30% Cu reinforcement and strength decreases with 40% reinforcement. The increased strength of Cu-epoxy composites can be attributed to smaller particle size (5 to 15μm), higher density (8gm/cc) and higher inherent strength (200MPa) of Cu fillers. For a given weight fraction, Cu particulates occupy less volume and each particle is well covered by resin to form a good interface. A good interface can effectively transfer load between matrix and reinforcement. The strength of hybrid (Al-Cu) composites also decreases with increased hybrid fillers; however, the strength of these composites is in between Al-epoxy and Cu-epoxy composites. At 40% filler loading, irrespective of filler type, strength degrades for all filled composites due to agglomeration of particles. Increased fillers reduce the interfacial adhesion due to particle to particle contact, which leads to non-effective transfer of load. Higher particle size makes adhesion poor at the interface leading to poor strength of the composites [10,11 and 12]. Further, dispersed particulates in epoxy have different moduli and poison's ratio from the resin, creates stress concentration and this magnifies the stress surrounding the particle leading to lowering of strength [13]. Figure 2 (b) presents the tensile modulus of epoxy composites. Cured epoxy exhibits a tensile modulus of 2.51GPa. Tensile modulus of Al-epoxy, Cu-epoxy and hybrid (Al-Cu)-epoxy increases with increased fillers. Modulus of Al-epoxy composites are higher than Cu-epoxy and hybrid (Al-Cu) epoxy due to more particles for a given weight fraction. Figures 3 to 8 compares tensile strength and modulus predicted by analytical equations with experiment. The values are shown in table 2.

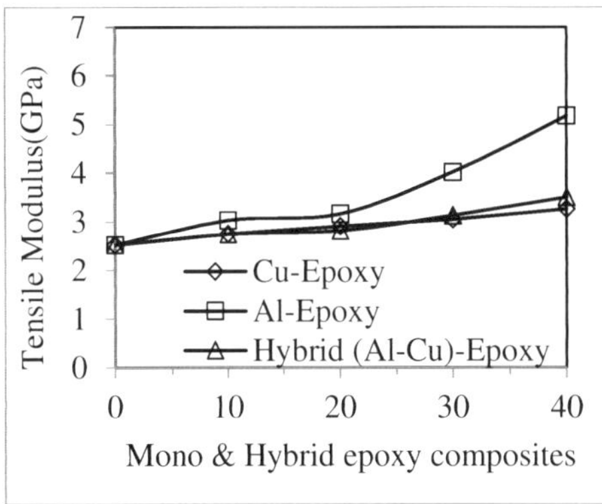

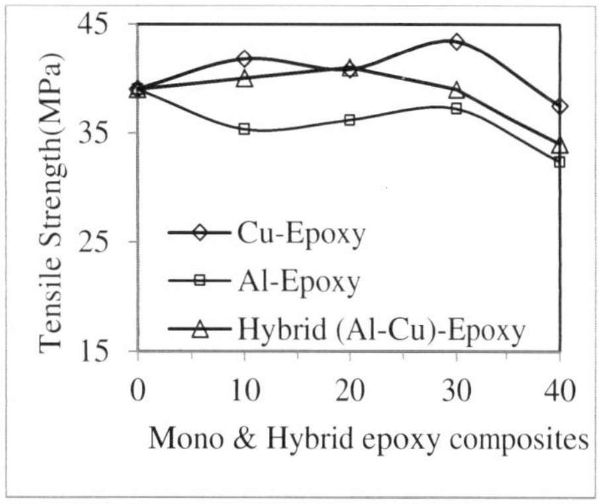

a) Tensile strength b) Tensile Modulus

Figure 2: Effect of fillers on tensile strength and tensile modulus of epoxy composites

4.2. Evaluation of predictive equations

Bigg's model is based on the adhesion between filler and matrix [14]. The value of b is taken as 1.1 (good adhesion) in Bigg's equation. The equation predicts decrease of strength increased volume fraction of fillers. The deviation is more with Al particulate fillers as volume fraction is more. The product b $(\varnothing)^{2/3}$in the equation increases with increased volume fraction. Bigg's model doesn't present the exactness of the experimental findings. Prediction of strength by Ross Murphy &Todd [15]is also not in line with experimental results. Increased volume fraction changes the strength marginally. Langley & Green Equation [15] predicts higher strength for Al-epoxy composites while for Cu-epoxy composites the strength is in line with experiment up to 30% Cu and show higher strength for 40% Cu. The predicted strength for hybrid (Al-Cu) composites is in line with experiment up to 20% hybrid fillers over estimates above 20% fillers. The value $(1-(\varnothing_f/\varnothing_m)^{-3/2}$ increases with increased volume fraction. The prediction by Wong's equation is reasonably good. The value of b in Wong's equation is taken as 0.5 for 10 and 20% fillers, 0.7 for 30 and 40% fillers [16].

4.3 New Equation: Most of the predictive models assume that the inclusions of fillers in a matrix, reduce the strength of composites. Furthermore, above models assume that composite strength cannot be higher than the strength of the matrix. But in some cases, increased fillers may increase the strength of the composites signifying that fillers can also contribute to composite strength by creating good interface. Hence the proposed empirical equation (a modified rule of mixture) considers the contribution of constituents i.e. filler and matrix on composite strength.

$$\sigma_c = \sigma_m W_m \pm Y \sigma_f W_f \tag{4}$$

Where, $\sigma_{c\ m,\ f}$ = tensile strength of composite, matrix and filler, $W_{m\ f}$ = weight fraction of matrix and

$$Y = \frac{1}{a + (n-1)d} \tag{5}$$

a and d are constants and their values are a=2, d = 0.1 and n represents the percentage of fillers by weight. The positive sign in the equation is to be considered when the inherent strength of filler is higher

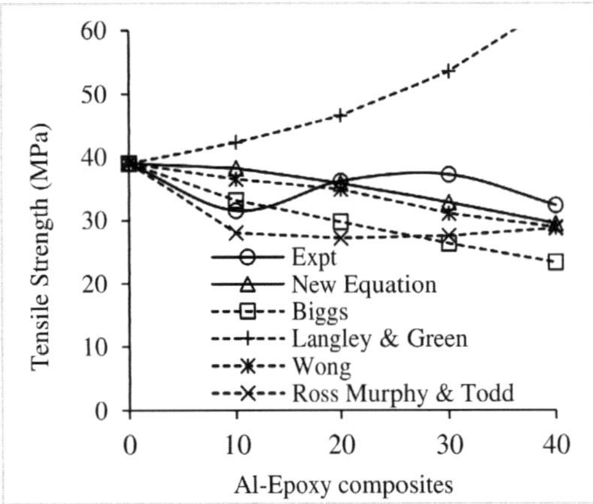

Figure 3 Tensile strength Al- Epoxy composites

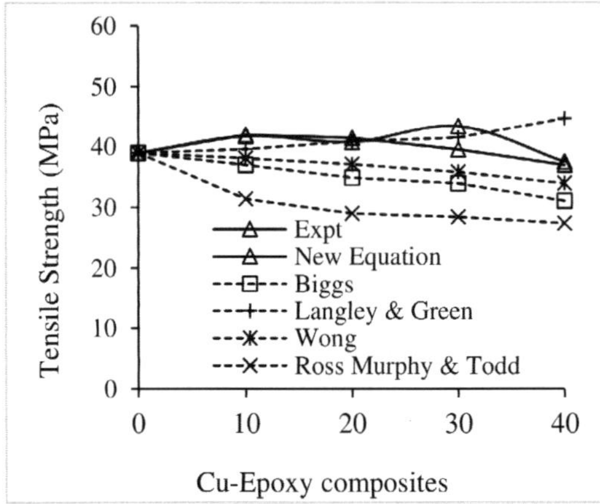

Figure 4 Tensile strength Cu- Epoxy composites

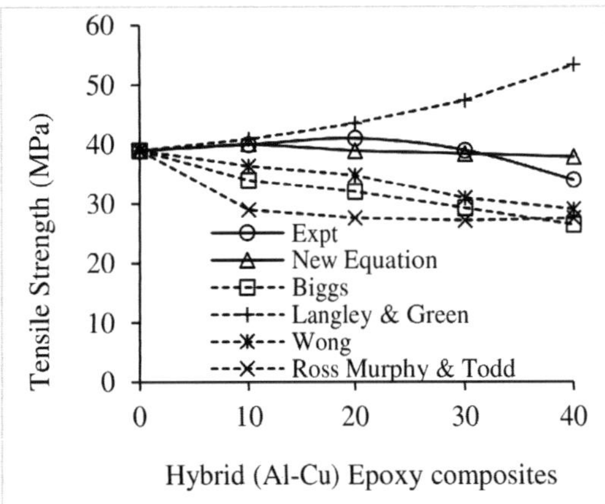

Fig 5 Tensile strength of Hybrid (Al-Cu)-epoxy composites

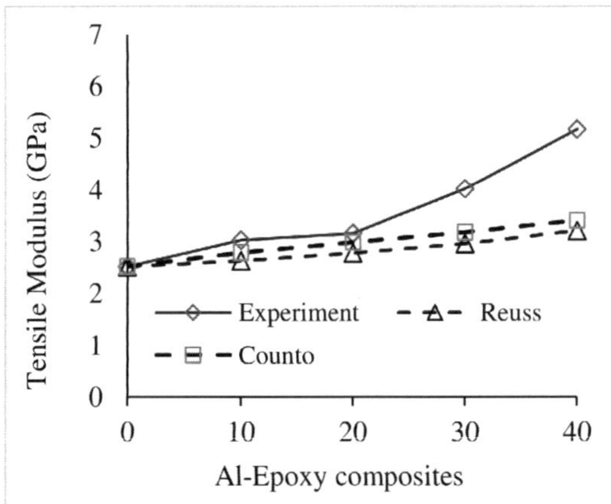

Fig 6 Tensile modulus of Al-epoxy composites

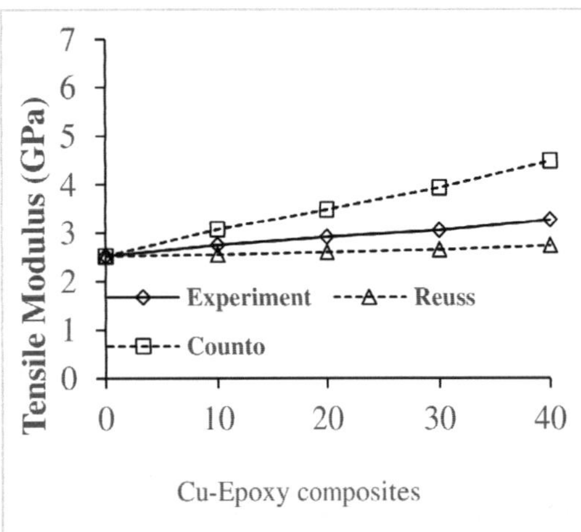

Fig7: Tensile modulus of Cu-Epoxy composites

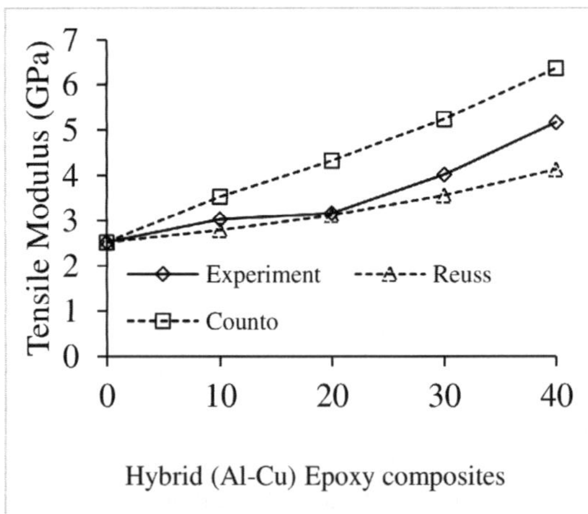

Fig 8: Tensile modulus of hybrid composites

IOP Conf. Series: Materials Science and Engineering 1189 (2021) 012003 doi:10.1088/1757-899X/1189/1/012003

than epoxy and the negative sign is to be considered for fillers that have lower inherent strength than epoxy. The prediction by new equation is in line with experiment values for Al-epoxy, Cu-epoxy and hybrid (Al-Cu)-epoxy composites. The deviation of new equation is minimum compared to other predictive models. Figures 6 to 8 compares the tensile modulus predicted by analytical equation and experiment. The Reuss equation predicts a marginal improvement in tensile modulus with increased fillers and the values are lower than experimental findings. Count's prediction is lower than experiment for Al-epoxy composites, higher than experiment for Cu–epoxy and hybrid (Al-Cu)-epoxy composites.

5. SEM of Fractured surfaces

The images of shown in figure 9(a-c) has the features of a brittle fracture such as crack, fan blade marks, hackle zones which are the characteristics of brittle fracture [17]. These images correspond to neat epoxy 5A5C-Ep and 10A10C-Ep composites. The images of 15A15C-Ep have the characteristics such as particle removal, void, and formation of cracks which could not sustain the applied load and makes the composites to land up with lower tensile strength.

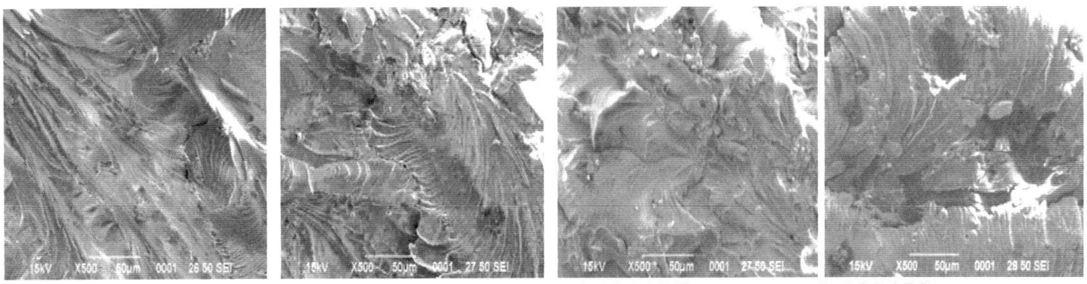

a) Epoxy b) 5A5C-epoxy c) 10A10C-epoxy d) 15A15C-epoxy

Fig 9 SEM of fractured surface of epoxy-mono and hybrid composites

Table 1Description of Models

Strength Prediction Models
Bigg's $\sigma_c = \sigma_p[(1-b(\phi^{2/3})]$ where σ cp m = tensile strength of the composite, matrix, b= 1.21 for poor adhesion between filler and matrix, b= 1.1 for good adhesion,ϕ = volume fraction of fillers Ross Murphy and Todd$\frac{\sigma_c}{\sigma_m} = \left(\frac{1-\emptyset^{\frac{1}{3}}}{1-\emptyset^{\frac{5}{2}}}\right)$ where, σ c, m = tensile strength of the composite and polymer matrix, ϕ = volume fraction of fillers Langley and Green$\frac{\sigma_c}{\sigma_m} = \left(1-\frac{\emptyset f}{\emptyset m}\right)^{-3/2}$ where, σ c, m = tensile strength of the composite and polymer matrix, ϕf, m = volume fraction of fillers and matrix. Wong $\sigma_c = \sigma_m\left(1 - bV_f^{\frac{2}{3}}\right)$ V_f = volume fraction of fillers
Modulus Prediction Models
Reuss$Ec = \frac{EmEf}{Ef(1-Vf)+EmVf}$ **Counto**$\frac{1}{Ec} = \frac{1-Vf^{\frac{1}{2}}}{Em} + \frac{1}{1-\frac{Vf^{\frac{1}{2}}}{Vf^{\frac{1}{2}}Em}+Ef}$ where, E cmf = Tensile modulus of composite, matrix and filler V_f = Volume fraction of reinforcement

Table2 Comparison of tensile strength & modulus of experiment with prediction models

	Tensile strength (MPa)						Tensile Modulus (GPa)		
	Expt	New Equation	Biggs	Langley & Green	Wong	Ross Murphy & Todd	Expt	Reuss	Counto
10A-Ep	31.55	38.20	33.18	42.29	36.50	28	3.03	2.63	2.79
20A-Ep	36.18	35.82	29.76	46.54	34.86	27.20	3.16	2.78	2.99
30A-Ep	37.22	32.81	26.36	53.55	31.07	27.57	4.02	2.96	3.18
40A-Ep	32.38	29.50	23.37	64.11	28.93	28.77	5.17	3.21	3.41
10C-Ep	41.78	42.00	37.01	39.60	38.09	31.38	2.75	2.55	3.07
20C-Ep	40.81	41.46	34.86	40.88	37.12	29.01	2.91	2.60	3.48
30C-Ep	43.40	39.54	33.98	41.57	35.81	28.42	3.05	2.65	3.93
40C-Ep	37.50	36.96	31.04	44.70	33.93	27.34	3.26	2.73	4.48
5A5C-Ep	40.00	40.10	33.91	40.91	36.35	28.98	2.74	2.79	3.53
10A10C-Ep	41.00	38.90	32.00	43.53	34.80	27.57	2.81	3.12	4.33
15C15A-Ep	39.00	38.44	29.27	47.33	30.95	27.18	3.13	3.56	5.24
20A20C-Ep	34.00	37.89	26.44	53.33	29.05	27.55	3.50	4.13	6.36

Summary

Micro particulate characteristics such as higher inherent strength, higher density and smaller size enhance the tensile strength of epoxy composites while filler characteristics such as low density, lower inherent strength and bigger size decrease the tensile strength of epoxy composites. Fillers like Cu which enhances the strength of epoxy can be called as positive fillers while Al fillers reduce the strength of epoxy and can be termed as negative fillers. The deviation of new equation with experimental data is found to be much lower than other predictive models considered in this study.

Reference:

[1] P.L. Teh, M. Mariatti, H.M. Akil, K.N. Seetharamu, A.N.R. Wagiman *Journal of Composite Materials,* Vol. 42, No. 2 (2008)

[2] H. V. Ramakrishna, S. Padma Priya and S. K. Rai, *Journal of Reinforced Plastics and Composites* 25 (2006) 227

[3] Laura M. Mc Grath, Richard S. Parnas, Saskia H. King, John L. Schroeder, Daniel. A, Fischer, Joseph L. and Lenhart *Polymer* 49 (2008) 999

[4] P. Firoozian, Hazizan Md. Akil and H.P.S. Abdul Khalil *Journal of Reinforced Plastics and Composites*, 29, (2010) 2368

[5] Long-Cheng Tang, Hui Zhang a, Stephan Sprenger, Lin Ye, Zhong Zhang *Composites Science and Technology 72* (2012) 558

[6] H Ku, J Epaarachchi M Trada and P Wong, *Journal of Reinforced Plastics and Composites, 32* (2013) 15

[7] Gaurav Agarwal, Amar Patnaik and Rajesh Kumar Sharma *International Journal of Advanced Structural Engineering* 2013

[8] ASTM Standards Designation: D638 Standard Test Method for Tensile Properties of Plastics

[9] Mohammed, M. (2011) Al-Nahrain *Journal for Engineering Sciences*, 14(2), 160

[10] Jonghwi Lee, Albert F. Yee, *Journal of Applied Polymer Science*,79, (2001)1371

[11] M. Sudheer, K. M. Subbaya, Dayananda Jawali, Thirumaleshwara Bhat *Journal of Minerals & Materials Characterization & Engineering*, 11, (2012) 193

[12] Amar Patnaik, Alok Satapathy, Sandhya rani Biswas *Malaysian Polymer Journal,* 5(2010), 37

[13] Qi Zhao and S. V. Hoa *Journal of Composite Materials* 41(2007), 201

[14] S.M. Adedayo and M.A. Onitiri *The West Indian Journal of Engineering* 35, (2012) 51

[15] Chrystel Loret, William J. Frith and Peter J. Fryer *Applied Rheology* 3 (2007) 1-13

[16] Athimoolam M.1 Dr. Moorthy T.V and Dr. Suresh babu A. *Polymers & Polymer Composites*, 24, (2016),539

[17] S.-Y. Fu Composites: *Composites Part B: Engineering* Volume 39, Issue 6, 2008, 933-961

Studies on Tensile Strength of Notched Woven Fabric Composites for Different Stacking Sequence

[1]B Latha Shankar, [2]G S Shivashankar, [3]D Thulasi Durai and [4]Dushyanth Kumar K R

[1]Associate Professor, Department of Industrial Engineering and Management
[2] Professor, Department of Mechanical Engineering, Siddaganga Institute of Technology, Tumakuru – 572 103, Karnataka, India
[3] Scientist, CSIR - National Aerospace Laboratories, Bangalore, Karnataka, India
[4] Associate Professor, Department of Mechanical Engineering, Malnad College of Engineering, Hassan, Karnataka, India

Keywords*:* Composite laminates, Stacking sequence, Point stress criterion, Notch strength.

Abstract. Present work emphasizes on the effect of stress concentration based on hole size, specimen width and stacking sequence on the fracture behavior of the woven-mat glass/epoxy reinforced composites both experimentally and numerically. In the present work three different stacking sequences: [45/-45], [0/90] and [30/60] were considered for the investigation. Laminates were fabricated using wet layup technique. Circular notches of six different diameters were considered for the analysis. Both notched and un-notched test specimens of all the configuration were subjected to tensile testing till fracture. Both experimental and analytical results show that stacking sequence has a significant influence on fracture stress in the presence of notch. Numerical results obtained using FEA were compared with experimental results. The obtained results showed that the notched fracture stresses were maximum for stacking sequence angles [+45/-45], and differ significantly from those of the [0/90] and [30/60]. The characteristic length (d_o) influenced by hole size and specimen width as per the point stress criterion (PSC).

1. Introduction

Composites with woven reinforcements are of special interest compared to their counterpart, unidirectional ones because of lesser cost of production, excellent out-of -plan properties, good drape ability and improved impact resistance [1,2]. Cut-outs or holes are to be made in composites parts, to assist their joining when used in structural applications. These holes made in composites negatively influence mechanical properties by creating stress concentrations and hence are of great importance for designers. Whitney and Nuismer derived point stress criterion(PSC) and average stress criterion(ASC) to predict the strength of composites with holes or cut-outs [3,4]. These models were further discussed while evaluating mechanical characteristics of composites with woven fabric and having discontinuities such as holes or notches [5,6]. Various models were proposed in literature to determine the influence of notches or holes on tensile strength of the woven composites. Waddoups et al [7] developed a model based on Linear Elastic Fracture Mechanics principle. As an alternative to this model, Whitney et al. [3, 4] developed a model to explain variation of tensile strength of structural

composites due to presence of holes or notches. Later this model was extended into three parameter based model by Pipes et.al [8, 9]. The three parameters considered were hole size, reference radius and notch sensitivity factor.
Later it was showed that above models developed for laminate composites could also be applied for calculating tensile strength of fabric composites with notch[9-11]. Finite element analysis was employed subsequently by Hwan et al to predict the notched strength of composite plates without using analytical solutions [12]. A combination of fracture and stress energies was used to formulate a phase-field model and was presented by Reinoso J et.al [13] for notched strength prediction. Recently an investigation was conducted to analyse the effect of stress concentration related various parameters on the mechanical properties [14].

1.1. Theoretical Background

As per Whitney et al. [3, 4], notched tensile strength of a composite laminate can be predicted by the PSC and ASC based two parameter model. This model majorly considers the stress distribution adjacent to the notch. According to PSC based model, it is assumed that failure occurs when the normal stress σ_y generated at some fixed distance (d_0) or the average stress σ_y generated over some fixed distance (a_0) near the notch will be equal to the strength (σ_0) of the un-notched material.

For a circular hole, the failure occurs as per the following Equations (1) & (2)

$$PSC=\sigma_y(x,0)\Big|_{x=R+do}=\sigma_o \tag{1}$$

$$ASC=\frac{1}{a_o}\int_{R}^{R+a_o}\sigma_y(x,0)\,dx=\sigma_o \tag{2}$$

Along the X-axis The normal stress (σ_y) developed ahead of the hole can approximately be expressed by Reference [15].

$$\sigma_y(x,0)=\frac{\sigma^{\infty}}{2}\left[2+\left(\frac{R}{x}\right)^2+3\left(\frac{R}{x}\right)^4-\left(K_T^{\infty}-3\right)\left\{5\left(\frac{R}{x}\right)^6-7\left(\frac{R}{x}\right)^8\right\}\right],\ x>R \tag{3}$$

Here, K_T^{∞} represents factor for stress concentration when an infinite orthotropic plate is considered. σ^{∞} represents uniform stress along the Y-axis for the similar plate containing a circular hole of radius R.

$$K_T^{\infty}=1+\left\{\frac{2}{a_{22}}\left(\sqrt{a_{11}\,a_{22}}-a_{12}+\frac{a_{11}a_{22}-a\frac{2}{12}}{2a_{66}}\right)\right\}^{\frac{1}{2}} \tag{4}$$

where a_{ij} denotes orthotropic in-plane stiffness of the laminate.

Substitution of equation (3) in (1), gives

$$\frac{\sigma_N^{\infty}}{\sigma_o}=\frac{2}{2+A_1^2+3A_1^4-\left(K_T^{\infty}-3\right)\left(5A_1^6-7A_1^{\infty}\right)} \tag{5}$$

$A_1=R/(R+d_0)$
σ_0= Tensile strength of the laminate without notch or hole.

σ_N^{∞} = Strength of an infinite width plate with notch.
A modification to the PSC was proposed by Karlak [16] in which d_o is a function of the hole radius (R),

$$d_0 = K_0\sqrt{R} \tag{6}$$

2. Experimental Analysis

Woven Glass/Epoxy laminates with 60% volume fraction was considered for investigation. They were fabricated using wet lay-up technique. The geometry & the dimensions of the straight-sided tension specimen with a hole in the middle are shown in Fig.7 All the static tension tests were conducted on UTM. Tensile loading was applied on the specimens till failure at room temperature. The experimental value were compared with FEA as shown in figure 1 to 6.The experimental results were having good agreement with Finite Element results. Table 1 shows the theoretical stress distribution around the notch.

3. Tensile Strength of Notched Laminates

Orthotropic Stress Concentration factor (K_T^{∞})

$$K_T^{\infty} = 1+\left\{\frac{2}{a_{22}}\left(\sqrt{a_{11}\,a_{22}} - a_{12} + \frac{a_{11}a_{22} - a_{12}^2}{2a_{66}}\right)\right\}^{\frac{1}{2}} = 4.35$$

3.1. Failure Strength of Notched Laminate in Tension

The relationship between applied tensile stress on notched laminate and unnotched laminate failure strength is given by

$$\frac{\sigma_N^{\infty}}{\sigma_o} = \frac{2}{2 + A_1^2 + 3A_1^4 - \left(K_T^{\infty} - 3\right)\left(5A_1^6 - 7A_1^{\infty}\right)}$$

σ_0= Unnotched tensile strength of the laminate, found experimentally=205 N/mm^2

σ_N^{∞} = 0.9134Xσ_0=.9134X205=187.25 N/mm^2

Failure strength of laminate=187.25 N/mm^2

3.2. *Failure Load of Laminate*

Applied load at failure or Failure load of laminate (N) = σ_N^{∞} X Area=187.25x197.6=37KN

Theoretical stress distribution is given by

$$\sigma_y(x,0) = \frac{\sigma^{\infty}}{2}\left[2 + \left(\frac{R}{x}\right)^2 + 3\left(\frac{R}{x}\right)^4 - \left(K_T^{\infty} - 3\right)\left\{5\left(\frac{R}{x}\right)^6 - 7\left(\frac{R}{x}\right)^8\right\}\right], x > R$$

R=4mm; K_T^{∞}=4.35; σ_N^{∞}=187.25 N/mm^2

Table 1: Theoretical Tensile Stress Distribution

Distance X(mm)	3	5	7	9	11	13	15	17	19
Theoretical Stress Distribution σ_y (MPa)	833.26	241.65	210.79	200.33	195.52	192.94	191.40	190.42	189.75
Normalized Stress $\sigma_y / \sigma_N^\infty$	4.45	1.290	1.125	1.070	1.044	1.030	1.022	1.017	1.013

3.3. Finite Element Analysis

Finite element analysis carried out on notched and unnotched laminates for stress analysis by using analysis software NISA Ver 12.0. The elements used were NKTP 32-3D composite shell element it has six degree of freedom, for the same geometrical configuration for both experimental and numerical analysis for comparing the obtained results. Few FEA results were shown in Figures 1 to 6. Comparisons between experimental test results and predictions of PSC and modified PSC is shown in figure 8. Relationship between notch strength and Characteristic length is shown in figure 9.

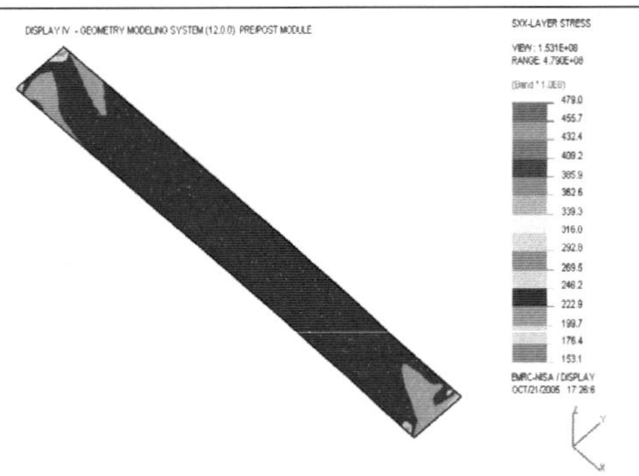

Figure 1: Un notched strength(30/60 stacking sequence)
Applied load 40.7KN
Failure stress: FEA 220 MPa
Experimental 198 MPa

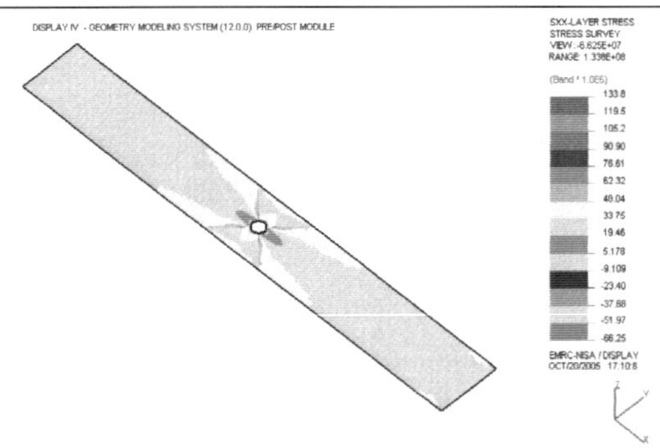

Figure 2: Stress distribution around the Notch (8mm dia, 0/90 stacking sequence)
Applied load 24.45 KN
Failure stress: FEA 134 MPa
Experimental 124 MPa

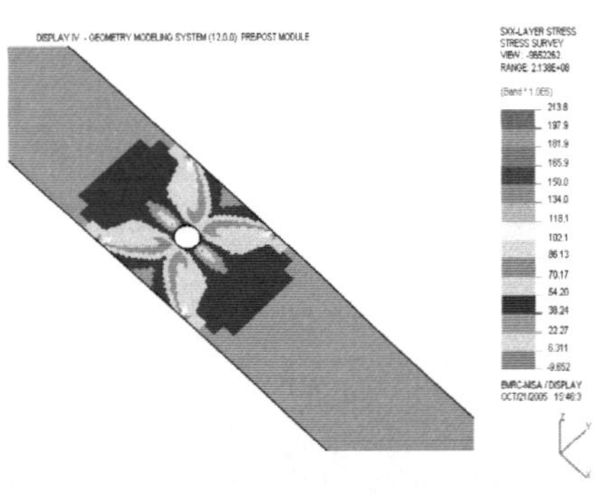

Figure 3: Stress distribution around the Notch(8mm dia, 45/-45 stacking sequence)
Applied load 29KN

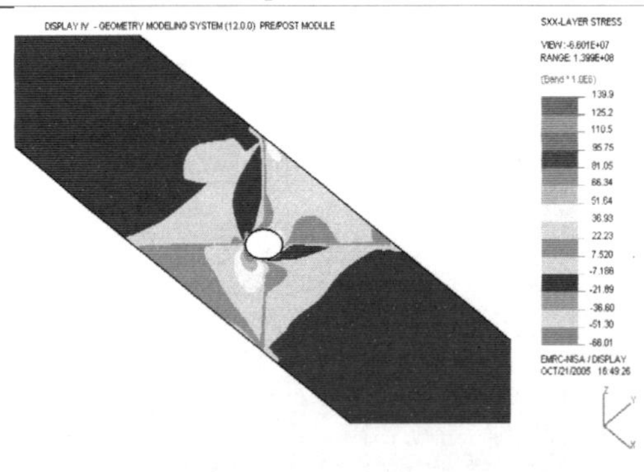

Figure 4: Stress distribution around the Notch(8 mm , 30/60 stacking sequence)
Applied load 26.6KN

IOP Conf. Series: Materials Science and Engineering 1189 (2021) 012004 doi:10.1088/1757-899X/1189/1/012004

Failure stress: FEA 150.5 MPa
Experimental 147 MPa

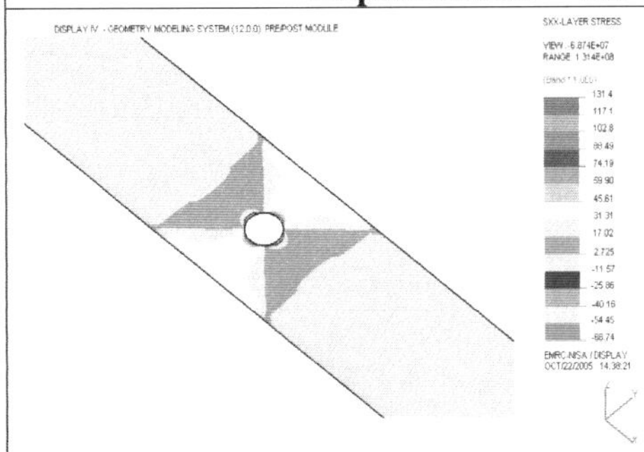

Figure 5: Stress distribution around the Notch (10 mm dia, 0/90 stacking sequence)
Applied load 20.7KN
Failure stress: FEA 110 MPa
Experimental 105 MPa

Failure stress: FEA 136 MPa
Experimental 135 MPa

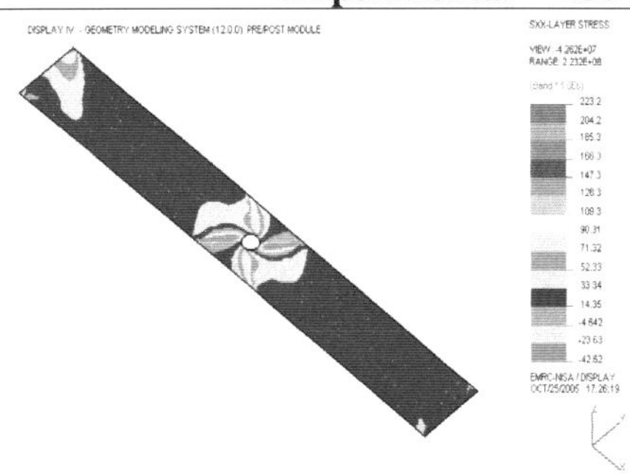

Figure 6: Stress distribution around the Notch(10 mm, 45/-45 stacking sequence)
Applied load 27.2KN
Failure stress: FEA 140 MPa
Experimental 138 MPa

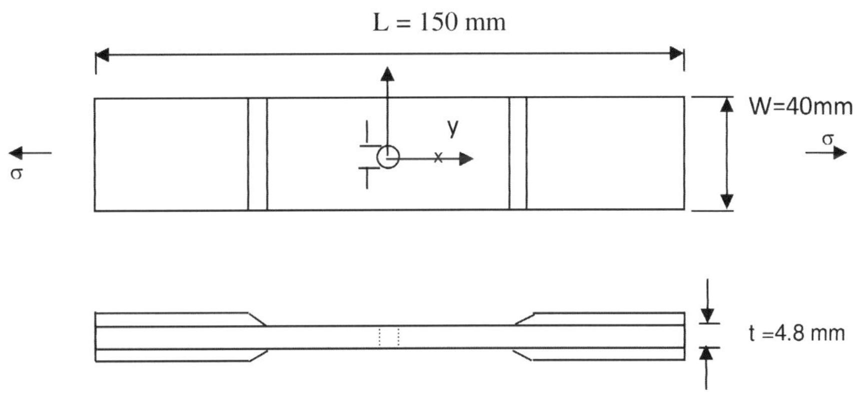

Figure 7.Tension test specimen

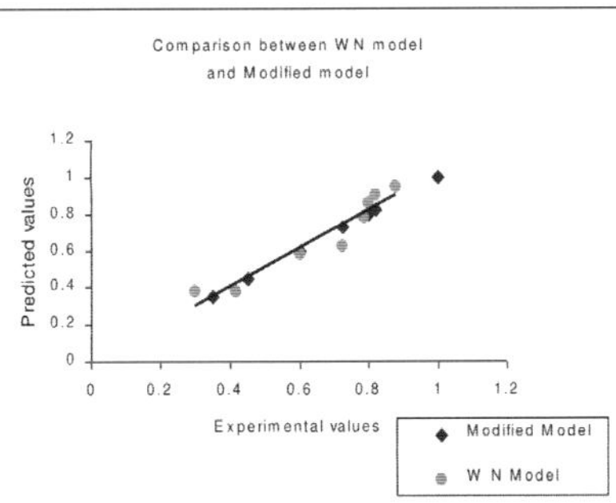

Figure 8. Comparisons between test results and predictions of PSC and modified PSC

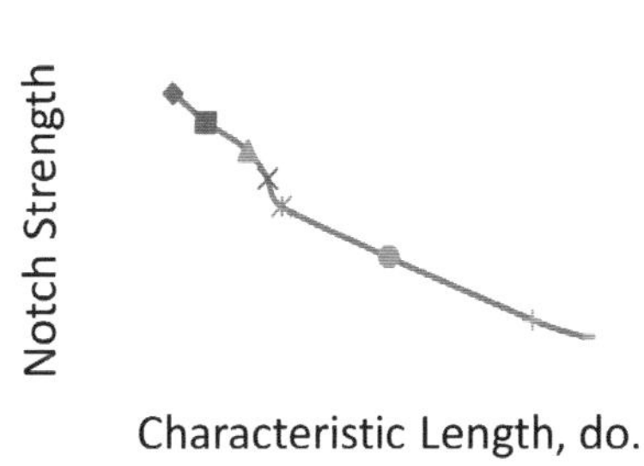

Figure 9. Relationship between notch strength and Characteristic length.

IOP Conf. Series: Materials Science and Engineering 1189 (2021) 012004 doi:10.1088/1757-899X/1189/1/012004

4. Conclusion

Based on the results of present study, the conclusions made are summarised below:

1. In the modified PSC the characteristic length (d_o) is denoted as a function of the geometry of the specimen (2R and W). It was found that the experimental results and the predictions of the modified PSC model tightly agree.

2. As the characteristic length (d_o) decreases an increase in (σ_N/σ_O) is observed.

3. There is a definite staking sequence dependence on the strength of laminates with holes.

4. The Von Mises stress value for laminate with hole is always higher than the laminate without hole.

5. The stress induced is maximum in 45/-45 stacking sequence and minimum in 30/60 stacking sequence.

5. References

1. N. Alif and L. Carlsson, "Failure Mechanisms of Woven Carbon and Glass Composites," in Composite Materials: Fatigue and Fracture (Sixth Volume), ed. E. Armanios (West Conshohocken, PA: ASTM International, (1997) 471-493.

2. Barbero, Ever J., P Lonetti and K. K. Sikkil. "Finite element continuum damage modeling of plain weave reinforced composites." Composites Part B: Engineering, vol.37, no.2, pp. 137-147, 2005.

3. Whitney JM, Nuismer RJ., Stress fracture criteria for laminated composites containing stress concentrations. J Compos Mater (1974) 8:253.

4. Nuismer RJ, Whitney JM., Uniaxial failure of composite laminates containing stress concentrations. Fracture mechanics of composites, ASTM STP 593(1975) P. 117.

5. Naik NK, Shembekar PS., Notch strength of fabric laminates: I. Prediction. Compos Sci Technol (1992) 44:pp1–12.

6. Kim JK, Kim DS, Takeda N., Notched strength and fracture criterion in fabric composite plates containing a circular hole. J Compos Mater (1995)

7. Waddoups, M.E., J.R.Eisenman and B.e.Kaminski., Macroscopic Fracture Mechanics of Advanced Composite Materials, Journal of Composite Materials(1971)Vol. 5, pp.446-454.

8. Pipes, R.B., J.W.Gillespie, Jr.and R.C. Wetherhold., Superposition of the Notched Strength of Composite Laminates." Polymer Engineering and Science(1979) Vol.19.pp.1151-1155.

9. Belmonte HMS, Manger CIC, Ogin SL, et al. Characterization and modelling of the notched tensile fracture of woven quasi-isotropic GFRP laminates. Comp Sci Technol (2001) 61: 585–597.

10. Toubal L, Karama M and Lorrain B. Stress concentration in a circular hole in composite plate. Compos Struct (2005) 68: 31–36.

11. Kazemahvazi K, Zenkert D and Burman M. Notch and strain rate sensitivity of non-crimp fabric composites. Comp Sci Technol (2009) 69: 793–800.

12. Hwan CL, Tsai KH, Chiu CH, et al. Strength prediction of braided composite plates with lateral holes by a finite element based point stress criterion. J Comp Mater (2013) 47(9): 1161–1168.

13. Reinoso, J., Arteiro, A., Paggi, M., & Camanho, P. P. "Strength prediction of notched thin ply laminates using finite fracture mechanics and the phase field approach." *Composites Science and Technology* 150 (2017): 205-216.

14. Fontes RS, Bezerra HA, Batista AC, Tinô SR, Aquino EM. "Failure theories and notch type effects on the mechanical properties of jute-glass hybrid composite laminates." Materials Research (2019)22(2).

15. G. P. Sendeckyj, Fracture Mechanics of Composites, Volume 593 of ASTM special technical publication:American Society for Testing and Materials.

16. Karlak, R. F. "Hole Effects in a Series of Symmetrical Laminates." In Proceeding of a Conference on Failure Modes in Composite (III), American Society for Metals, Chicago(1977).

IOP Conf. Series: Materials Science and Engineering 1189 (2021) 012005 doi:10.1088/1757-899X/1189/1/012005

Influence of Heat Treatment on Stress Intensity Factor of Nano Boron Carbide Hybrid Al2219 Composites with MoS_2

[1*]Sagar B Sansuddi, [2*]G S Shivashankar,[3]B Latha Shankar and [4]Dushyanth Kumar K R

[1]P.G Student
[2]Professor
*Department of Mechanical Engineering, Siddaganga Institute of Technology, Tumakuru – 572 103, Karnataka, India
[3]Associate Professor, Department of Industrial Engineering and Management Siddaganga Institute of Technology, Tumakuru – 572 103, Karnataka, India
[4]Associate Professor, Department of Mechanical Engineering,Malnad College of Engineering, Hassan,Karnataka,India

Keywords: Nano B_4C, MoS_2, Stir casting, T6 type heat treatment, Stress intensity factor

Abstract. The current work emphasizes on study of influence of heat treatment on Stress Intensity factor (K_{IC}) of Al alloy matrix hybrid composites. Hybrid composites with a constant Nano B_4C content of 3 weight % and MoS_2 of 4 weight % as reinforcements to Al2219 alloy matrix are fabricated by using two step stir casting technique and tested for Stress Intensity factor and fracture toughness as per ASTM standards. One set of composites are subjected to T6 type heat treatment. Compact Tension (CT) Specimen were prepared using wire EDM process and tested in BISS Axial Universal Testing Machine to find Stress Intensity factor (K_{IC}). From this study stability of crack propagation under mode-I failure is analyzed. Results show generous increment in energy required to open the crack and crack propagation is delayed in heat treated composites when compared to non-heat treated composites and furthermore improvement in fracture toughness of the heat treated specimen over non-heat treated specimen.

1. Introduction

Aluminium hybrid alloys are being the predominant material used in high temperature applications using nano boron carbide with molybdenum Disulphide as filler material. Greater part of engineering machine parts and structures are failed by fatigue and fracture cause issues [1]. Biggest challenge is to understand how materials fail and how crack begin and propagate, how it can be controlled and how to prevent such failures. Aluminium composites have most advantages in aircrafts, rockets and submarines. In aluminium metal matrix composites, ceramic materials are added as a reinforcing material which imparts better physical and mechanical properties which are impossible to get with monolithic alloys [2]. Usually researchers adopt molten metal stir casting technique or powder metallurgy for fabrication of ceramic-reinforced aluminium matrix composites [3]. Composite with B_4C content of 3 weight % and MoS_2 of 4 weight % as reinforcements to Al2219 alloy matrix yielded best results [4]. Al 2219 matrix composite is very suitable to age hardening irrespective of lower or

higher aging temperature. Clearly the composites show accelerated rate of aging kinetics as compared to unreinforced matrix alloy [5].
Sufficient investigation has not been accounted to come across the fracture toughness of heat treated nano B_4C hybrid composites. In the present work, the nano B_4C hybrid composite was fabricated by using two step stir casting technique. The aim of this study is to evaluate the influence of heat treatment on Stress intensity factor (Fracture toughness) of the nano B_4C hybrid composites. In this study, hard particle nano B_4C and dry solid lubricant MoS_2 are used as reinforcement for matrix material Al2219 to investigate the stress intensity factor of heat treated hybrid composites on Mode-I failure.

2. Synthesis of Hybrid Composite

2.1. Matrix and Reinforcement material
Table 1 shows the nominal composition of matrix material (Al 2219) and Table 2 shows the reinforcements used.

Table 1: Composition (Wt. %) of Al2219 alloy

Element	Mg	Si	Cu	Zr	Fe	Zn	Ti	V	Mn	Al
%by Wt	0.02	0.2	6.3	0.18	0.3	0.1	0.06	0.1	0.3	Base

Table 2: Composition (Wt. %) of Reinforcement

Reinforcement	Wt%
Nano B_4C	3 Wt%
MoS_2	4 Wt%

2.2. Stir casting technique
Stir casting technique is the Liquid Metallurgical Technique used to synthesis the hybrid aluminium composites with two step addition of reinforcement. Resistance furnace used for stir casting process is shown in figure 1. Matrix material Al2219 alloy was melted in graphite crucible, the mixture of Nano B_4C with MoS_2 were preheated at the temperature of 300^0C before adding to melt. In order to eliminate absorbed gases present in melt a tablet of degassing agent is introduced into the melt and K_2TiF_6 is added with reinforcement as wetting agent. At the melt temperature of 750^0C the preheated mixture of reinforcement were added into the Al2219 melt in step of two, because to avoid agglomeration. Zirconia coated stainless steel stirrer is used for stirring the melt, stirring is done for 5-6 min at 600 rpm spindle speed. The melt was poured into the preheated die and the obtained cast product is shown in figure 2.

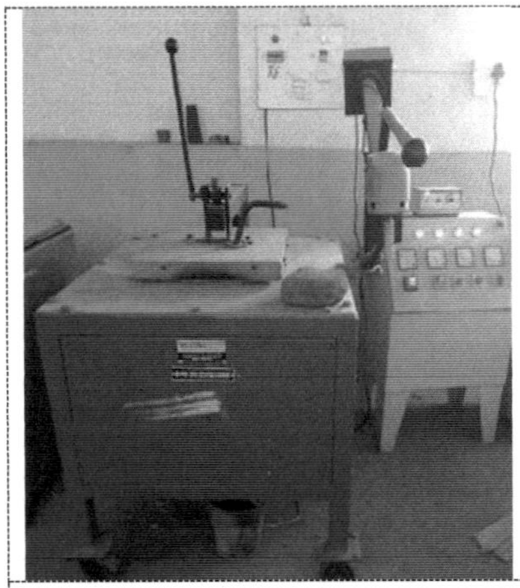

Figure 1. Resistance furnace

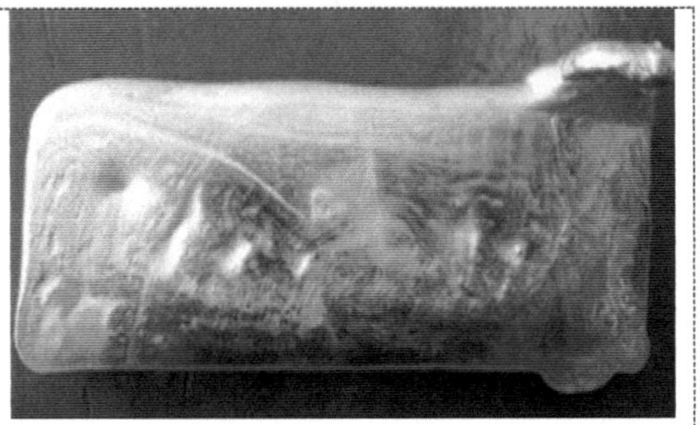

Figure 2. Prepared composite of Al2219-B_4C-MoS_2.

2.3. T6 Heat Treatment (Age hardening treatment)
Heat treatment of selected samples of the composites produced was carried out using Muffle furnace as shown in figure 3. The samples were first solution-treated in the furnace at 530°C for duration of 2 hrs, followed by water quenching. Then, ageing was carried out at 180 °C for a period of 6 hrs, followed by water quenching.

Figure 3. Muffle furnace used for heat treatment process

3. Experimentation

3.1. Specimen preparation
For fracture toughness (K_IC) test, Compact Tension (CT) specimen is prepared as dimensions are conformity to ASTM E-647 standard. The CT specimen prepared is shown in figure 4. A 7.6 mm of notch length is cut for pre-crack by using wire EDM shown in figure 5, so as to obtain a/w ratio of 0.3. Hole of 6.35 mm diameter is drilled by drilling process. Dimensions of CT specimen are shown in table 3.

Figure 4. CT specimen

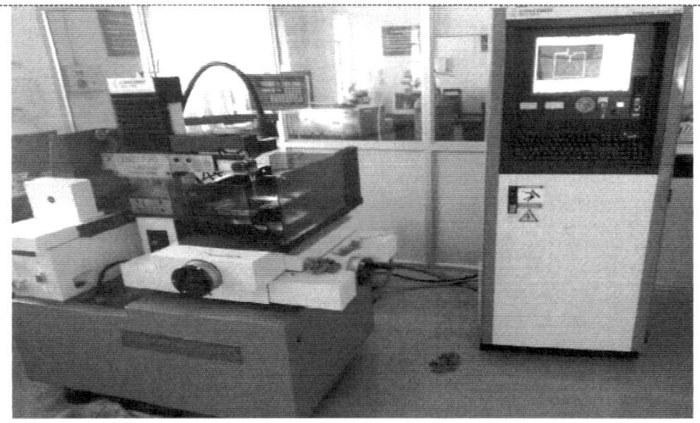

Figure 5. Wire EDM used for Notch Preparation

Table 3: Dimensions of CT specimen

Factor	Dimension(mm)
Total length	31.75
Total width	30.48
W	25.5
Thickness	6.35
Hole diameter	6.35
a(initial)	7.6

3.2. Experimental Procedure

The compact tension (CT) specimen is positioned in fixture in Axial universal testing machine as shown in figure 6. Fatigue crack growth according to ASTM-E-1820 standard following the equation (1) for fatigue loading.

$$P_f = \frac{0.4Bb_o^2\ \sigma_y}{(2W+a_o)} \tag{1}$$

$$\sigma_y = \frac{\sigma_{YS}+\sigma_{TS}}{2} \tag{2}$$

$$b_o = W - a_o \tag{3}$$

where a_o is initial crack length, W is the length from centre of hole grove to initial notch length, B is thickness of material. σ_y and b_o are obtained as shown in equation (2) and (3). All the inputs are given at the beginning of test such as standard dimension of specimen, properties of the material and fatigue pre-cracking load (P_f) is applied to produce pre-crack, then from this point material is loaded for K_IC test condition i.e., tension fatigue load is applied until the fracture occurs as shown in figure 7 and all the results and load v/s COD graph is formed and same procedure is repeated for different specimen.

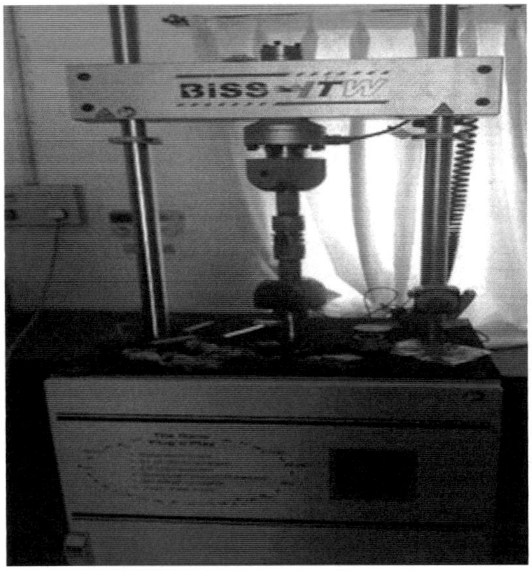

Figure 6. Axial Universal testing machine

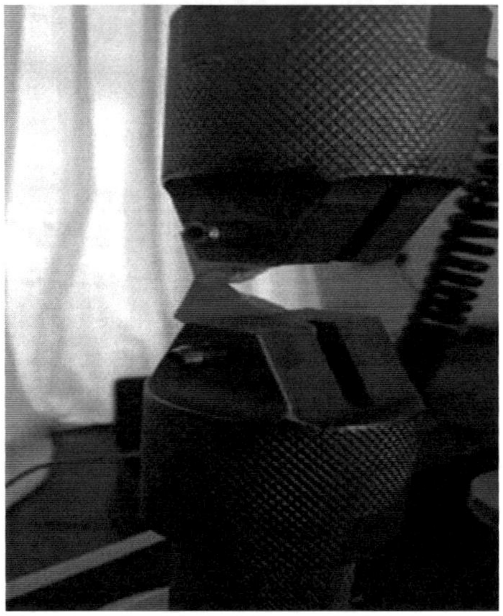

Figure 7. Fracture of specimen under fatigue loading condition

4. Results and discussion

The Load v/s COD graphs clearly show the load required to open the crack. In figure 8(a), non-heat treated Al2219 as cast requires 2.01 KN of load to open up the crack. From figure 8(b), the heat treated Al2219 requires little more load to open up the crack than non-heat treated alloy. From figure 8(c), non-heat treated composite of Al2219-3Wt% B_4C and MoS_2 of 4Wt% requires more load than as cast alloy due to the addition of B_4C and MoS_2 to Al2219 alloy. In the figure 8(d) heat treated composite of Al2219-3Wt% B_4C and MoS_2 of 4Wt% requires higher load than non-heat treated composite. Thus heat treated composite specimen requires higher load to open up the crack compared to other specimen prepared. Comparison of Stress Intensity Factor of various samples indicated in the figure 9.The results clearly indicates that heat treated hybrid alloy possess higher stress intensity factor.

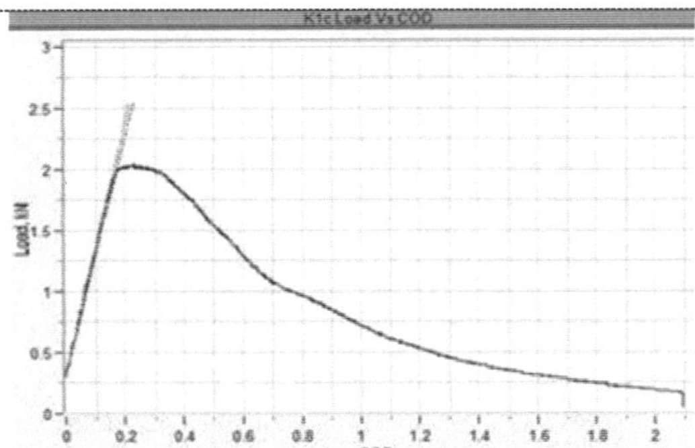

Figure 8 (a) Al2219

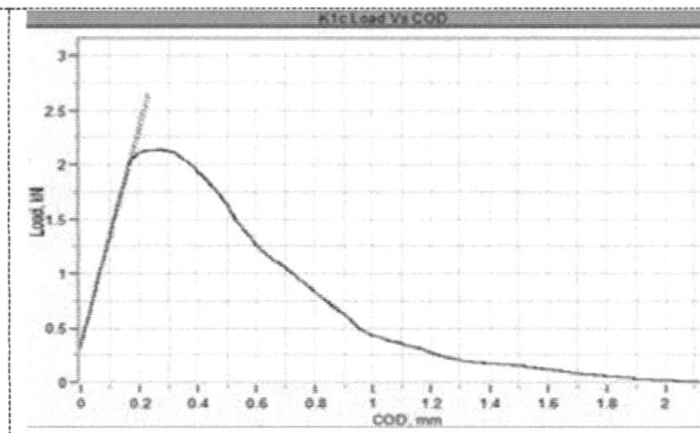

Figure 8 (b) Al2219-heat treated

IOP Conf. Series: Materials Science and Engineering 1189 (2021) 012005 doi:10.1088/1757-899X/1189/1/012005

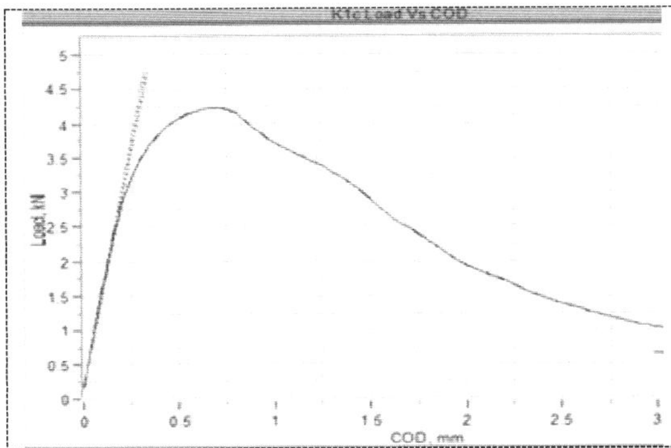

Figure 8 (c) Al2219-3%B_4C-4%MoS_2

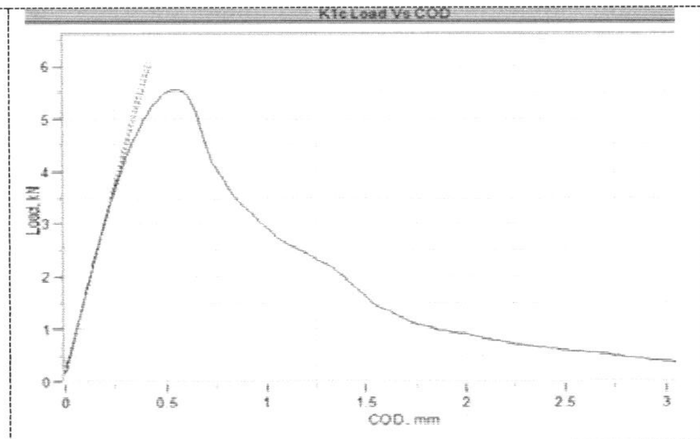

Figure 8 (d) Al2219-heat treated-3%B_4C-4%MoS_2–Heat treated

Table 4: Validation of K_{IC}

Material	P_{max}/P_Q	K_{IC}(MPa $m^{0.5}$)	%increase in K_{IC}
Al2219	1.051	21.62	-
Al2219-Heat treated	1.059	23.56	8.9
Al2219-3%B_4C-4%MoS_2	1.073	26.01	20.3
Al2219-3%B_4C-4%MoS_2–Heat treated	1.093	28.12	30

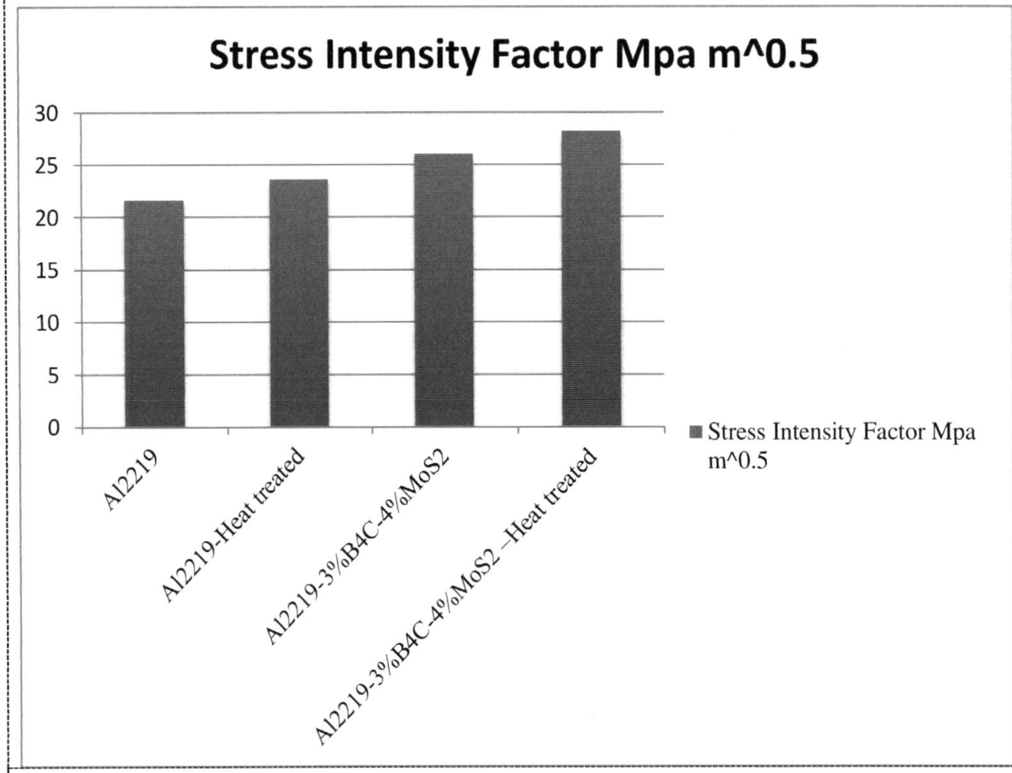

Figure 9. Comparison of Stress Intensity Factor of samples

The addition of reinforcement to hybrid composite reduces the inter-particle arrangements and thus creating obstacle to the movement of dislocation and thus addition of B_4C and MoS_2 to aluminium matrix creates higher resistance to open up the crack front thereby increasing the load bearing capacity and also increase in stress intensity factor. Further the load bearing capacity and stress intensity factor of heat treated cast alloy and composites is much more than the non-heat treated as shown in table 4.

5. Conclusion

This work makes an attempt to find the Stress Intensity Factor of heat treated and non-heat treated Nano Boron Carbide Hybrid Composites under mode I failure. The composites were prepared with uniform distribution of B_4C and MoS_2 which resulted in increase in bearing load to open the crack front. When compared to as-cast alloy and non-heat treated composites, the heat treated composite Al2219- 3wt%B_4C- 4wt%MoS_2 has higher Stress intensity factor. Thus the heat treatment of composites resulted in better fracture toughness.

6. References

[1] N. G. Siddesh Kumar, G. S. Shiva Shankar , S. Basavarajappa "Some studies on mechanical and machining characteristics of Al2219/n-B4C/MoS2 nano-hybrid metal matrix composites" Elsevier Journal of Measurement 107 (2017) 1–11

[2] A.K.M.Asif Iqbal, Dewan Muhammad Nuruzzaman, Effect of the Reinforcement on the Mechanical Properties of Aluminium Matrix Composite: A Review by, International Journal of Applied Engineering Research ISSN 0973-4562 Volume 11, Number 21 (2016)

[3] Dipti Kanta Das, Purna Chandra Mishra, Saranjit Singh, Swati Pattanaik Fabrication and heat treatment of ceramic reinforced aluminium matrix composites – a review by, International Journal of Mechanical and Materials Engineering(2014), 1:6

[4] Niranjan D. B., G. S. Shiva Shankar and Latha B. Shankar -Study on fracture behaviour of Hybrid Aluminium Composite- MATEC Web of Conferences 144, 02012 (2018)
[5] Gowrishankar.M.C, Sathyashankara Sharma, B.K.Pavan, „Mechanical characterization of heat treated Al2219 hybrid compositesr, Journal of Mechanical Engineering and Sciences ISSN (Print): 2289-4659; e-ISSN: 2231-8380 Volume 13, Issue 1, pp. 4380-4389, March (2019)

IOP Conf. Series: Materials Science and Engineering 1189 (2021) 012006 doi:10.1088/1757-899X/1189/1/012006

Experimental study of cooling characteristics to analyse quenching phenomenon of non-heat treatable material with different quenchants

M C Yogesh Gowda[1*], H M Sathisha[2], L Bharath Raj [3], and T S Prasanna Kumar[4]

[1*, 2, 3] Department of Mechanical Engineering, The National Institute of Engineering, Mysuru, Karnataka-570008, India.
[4]Thermet solutions Pvt. Ltd Bangalore, Karnataka-560076, India
E-mail: [*1]yogeshgowdamc@gmail.com, [2] sathishahm67@gmail.com, [3]bharathraj@nie.ac.in, [4]tsp@thermetsolutions.com

Abstract: Quenching is one of the commonly used process to improve mechanical properties for any materials. The quality of quenching and efficiency are usually measured by using various quench probes and get the cooling curves. The main parameters which are affecting the cooling characteristics are, geometry, quenchant type and materials grade, etc. A mathematical model that can be employed to get heat flux and cooling curves for a different materials/cooling medium combination. Present model coupled with nonlinear unsteady heat transfer with phase change, which leads to cooling curves for different materials/quenchant. Experiments are conducted in the standard laboratory condition to determine heat flux and surface temperature at the mid-plane of a 10 mm diameter by 70 mm long specimen of Stainless Steel material. InverseSOLVER software is used to analyse the experimental data to obtain cooling curves. This specimen is cooled with different quenching medium, such as, water, vegetable oil and different grade of lubricating oils. The experimental results are analysed for different quenchants using Inverse solver. The experimental results indicated that cooling curves, cooling rate, and the surface heat flux vary with quenchants.

1. Introduction

Immersion quenching is commonly used in heat treatment processes for obtaining martensitic and bainitic steels. The probes made are of standard material example stainless steel various grades of stainless steel can be used the quenching properties vary according to the grade therefore a probe made of standard material is used in the process for all the quenchants. A cooling curve is the graph obtained by plotting the measured temperature of materials with respect to time at geometric centre of cylindrical specimen when it is subjected to quenching process.

Most of the researchers have done work on quenching process but they have studied the material properties there are very few works on the thermodynamic properties so this is our field of interest.

IOP Conf. Series: Materials Science and Engineering 1189 (2021) 012006 doi:10.1088/1757-899X/1189/1/012006

The estimation of cooling rate, hardness, microstructure distribution for steel grade material during quenching process is done with the help of reference quench probe [1]. The measuring equipment was specially developed in order to compute the cooling rate of different quenchants [2]. Standard reference quench probes are used to determine the quality of quenchant. The important parameters like steel grade, bath temperature, thickness are effects the cooling curves. Different phase change phenomena as a function of temperature were determined using JMatPro software [3]. The polymer solution, water brine solution, and Mineral oil are employed as quenchantants of different thermo-physical properties, The standard ISO /DIS 9550 quench probe is employed to compute the cooling rate [4]. The quenching performance was evaluating heat transfer coefficient variation at different times after it is subjected to to an accelerated thermal-oxidation aging test [5]. The microstructure of quenched specimens is studied by optical microscope [6]. The cooling rate of quenchant is analysed of commercial quench oil MT650 [7]. To determine time dependent heat flux of stainless steel, function specification method is used at unknown boundary. The present formulation did not consider the magnitude of latent heat due to austenite decomposition

The objective of this paper is to compute the heat flux variation of stainless steel, to analyze the cooling rate, to determine the convective heat transfer coefficient during Quenching processes and to understand the physics of boiling during Quenching process.

2. Material selection

Experiments are performed for stainless steel specimen and quenchants as water, vegetable oil, Castrol 4T and 2T oils at room temperature and their corresponding chemical composition, properties of material and Quenchants as shown in Table.1, 2 and 3 below.

Table 1. Chemical Composition of Stainless Steel (304)

Composition	C	Mn	Si	P	S	Cr	Mo	Ni	N
%	0.08	2.0	0.75	0.045	0.030	18.0±20.0	-	8.0-10.5	0.10

Table 2. Properties of Stainless Steel Material.

Density	Thermal Conductivity	Melting range
8.03 g/Cm^3	16.3 W/m-K	1399-1454^{0}C

Table 3. Properties of Quenchants

Water		Sunflower Oil		Castrol Activ 4T Oil		Castrol Go 2T Oil	
Density (kg m-^{3})	1000	Density (kg m-^{3})	917	Density at 15 ^{0}C(g/ml)	0.8913	Density at 15 ^{0}C(g/ml)	0.8737
Viscosity at 25°C (cP)	0.890	Viscosity at 40 °C (cP)	28.56	Kinematic Viscosity 40^{0}C(mm²/s)	122.80	Kinematic Viscosity 40^{0}C (mm²/s)	51.9
Thermal Conductivity at 22 ^{0} C W/m-K	0.606	Flash point (°C)	314	Flash Point(°C)	>185	Flash Point(°C)	119

3 Experimental procedure

The Unique quench probe was used for the experiment. Which includes Electric resistance Furnace, probe, K-Type thermocouple, Quenchant, Data acquisition system, quenchant vessel and the vessel holding stand as shown in Figure. 1. The stainless steel specimen of 10 mm diameter and 70 mm long and this specimen is heated up to 850^0 C in electric resistance. The K-type Thermocouple is fitted in the probe further connected to data acquisition system. The vessel contains the quenchant having volume of 2 litres when the specimen is heated to the required temperatures it is dipped into the quenchant and the temperature changes with respect to time are observed and recorded with the help of data acquisition system. Also different quenchants are used for the experiment such as, distilled water, vegetable oil, two varieties of engine oil.

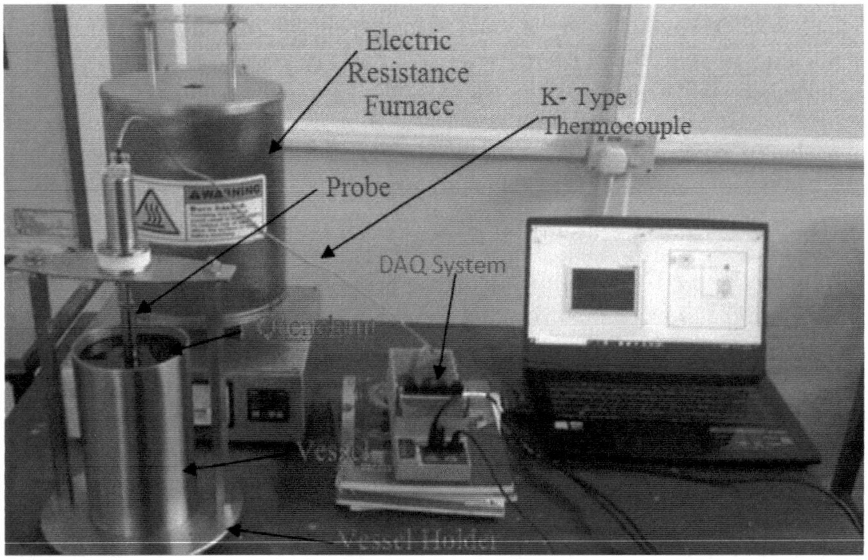

Figure.1 An experimental overview to measure the Temperature-time data during quenching

3.1 Inverse Heat Conduction Including Source Term

The boiling heat transfer phenomena during quenching process was analysed for cylindrical geometry using cylindrical coordinate system; assuming symmetry with respect to axis, heat flux at the boundary treated as an unknown parameter. The heat transfer variation in the specimen while quenching was analyzed by the following equations:

$$\frac{k}{r}\frac{\partial}{\partial r}\left(r\frac{\partial T(r,z,t)}{\partial r}\right) + k\frac{\partial T}{\partial z}\left(\frac{\partial T(r,z,t)}{\partial z}\right) + \dot{q} = \rho c\frac{\partial T(r,z,t)}{\partial t} \quad (1)$$

Initial conditions are, at t=0, T (r,z,t)=T $_{soak}$,

The boundary conditions are,

$$-k\left(\frac{\partial T}{\partial z}\right) = 0 \text{ at } S_{2,} \quad -k\left(\frac{\partial T}{\partial r}\right) = q(t) \text{ at } S_1; \quad T\ (r,z,t) = T_{max}\ (t) \text{ at } r=r_0;\ t= 0< t \leq t_{end.}$$

Where

C = Specific heat (J/kg K),K =Thermal conductivity (W/m K) ,r=Radial coordinate (m),S= Objective function (K_2), s= No. of thermocouples (-) , T = Temperature (K), t = Time (s) , z = Axial coordinate (m), ρ = Density (kg/m^3).

4. Results and discussion

4.1 Cooling Curve Analysis

The temperature variation is recorded by thermocouple, which is fixed at centre of the stainless steel specimen. These temperature variations with time data, recorded during quenching, are used as input to the Inverse Solver. From Figure. 2 it observed that the actual cooling path of the sample is decrease with temperature with respect to time. There are intersecting points at 295^0 C for water and vegetable oils, 525^0 C for water, lubricating 2T and 4T oil.

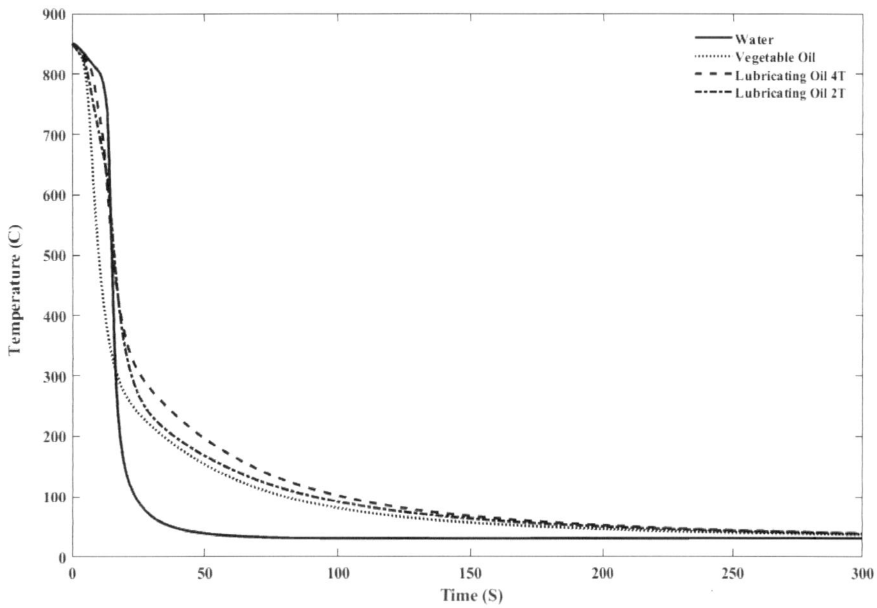

Figure.2 Cooling curves for different quenchants

4.2 Computed Surface Cooling Rate

The variation of cooling rate for Stainless steel material with different quenchants as shown in Figure.3 During heating condition, the material is completely austenitic having same thermal characteristics. Once austenite decomposition starts, different magnitude of latent heat will be generated in the specimen. The thermal properties also differ across the specimen cross section. The total effect of this is observed on the cooling rate curves as being distinct for the specimen below the Ae3 (Temperature at which α ferrite austenite can exits equilibrium) temperature. The maximum cooling rates variation for this specimen are at 170^0 C/s, 75 ^{0}C/s, 59^0 C/s, 55^0 C/s, at a temperature of 425^0C , 655^0C, 582^0C, 496^0C for distilled water, vegetable oil, lubricating 4T oil and 2T oil respectively.

4.3 Surface Heat Flux Curve Analysis

Figure 4 shows the surface heat flux was determined by using Equation 1. The stainless steel specimen is quenched in distilled water, vegetable oil, lubricating oil 4T and 2T. It is observed that the physics of the heat flux curves follows the same nature of the surface cooling rate curves in all the cases. The rate of heat energy removed from the solid is equal to the heat gain by the fluid during quenching, this is result of energy balance. Heat transfer within the solid specimen is purely by conduction, whereas heat transfer process between fluid and solid surface by convection. Here convection heat transfer process is extremely complicated due to two phase flow heat transfer. So there is boiling heat transfer which includes various regimes such as, nucleate boiling, vapor phase and the convection phase with time. This phenomenon of heat transfer depending on the surface temperature. The stainless steel material releases the heat with respect to time is highly complicated and non linear way this is because of phase transformation, also which is depending on cooling rate. The maximum heat fluxes are 19.05 KW/m^2 at 349.8 ^{0}C, 26.06 KW/m^2 at 519^{0}C, 17.11 KW/m^2 at 478.5^{0}C, 16.79 KW/m^2 at 469.2 ^{0}C for water, vegetable oil, lubricating 2T oil, and lubricating 4T oil. Among these the maximum heat flux occurs when stainless steel probe is quenched in vegetable oil during quenching.

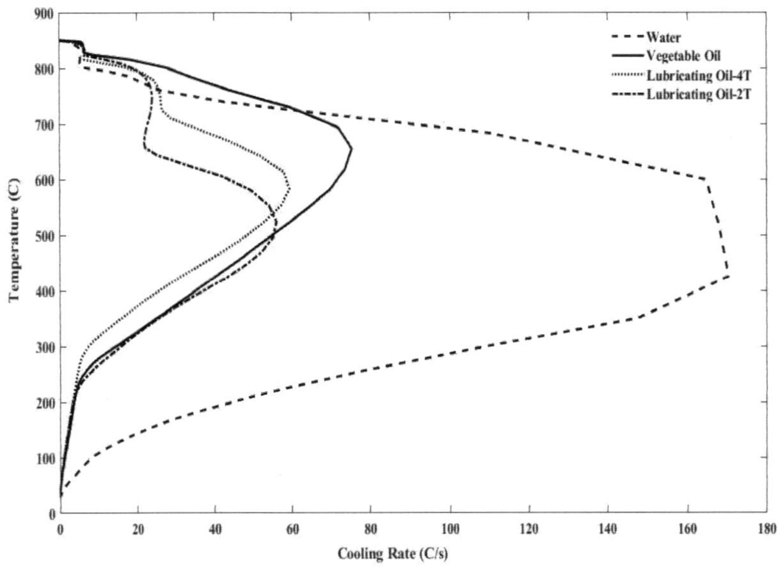

Figure.3 Cooling Rate for different quenchants

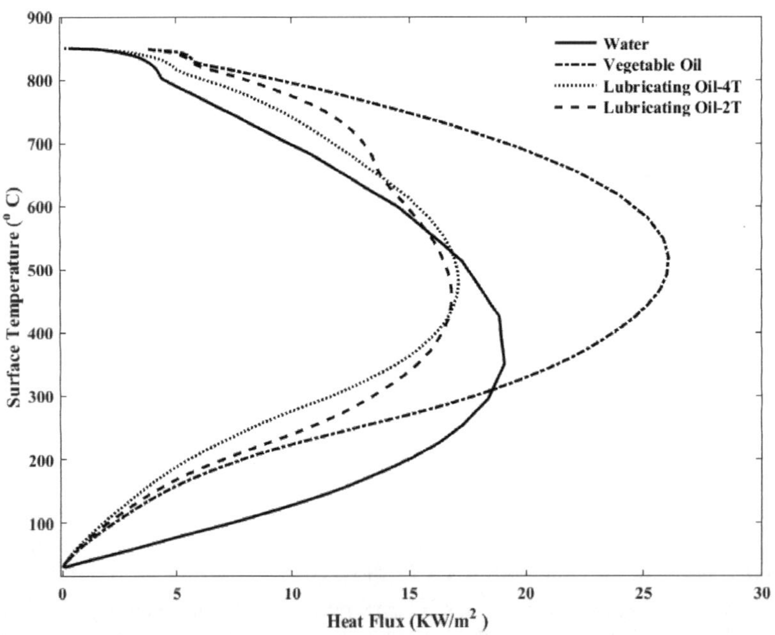

Figure.4 Heat flux variation for different quenchants.

5. Conclusions

Experiments are conducted to analyse cooling rate and heat flux during quenching of stainless steel specimen with various quenchants. Using this data with coupled inverse heat transfer solver is used to determine heat flux and surface cooling rates. The results show that the cooling curves depending on the quenchants. The cooling curves are cross over each other. This indicates that there is an identical heat transfer to be takes place. The magnitude of heat flux also varies significantly for different quenchants. The maximum cooling rates for the specimen 170 ^{0} C/s, 75 ^{0} C/s, 59 ^{0} C/s, 55^0 C/s, at a temperature of 425 ^{0}C , 655^{0}C, 582^{0}C, 496^{0}C for distilled water, vegetable oil, lubricating oil 4T and 2T oil respectively. At higher temperature there is a film boiling takes place. Due to this the heat flux is to be minimum. The convective heat transfer coefficient is low because the energy absorbed by the quenchants is slow. In nucleate boiling region there is increase in heat flux and convective heat transfer coefficient is increased due to decrease in temperature. Cooling rate decreases at the lower temperature the heat flux is variation of stainless steel materials decreased. The transfer through natural convection only.

Present work is able to analyse complicated boiling heat transfer phenomena for transient heat treatment process.

References

[1] Prasanna Kumar T. S, Hernandez-Morales B., and Totten, G. E.,2014, Reference QuenchProbe"—An Alternative Probe Design for In-Situ Estimation of Cooling Rates, Heat Flux, and Hardenability During Immersion Quenching of Hardenable Steels, *Materials Performance and Characterization* January (2015)

[2] Elmi Hosseini, S. R., Zabett, A., and Li, Zhuguo 2014, Cooling Curve Analysis Of Heat Treating Oils And Correlation With Harden And Micro-structure Of Low Carbon Steel 2014, *Materials Performance and Characterization*, Vol. **3,** No. 4, 2014, pp. 427–445.

[3] T.S. Prasanna Kumar, 2013, Influence of Steel Grade on Surface Cooling Rates and Heat Flux during Quenching, Journal of Materials Engineering and Performance. *ASM International* JMPEG (2013) 22:1848-1854

[4] G. Ramesh and K. Narayan Prabhu, 2014, Effect of thermal conductivity and viscosity on cooling performance of liquid quench media, *International Heat Treatment and Surface Engineering*, 8:1, 24-28.

[5] Ester Carvalhode Souza, Lauralice C.F. Canale, G.Sa´ nchez Sarmiento, Eliana Agaliotis, Juan C. Carrara, Diego S. Schicchi, and George E. Totten,2013, Heat Transfer Properties of a Series of Oxidized and Unoxidized Vegetable Oils in Comparison with Petroleum Oil-Based Quenchants, *Journal of Materials Engineering and Performance* Volume **22(7)** July 2013—1871.

[6] J. K. Odusote, T. K. Ajiboye and A. B. Rabiu, 2015, Effect of Vegetable oil Quennchants On The Properties Of Aluminium During Solution Heat Treatment. *Journal of Mechanical Engineering and Sciences* (JMES); Volume **8,** pp. 1343-1350, June 2015 ISSN (Print): 2289-4659; e-ISSN: 2231-8380..

[7] AtulShelar, Kiran Wale, Manoj Muchalambe and SurajPattekar, 2019, Study of Mechanism of Quenching & Calculate the Cooling Rate of Quenching Oils by Quenching Instrument. *International Research Journal of Engineering and Technology* (IRJET) volume **05** May (2015) -ISSN: 2395-0056.

Acknowledgment

We are thankful to author T S Prasanna Kumar Thermet solutions Bangalore, Karnataka, India, for guidance and support is help us coordinate the project

IOP Conf. Series: Materials Science and Engineering 1189 (2021) 012007 doi:10.1088/1757-899X/1189/1/012007

Wear Behavior of Aluminum-TiB_2 Metal Matrix Composites using Taguchi Method.

Dr.Manjunatha H.S[1], Satish V T[2], Dr.G.Mallesh[3]
*[1]Assistant Professor, Department of Mechanical Engineering, ATMECE, Mysore, Karnataka, India
*[2] Assistant Professor, Department of Mechanical Engineering, MCE, Hassan, Karnataka, India
*[3] Professor, Department of Mechanical Engineering, SJCE, Mysore, Karnataka, India

E-mail: manjuhs.hsn@gmail.com

Abstract. In the present day engineering design and growth activities many Engineers and Scientists are striving hard to develop novel and better engineering materials, which accomplishes high strength, light weight and energy efficient materials since the problems of environment and energy are major threshold areas. In this paper attempts are made to study the abrasive wear loss and coefficient of friction of the Al 6061 reinforced with varying weight percentage of TiB_2 composites by in situ process was investigated. The wear test was conducted using pin-on-disc wear testing apparatus (POD) as per the design of experiments (DOE).Taguchi Technique, Analysis of Variance (ANOVA) and Regression Analysis were adopted to optimize the wear loss. Further, Scanning Electron Microscopy images of worn out surface are carried to study the wear behaviour. Wear loss and Co-efficient of friction decreases with increase in the wt. % of TiB_2 particles and noticed form the ANOVA and regression analyses that reinforcement plays a crucial role in the development of Al 6061-TiB_2 composite system.

1. Introduction

An aluminium matrix composite plays very important role in the present day engineering applications. Because conventional materials have many limitations in achieving required strength, stiffness, toughness, corrosion and wear resistance. The development of new materials is growing day by day to replace the conventional materials in the field of automobile, marine engineering, aerospace etc., Hence, composite material are found to be an alternative materials. Fabrication of metal matrix composites has several problems like poor wettability between matrix and reinforcement, porosity and non homogeneous distribution of reinforcement.

Aluminium 6061 is the mostly used 6xxx series which offer good mechanical, corrosion and wear resistance. Among various reinforcements titanium diboride is an exceptional ceramic material due to its high strength, hardness, superior wear resistance and good thermal stability and more prominently it does not react with molten aluminium [1, 2]. The aluminium 6061 embedded with different ceramic particles can be synthesized using various manufacturing methods like stir casting, in situation stir casting process, powder metallurgy, squeeze casting etc.,[3]. In situ stir casting route has been found to be one of the most important technique in the synthesis of aluminium reinforced with TiB_2 due to the fact that this process is thermodynamically stable, uniform distribution of reinforcement in the

IOP Conf. Series: Materials Science and Engineering 1189 (2021) 012007 doi:10.1088/1757-899X/1189/1/012007

matrix and economical. Composites materials are developed using an exothermic reaction route with addition $K2TiF_6$ and KBF_4 salts [3, 4]

Wear is a specific property of substance which is denoted by removal or loss of the material from surface due to relative motion between a surface and the contacting substance. Wear test can be performed in dry as well as in lubricating conditions.

In the present paper dry sliding wear test has been carried out on Al 6061 reinforced with 3 to 12 wt. % of TiB_2 composites by using Pin-on-Disk wear test apparatus with normal loads of 10-50 N, sliding distance 500 m to 2500 m in the step of 500 m and sliding velocity 1 m/s to 5 m/s to know the influence of applied load, reinforcement, sliding speed and sliding distance on wear rate. The worn out surface morphology was studied using Scanning Electron Microscope.

2. Optimization technique

Optimization is a method for maximize or minimize a real function by choosing the input functions. In general, optimization is a process of finding the "best available" value among the variety of objective functions and domains. Design of Experiments (DOE), Taguchi Technique, Analysis of Variance (ANOVA) and Regression Analysis are adopted in this paper [6]

3. Experimental Procedure

Aluminium 6061 reinforced with TiB_2 by varying 3 to 12 Wt % are used to develop composite materials to study the wear characteristics of the composites synthesized using In-situ technique. In this process, the reinforcement phase is formed in situation, by adding halide salts to form required combination of composites.

Table 1. Chemical Composition of Al 6061 alloy

Element	Mg	Fe	Si	Cu	Mn	V	Ti	Al
Weight %	1.08	0.17	0.63	0.32	0.52	0.01	0.02	Balance

Al 6061 rods were weighed and placed in a coated graphite crucible and heated by means of an electrical furnace up to 850°C. The chemical composition of Al 6061 aluminium alloy is presented in Table 1. A coating was applied inside the crucible to avoid contamination. The temperature of the molten aluminium was maintained at 850^0C. The measured quantities of inorganic salts like K_2TiF_6 and KBF_4 were added into the molten aluminium and thoroughly stirred using a graphite stirrer at intervals of every 10 minutes. A total reaction holding time for about 45 min to 1 hr to form titanium diboride and then 1.5 Wt % of magnesium was added and stirring was continued so that the TiB_2 and Mg particles are completely dispersed in to the molten alloy [15, 16, and 18]. After solidification process castings were machined as per ASTM standards.

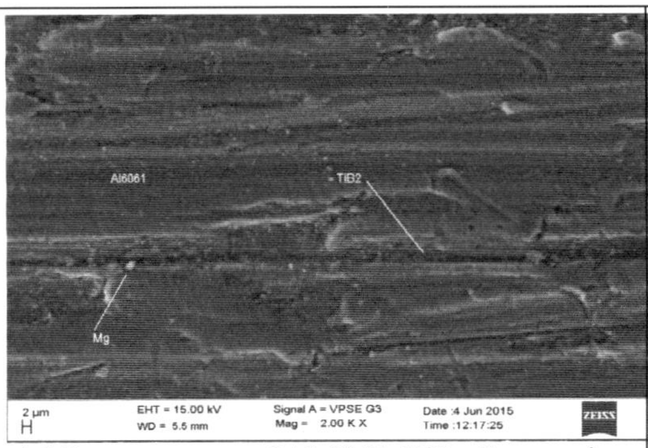

Fig 1. Microstructure of Al 6061-TiB_2 Composites

Fig 2. POD Apparatus

IOP Conf. Series: Materials Science and Engineering 1189 (2021) 012007 doi:10.1088/1757-899X/1189/1/012007

The dry sliding wear test was carried out using pin-on disc apparatus as shown in Fig.2. The wear test was conducted as per ASTM G-99 standards. The dimensions of test specimens are 10mm×10mm×50mm. Results obtained from the Pin-on-disc experiments are used to conduct statistical using MINITAB software. After conducting the tests, the worn out surfaces were observed by scanning electron microscope to examine the wear behaviour of the composite materials.

Plan of Experiments

The experimental plan is to formulate by considering parameters (factors) and levels based on the Taguchi technique. The levels of variables chosen for experiments are planned in Table 2.

Table 2. Parameters and their levels

Level	Load(N)	SV (m/s)	Sliding Distance (m)	Wt% of TiB_2
1	10	0.5	300	0
2	20	1	600	3
3	30	1.5	900	6
4	40	2	1200	9
5	50	2.5	1500	12

Dry sliding wear test was conducted by varying load, sliding velocity, sliding distance and wt. % of TiB_2 at five levels. According to the rule, the degree of freedom for an orthogonal array should be greater than or equal to sum of wear parameters. In this study L25 Orthogonal array was selected.
The Signal to Noise (S/N) ratio, was used to condense numerous data points within a trial to establish different characteristics. In this process "Smaller the Better" characteristic was selected to study the dry sliding wear resistance. The S/N ratio of wear rate through "Smaller the Better" characteristic is given by the equation as follows

$$S/N = -10 \log [1/n (\Sigma y2)] \quad (1)$$

Where, n = number of observations and y = observed data.

4. Results and Discussion

Wear rate analysis

The experiments were conducted by varying the Reinforcement percentage, applied Load, Sliding velocity and Sliding distance to determine wear loss and coefficient of friction. The objective of this study is to find out the important factors influence on wear process to achieve the minimum wear rate.

Table 3. Results of L25 array for Al 6061/TiB_2 composites

Sl.No	Wt % of R	Load (N)	SV (m/s)	SD (m)	Wear Loss (gm)
1	0	10	0.5	300	0.0093
2	0	20	1	600	0.0111

3	0	30	1.5	900	0.0127
4	0	40	2	1200	0.0140
5	0	50	2.5	1500	0.0151
6	3	10	1	900	0.0082
7	3	20	1.5	1200	0.0086
8	3	30	2	1500	0.0087
9	3	40	2.5	300	0.0089
10	3	50	0.5	600	0.0091
11	6	10	1.5	1500	0.007
12	6	20	2	300	0.0071
13	6	30	2.5	600	0.0074
14	6	40	0.5	900	0.0078
15	6	50	1	1200	0.0081
16	9	10	2	600	0.0061
17	9	20	2.5	900	0.0064
18	9	30	0.5	1200	0.0068
19	9	40	1	1500	0.007
20	9	50	1.5	300	0.0071
21	12	10	2.5	1200	0.0048
22	12	20	0.5	1500	0.0051
23	12	30	1	300	0.0053
24	12	40	1.5	600	0.0055
25	12	50	2	900	0.0057

The Influence of process parameters such as applied load, percentage reinforcement, sliding speed and sliding distance on wear has been analyzed. The ranking of process parameter by signal to noise ratios obtained for different parameter levels for wear are given in the table 4 and table 5.

Table 4. Response table for means for wear

Level	Wt % Reinforcement	Load	SV	SD
1	0.01244	0.00622	0.00762	0.00754
2	0.0087	0.00766	0.00794	0.00784
3	0.00748	0.00818	0.00818	0.00816
4	0.00668	0.00864	0.00832	0.0076

5	0.00442	0.00902	0.00766	0.00858
Delta	0.00802	0.0028	0.0007	0.00104
Rank	**1**	**2**	**4**	**3**

Table 5. Response table for S/N ratio for wear

Level	Wt % Reinforcement	Load	SV	SD
1	38.23	47.22	42.56	42.62
2	41.21	42.62	42.25	42.4
3	42.54	42.12	42.1	42.11
4	43.52	41.71	42.09	45.99
5	49.56	41.39	46.06	41.93
Delta	11.33	5.84	3.97	4.05
Rank	**1**	**2**	**4**	**3**

It is observed from the above table 4 and 5 the control factors are statistically significant in the SNR for wear rate. Percentage reinforcement is the dominant factor on the wear rate followed by load applied, sliding distance and sliding velocity.

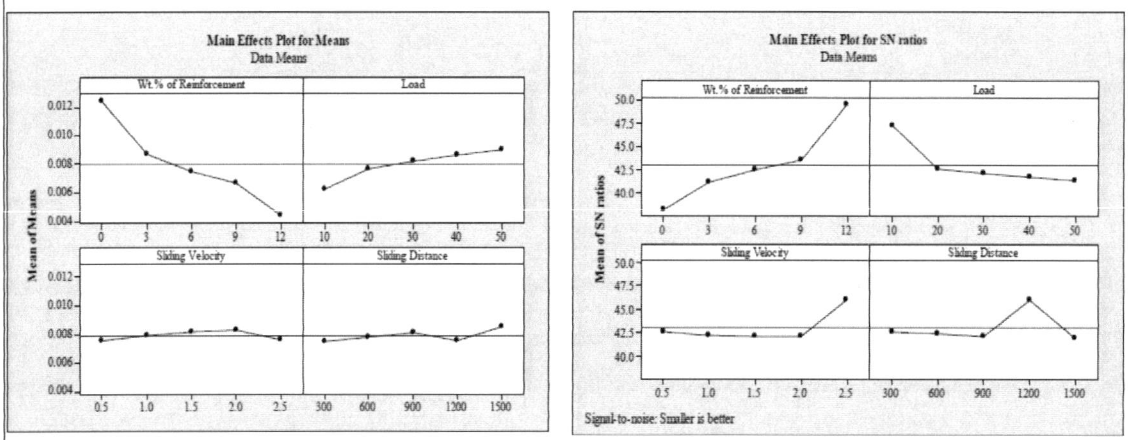

Fig 3. Main effects for plot for Al 6061-TiB_2 composites (a) Means Wear Rate (b) SN Ratio –Wear Rate

The main effect plot for means, Signal to noise ratio for the wear effect plots between the process parameters are shown in Figure.3. The significance of each factor is found out from the inclination of the main effect plot. A parameter for which the line has the maximum inclination will have the most significance. It is clear from the main effect plot that percentage reinforcement is the most significant wear parameter while applied load, sliding distance and sliding speed also has considerable effect [10]

ANALYSIS OF VARIANCE (ANOVA)

Table 6. ANOVA Results

Source	Df	Seq.SS	Adj.SS	Adj.MS	F	P	Pr.%
Wt% Reinforcement	4	347.5	347.5	86.88	5.07	0.025	48.57
Load	4	115.17	115.17	28.79	1.68	0.247	16.09
SV	4	58.79	58.79	14.7	0.86	0.528	8.21
SD	4	56.76	56.76	14.19	0.83	0.543	7.93
Error	8	137.19	137.19	17.15			19.17
Total	24	715.4					100

To find out the significance of each process parameters and their relations on the system response, an efficient statistical tool ANOVA is used. In this study, analysis of variance is applied to assess the significance parameters and their interactions on the wear behaviour of Al 6061-TiB_2 composite materials. Table 6 shows the ANOVA results with their wear performance of the composite materials developed. Percentage of contribution (Pr. %) of Wt. % of Reinforcement (48.57%), load (16.09%), sliding speed (8.21%) and sliding distance (7.93 %) are the main controlling factors on the wear behaviour.[7]

5. Multiple linear regression equation

The multiple linear regression equation is developed using MINITAB software. This developed equation gives the relation between predicted and independent variables using linear equation to the measured data.

Wear rate = 0.00889-0.000602 Wt. % of Reinforcement +0.000066 Load + 0.000090 Sliding Velocity +0.000001SlidingDistance. (2)

6. Confirmation Experimentation

Finally, confirmation test was conducted for the composite materials to check the accuracy of the equation developed with predefined experimental parameters as shown in table 7 and table 8.

Figure 4 shows the microstructure of the worn out surfaces of the Al 6061-TiB_2 composite materials. Grooves were formed by the reinforcing particles of titanium diboride.

Table 7. Confirmation Test values

Trails	% Reinforcement	Load	SV	SD
1	9	5	0.25	250
2	9	15	0.75	500

Table: 8 Comparisons of Confirmation test results

Trails	Experimental wear rate.	Regression wear rate.	% Error
1	0.0041	0.00407	0.60
2	0.00493	0.00503	2.02

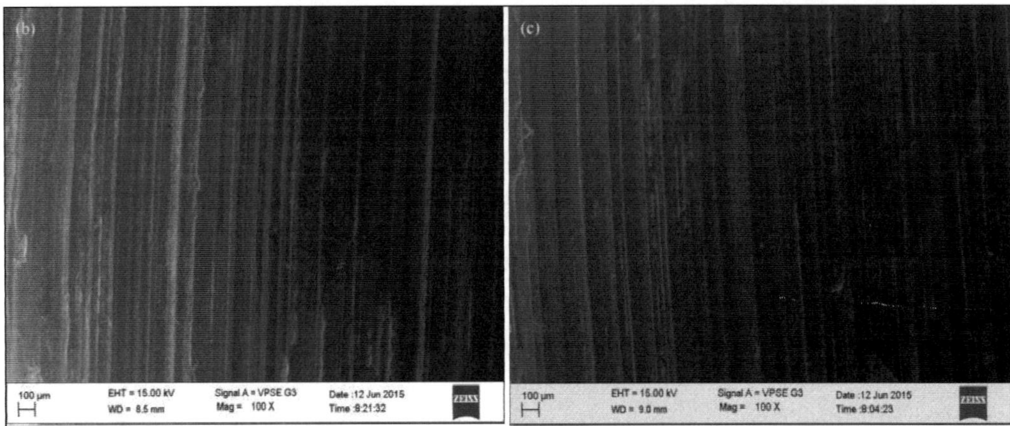

Fig: 4 SEM Images of Worn out Surfaces

7. Conclusions

- Wear performance of Al6061-TiB_2 composites were conducted using Pin-On-Disc equipment as per the design of experiments using L25 orthogonal array.
- Wear loss decreases with increase in wt % of Reinforcement.
- Reinforcement is the highest influencing factor on wear loss followed by load, sliding distance and speed.
- Linear Regression equation was developed which is good agreement with confirmation wear test results SEM images shows uniform tracks in the direction of rotation of the disc which clearly indicates the uniform distribution of the particles in the matrix.

8. References

[1] Suresh, S., N. Shenbaga Vinayaga Moorthi, S. C. Vettivel, and N. Selvakumar. "Mechanical behavior and wear prediction of stir cast Al–TiB_2 composites using response surface methodology." Materials & Design 59 (2014): 383-396.

[2] Ramesh, C.S., S.Pramod, and R.Keshavamurthy."A study on microstructure and mechanical properties of Al 6061–TiB_2 in-situ composites." Materials Science and Engineering: A 528, no. 12 (2011): 4125-4132.

[3] Mondal, D.P., J.H.A.Nidhi, Anshul Badkul, and S. Das. "Effect of Al–TiB master alloy addition on microstructure, wear and compressive deformation behavior of aluminum alloys." Transactions of Nonferrous Metals Society of China 22, no. 5 (2012): 1001-1011.

[4] Rajan, HB Michael, S. Ramabalan, I. Dinaharan, and S. J. Vijay. "Synthesis and characterization of in situ formed titanium diboride particulate reinforced AA7075 aluminum alloy cast composites." Materials & Design 44 (2013): 438-445.

[5] Aluminium Composites: Review and Discussion, Wear, 189, pp. 1-19.

[6] Taguchi G. 1990. Introduction to quality engineering, Asian productivity organization.

[7] Basvarajappa, S., Chandramohan, G., 2005. Wear Studies on metal matrix composites: a Taguchi Approach. Journal of Material Science Technology 21 (6), 845-850

[8] Surappa, M. K., 2003, Aluminium Matrix composites: challenges & opportunities. In: Sadhana 28 (1-2), 319-334.

[9] Chawla N and Chawla K 2006 *Metal Matrix Composites* Springer New York

[10] Niranjan, K., and P. R. Lakshminarayanan. "Dry sliding wear behaviour of in situ Al–TiB_2 composites." Materials & Design 47 (2013): 167-173

Investigation on the Influence of Nano Filler on Tribological Behaviour of Glass-Nylon fiber reinforced epoxy based Hybrid Composites

Dilip Kumar K[1], Shantharaja M[2], Santhosh G[3]

[1]Research Scholar, Dept. of Mechanical Engineering, UVCE, Bangalore University, Bangalore, Karnataka, India – 560001.

[1, 3]Assistant Professor, Dept. of Mechanical Engineering, NMAM Institute of Technology, Nitte, Karnataka, India -574110/VTU belgavi Karnataka, India -590018.

[2]Associate Professor, Dept. of Mechanical Engineering, UVCE, Bangalore University, Bangalore, Karnataka, India – 560001.

E-mail: dilk333@nitte.edu.in

Abstract. From the last few decades in engineering applications the polymer matrix composites (PMC's) were used over conventional engineering materials due to their improved properties and enhanced performance. The researchers are encouraged to perform comprehensive research work on such materials. The tribological behaviour of a glass-nylon fiber reinforced epoxy based hybrid composite material filled with alumina nanoparticles in three different compositions of 0.5%, 1.0 % and 2.0% have been investigated thoroughly using different load and speed conditions. Effect of Alumina nano particles on tribological properties like wear rate, wear loss and coefficient of friction on the Glass-Nylon fiber epoxy based hybrid composite has been studied. The experimental results have showed that hybrid composite with an increase in the percentage of nano filler enhances the tribological property up to 1.0% of nano filler and then shows a declining nature.

Keywords: Hybrid Polymer matrix composites, Nano Fillers, Alumina, Wear.

1. Introduction

Special materials are in great demand for innovative applications in engineering, leading to new inventions in the field of composites. A new trend in engineering has begun to optimize individual properties in order to understand a particular material. The current importance in research area is more on thermoset composites as it is superior in terms of engineering properties and applications. Due to ease in manufacturing and its properties like higher strength and durability, composite materials are gaining importance [1, 2]. Polymers are transformed to achieve high strength to weight ratio and low density, different types of fillers and fibers are reinforced into it [3, 4]. Moreover manufacturing industries, including aerospace, automobile, ships, infrastructure, and petroleum are focussing more on fabric-reinforced polymeric composites [5-7].

The improvement in the performance of composites is observed due to micro or nano filler inclusion in polymer matrix. The reinforcement of filler is not only to reduce the cost of the material but also to

IOP Conf. Series: Materials Science and Engineering 1189 (2021) 012008 doi:10.1088/1757-899X/1189/1/012008

enhance the mechanical, thermal and tribological properties [8]. Alumina is recognised as an important ceramic material used in the fields of aerospace, automotive, packaging industry etc., as it has high strength, excellent corrosion resistance, low thermal conductivity, and better electrical insulation properties [9]. The "pin on disk machine" has been used to study the wear and frictional properties of epoxy based polymer composites under various load and speed conditions [10-12]. An effort is made to enhance the wear characteristics of a hybrid thermoset composite by reinforcing the nano alumina filler under variable parameters [13, 14].

2. Materials and Preparation of Hybrid Samples

The α-Al2O3 nanopowders were purchased from Platonic Nanotech Private Limited, Jharkhand with an average particle size of 30 - 50nm. The type of epoxy resin used in the present investigation is Araldite LY-556 together with HY 951 hardner. The wear test was conducted on three types of samples with the variation of nano Al2O3 percentage i.e 0.5%, 1.0% and 2.0% and the Glass fiber and Nylon fiber of 15% each as per the results from tensile, flexural and modulus test results. The effect of 0.25 % nano Al2O3 on tensile, flexural and modulus test results found to be comparatively lower; hence only 0.5 % to 2.0 % variation is considered for wear test [15]. To avoid the adhesion of composite to mold, a release coating is applied on the bottom and inner mold surface. The prepared sample is kept to cure at ambient temperature for one day. Later, the post-curing has been done at a temperature of 100 ± 5°C for 2 hours in a hot air circulated furnace.

3. Experimental Procedure

Two-body abrasive wear test was conducted using a Pin on Disk Apparatus as per ASTM G99, Test Method [16, 17]. The sample of 10mm×10mm×4mm was attached to a holding pin at one surface, which was fabricated manually. Before the test the sample surface was cleaned using a soft paper that is soaked in acetone. Abrasive paper of 320 grit size was pasted on rotating disc. The composite specimens were abraded against a metal disc at a load of 20 N, 40 N and 60 N for 10 minutes at speeds of 1200 RPM, 1500 RPM, and 1800 RPM on a track of 100 mm diameter. Wear volume (ΔV) and specific wear rate (Ks) were calculated using the following equations Eq.1 and Eq.2 :

$$\Delta V = \frac{\Delta m}{\rho} \qquad (mm^3) \tag{1}$$

$$K_S = \frac{\Delta V}{Ld} \quad (mm^3/N\text{-}m) \tag{2}$$

Where, Δm is the loss in mass gms, ρ is the density of the test material in g/mm3, ΔV is the volume loss in mm3, L is the load in N, and d is the sliding distance in meters. The sample density was calculated by measuring its mass individually and dividing the volume.

4. Results and Discussion

The tribological study is done using three parameters, COF, Wear loss, and specific wear rate for various nano filler percentage, different speeds and load conditions. The hybrid nano composites with various nano filler percentage (0.5%, 1.0%, and 2.0%) at different sliding speeds (1200, 1500, and 1800 RPM) under dry sliding conditions is used to study the variation in different parameters.

4.1. Co-efficient of Friction

From the graphs (Fig.1), it is observed that COF decreases with percentage increment in filler content for lower loadings, but increases from 1.0 % to 2.0% filler addition. The increase in applied load leads to the temperature rise at the interface. Also, it is observed that at 1800 RPM has low COF than that noticed at other speeds. Thus, under dry sliding conditions, the COF decrease with lower applied load and speed leads to higher temperatures generated at the interface. The temperature increases as the filler content increased above 1.0% due to agglomeration of alumina in the epoxy and may be related to a weaker filler matrix interface and/or non-homogeneous filler distribution. Hence the increase in the friction coefficient noticed in the results. This heat transfer, loosen the fibers in the matrix and shear easily due to higher axial thrust.

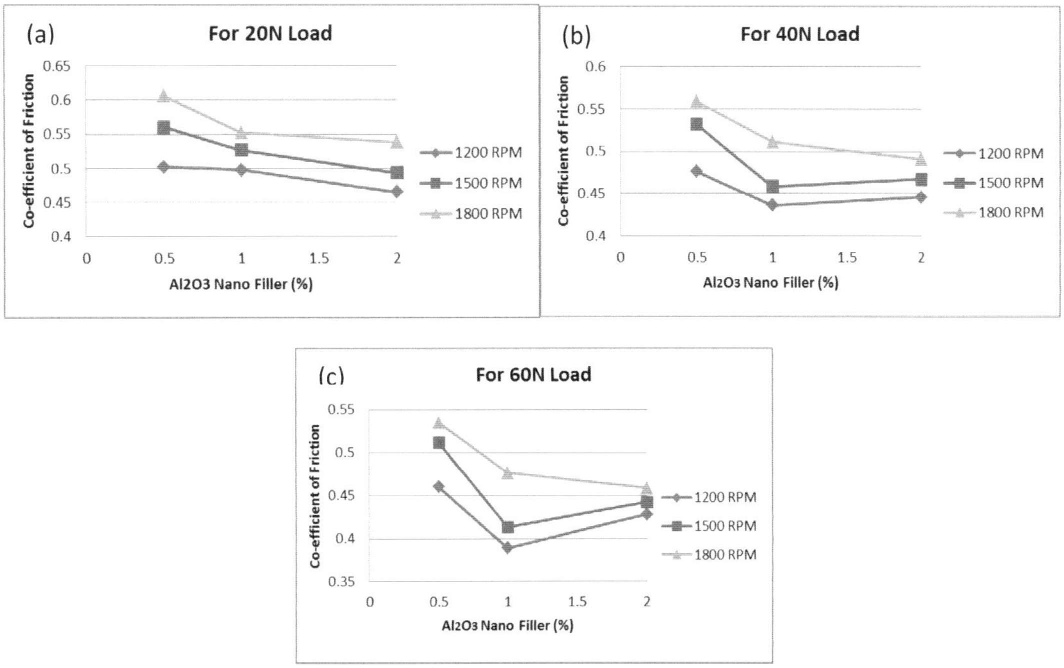

Figure 1. COF of hybrid nano composites at 20 N (a), 40 N (b) & 60 N (c) Load.

4.2. Specific Wear Rate

The result shows (fig.2) irrespective of the sample type a high rate of decrease in specific wear rate with an increase in applied load initially. Beyond 1.0% nano filler addition there no drastic decrease in the specific wear rate noticed. This behaviour of the hybrid composite may be due to once the initial abrasion of epoxy matrix, the fibers which are exposed, and some of the worn particles clog to the abrasive paper, and further wear loss is been restricted [18]. As the load is increased further, the micro cracking of resins, as well as the micro cutting process of fibers in to tiny sections may also increase proportionally. But the existence of debris on the rubbing surface reduces additional abrasion. Hence the wear rate reduction is witnessed with load increment.

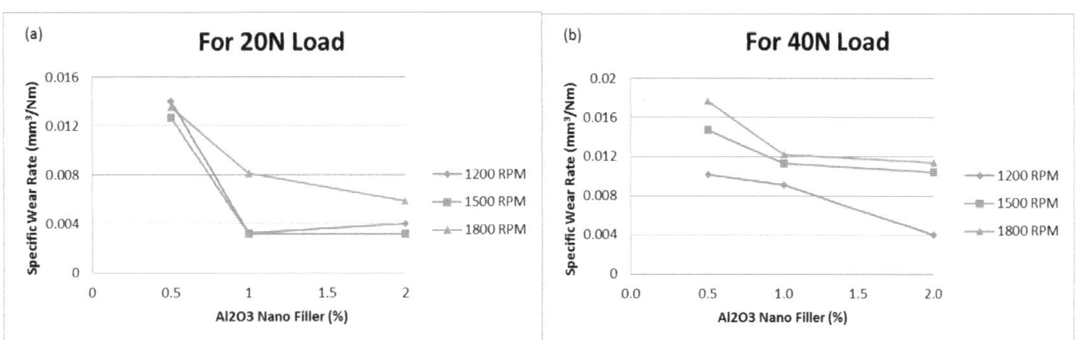

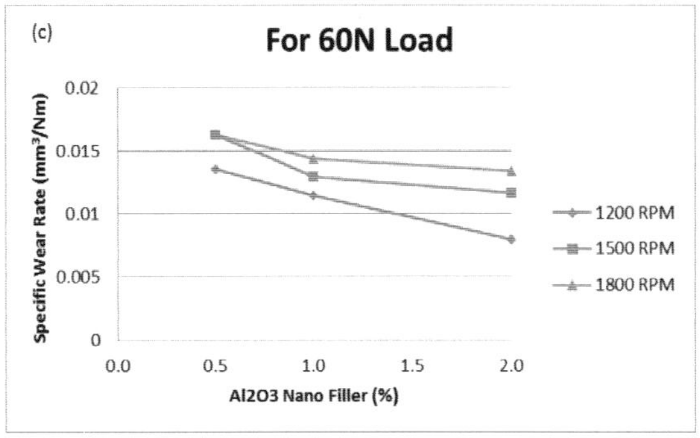

Figure 2. Specific Wear Rate of hybrid nano composite at 20 N (a), 40 N (b) & 60 N (c) Load

4.3. Wear Loss

It is obvious from the sample results (fig.3) that the hybrid nano composites wear decreases with an increase in nano filler percentage initially irrespective of the sample composition. It is also observed with an increase in speed, the wear loss was higher compared to lower speeds. Beyond 1.0% nano filler addition, not a considerable amount of decrease in wear loss is noticed. This behaviour of the hybrid nano composite is due to after the primary abrasion of the epoxy matrix, the fibers which are open to the surface and some worn particles clog to the abrasive paper, and further wear loss will slow down. The fiber debris is observed at the surface of the samples attributed to the twisting forces acting between the fibers efficiently contribute to higher wear resistance at higher load and speed. This will result in clogging of nano filler as well as worn out fibers. The wear loss change follows almost shows a linear trend at higher load. This behaviour is due to, when the material is in touch with the abrasive surface; the wear loss is high. As load and filler content increased, some of the worn particles clog on the abrasive surface, and further wear loss will slow down [18].

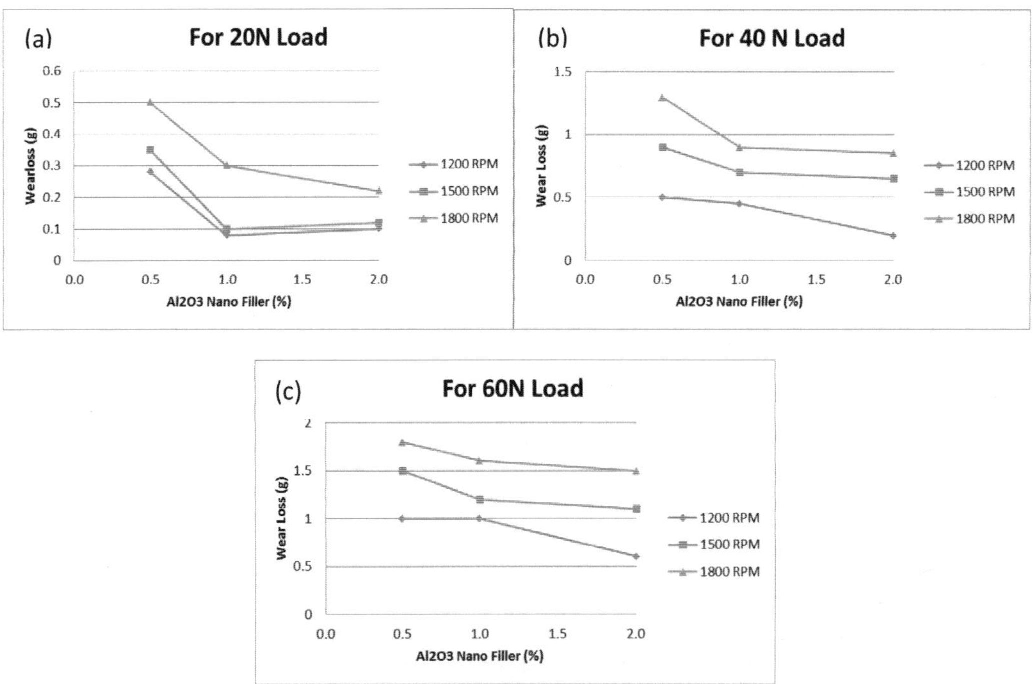

Figure 3. Wear Loss of hybrid composites at 20 N (a), 40 N (b) & 60 N (c) Load

5. Conclusion

The prepared hybrid nano epoxy based composites exhibit better anti-frictional behaviour with the reinforcement of aluminum nano particles. This behaviour is observed up to 1.0% of Nano filler addition. The COF of the nano hybrid epoxy based composites at 1800 RPM shows lower value compared to other tested speeds, indicating better nano filler interaction between the matrix and fibers. This behaviour of Nano composite is further attributed to a decrease in wear loss and specific wear rate, due to particles clog into the abrasive surface.

6. References

[1] Doreswamy D, Shivamurthy B, Anjaiah D and Sharma NY 2015 An investigation of abrasive water jet machining on graphite/glass/epoxy composite. *International Journal of Manufacturing Engineering*. 2015.

[2] Feng J, Venna SR and Hopkinson DP. 2016 Interactions at the interface of polymer matrix-filler particle composites. *Polymer.* **103** 189–95.

[3] Ragab KA, Abdel-Karim R, Farag S, El-Raghy SM and Ahmed HA. 2010. Influence of SiC, SiO2 and graphite on corrosive wear of bronze composites subjected to acid rain. *Tribology* International. **43(3).** 594–601.

[4] Sudeepan J, Kumar K, Barman TK and Sahoo P. 2014 Study of friction and wear of ABS/Zno polymer composite using Taguchi technique. *Procedia materials science.* **6** 391–400.

[5] Sampathkumaran P, Seetharamu S, Vynatheya S, Murali A, Kumar RK. 2000 SEM observations of the effects of velocity and load on the sliding wear characteristics of glass fabric–epoxy composites with different fillers. *Wear.* **237(1)** 20–27.

[6] Pihtili H and Tosun N. 2002 Investigation of the wear behaviour of a glass-fibre-reinforced composite and plain polyester resin. *Composites Science and Technology*. **62(3)** 367–70.

[7] Chand N, Naik A and Neogi S. 2000 Three-body abrasive wear of short glass fibre polyester composite. *Wear.* **242(1–2)** 38–46.

[8] Nair RP, Griffin D and Randall NX. 2009 The use of the pin-on-disk tribology test method to study three unique industrial applications. *Wear.* **267(5–8)** 823–27.

[9] Muhammed OS, Saleh HR and Alwan HL. 2009 Using of Taguchi method to optimize the casting of Al–Si/Al2O3 composites. *Engineering and Technology Journal*. **27(6)** 1143–50.

[10] Ekka KK and Chauhan SR. 2015 Effect of different reinforcements on sliding wear of aluminium matrix composites using Taguchi design of experimental technique. *Indian journal of engineering and material sciences.* **22** 195-202.

[11] Raju BR, Kanthraj BS, Suresha B and Swamy RP. 2013 Three-body abrasive wear behaviour of silicon carbide filled glass-fabric reinforced epoxy composites using Taguchi method. *Advances in polymer science and technology*. **3(3)**

[12] Mishra V and Biswas S. 2016 Three-body abrasive wear behavior of short jute fiber reinforced epoxy composites. *Polymer Composites*. **37(1)** 270–78.

[13] Mazza L, Trivella A, Grassi R and Malucelli G. 2015 A comparison of the relative friction and wear responses of PTFE and a PTFE-based composite when tested using three different types of sliding wear machines. *Tribology International.* **90** 15–21.

[14] Nirmal U, Hashim J and Lau STW. 2011 Testing methods in tribology of polymeric composites. *International Journal of Mechanical and Materials Engineering.* **6(3)** 367–73.

Characterisation of Aluminium Metal Matrix Composites reinforced with SiC (Carborundum) & Al2O3 (Corundum)

Gangaraj[1] Manjunath K.[2]

[1] Senior Grade Lecturer, Dept. of Mechanical Engg, Rajeev Polytechnic, Hassan -573201, India,

[2]Asstistant. Professor, Dept. of Mechanical Engg, Government Engineering College, Hassan -573201, India.

Abstract. Aluminium alloys are widely used in aerospace and automotive industries due to their advantages of physical and mechanical properties. This paper presents a study on the performance of stir cast SiC & Al2O3 reinforced Aluminium metal matrix composite materials. The particulate reinforcement of SiC & Al2O3 is in the ration of 3:2. The mechanical characterization of AMMC indicates that the composite exhibit improved physical and mechanical properties such as low coefficient of thermal expansion, high ultimate tensile strength, high impact strength and hardness. Through scanning electron microscope, the microstructure of AMMC & distribution of SiC & Al2O3 particles in Aluminium alloy (Al 1350) is uniform. By experimental analysis, it is observed that AMMC with 15% of particulate reinforcement of SiC & Al2O3 with a ration of 3:2 (9% of SiC & 6% of Al2O3) is the optimum mixture for AMMC to obtain better mechanical characteristics. This AMMC has good potential for the applications in automotive industry to manufacture engine components.

1. Introduction

Composites are materials in which two phases are combined with strong interfaces between the base metal & reinforcement. They usually consist of a continuous phase called the matrix and discontinuous phase in the form of fibers or particles called the reinforcement. The reinforcing component is distributed in the continuous or matrix component. When the matrix is a metal, the composite is termed as metal matrix composite (MMC). The matrix holds the reinforcement to form desired shape while the reinforcement improves overall mechanical properties of the matrix. Composite materials have high stiffness, high strength, low weight / density, high temperature stability, high electrical and thermal conductivity, corrosion resistance and improved wear resistance. In this paper, aluminium metal matrix composites with silicon carbide (SiC) and Aluminium Oxide (Al2O3) as reinforcement particles in the ratio 3:2 are studied to evaluate their physical and mechanical properties. Aluminium 1350 alloy is used as matrix which has 99.5% aluminium, 0.4% iron, 0.1% silicon, 0.05% copper, 0.05% Zinc, 0.05% Tin and 0.01% Manganese. The mechanical properties of aluminium metal matrix composites are affected by the volume fraction of the reinforced SiC & Al2O3 particles.

2. Problem Formulation

Through the analysis of literature survey, it is hypothesized that AMMC with SiC & Al2O3 particle reinforcement could be reliable materials to replace grey cast iron in the automotive industry. To improve the mechanical properties of AMMC the volume fraction of the reinforcement particles are varied compared to previous developments.

3. Objectives of Work

To reduce the weight of AMMC & to increase the tensile and yield strength with minimum ductility and toughness.
To increase creep resistance at higher temperatures compared to other alloys.
To improve corrosion resistance property.
To increase modulus of elasticity during tensile and compressive load.
To reduce thermal elongation of the material.

4. Methodology

The methodology used for stir cast of SiC & Al2O3 reinforced Aluminium metal matrix composite materials is shown in figure 1.

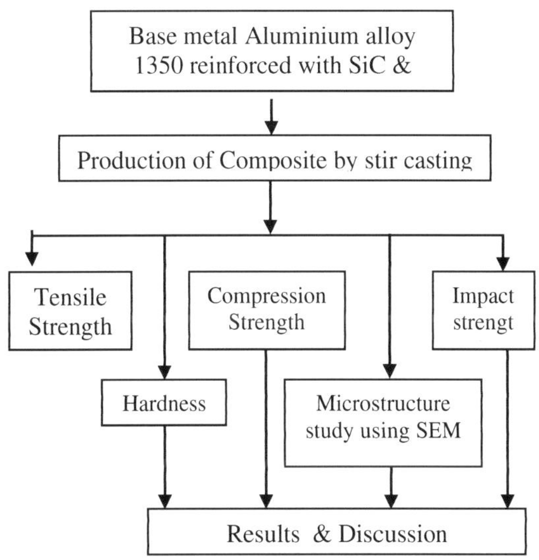

Figure 1. Flow chart of Methodology

5. Stir Casting

Stir casting is a liquid state method of composite material fabrication in which a dispersed phase is mixed with a molten matrix metal by means of mechanical stirring. The aluminium alloy 1350 in the form of wire,used as base metal is melted in the crucible by heating it in a muffle furnace at 7500C for 3 hours. The SiC particles and Al2O3 particles are preheated at 10000c for one hour to make their surfaces oxidized. The different specimens for tensile test, compressive test, impact test, hardness test and for microscopic study through SEM as per ASTM standards prepared with different composition and percentage of particulate reinforcement by weight in the ratio 3:2 as table 1 and shown in figure 2,3,4, 5 and 6 respectively.

IOP Conf. Series: Materials Science and Engineering 1189 (2021) 012009 doi:10.1088/1757-899X/1189/1/012009

Table 1. Composition and Percentage of Particulate Reinforcement

Composition	Aluminium Alloy 1350 %	SiC %	Al2O3 %	Total particulate SiC&Al2O3 reinforcement %
Pure aluminium alloy 1350	100	0	0	0
Al.alloy 1350 with 3 % SiC & 2% Al2O3	95	3	2	5
Al.alloy 1350 with 6 % SiC & 4% Al2O3	90	6	4	10
Al.alloy 1350 with 9 % SiC & 8% Al2O3	85	9	6	15
Al.alloy 1350 with 12 % SiC & 8% Al2O3	80	12	8	20
Al.alloy 1350 with 15 % SiC & 10% Al2O3	75	15	10	25

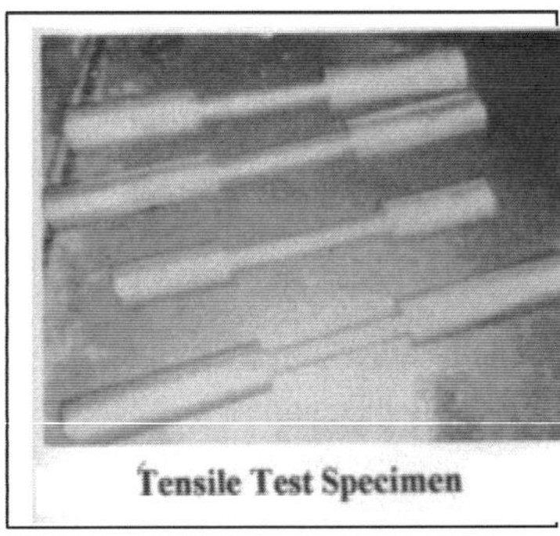

Figure 2. Tensile Test Specimens

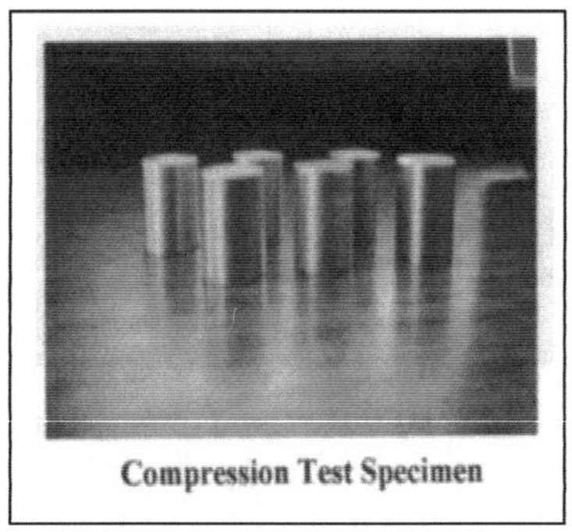

Figure 3. Compression Test Specimens

Figure 4. Hardness Test Specimens

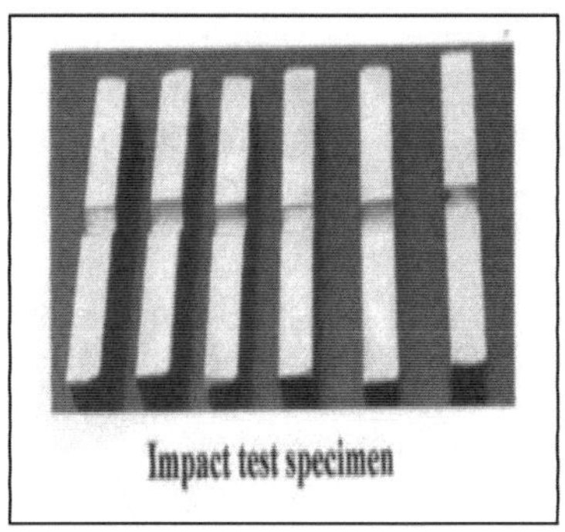

Figure 5. Impact Test Specimens

Figure 6. Diffrent Samples for SEM

6. Experimental Results

6.1 Tensile Test results

Experimentally obtained tensile test for different particulate reinforcement percentage in AMMC is carried out and results are shown in table 2.

Table 2. Tensile Test Results for Different Composition

Al.alloy 1350	Ultimate tensile strength N/mm^2 (MPa)	Percentage Elongation	Total percentage Particulate reinforcement ($SiC + Al_2O_3$) (3:2)
Pure Al. alloy 1350	83.10	16.37	0
Al. alloy 1350+ 3 % SiC + 2% Al_2O_3	94.74	27.1	5
Al. alloy 1350 +6 % SiC + 4% Al_2O_3	83.15	11.87	10
Al. alloy 1350 +9 % SiC + 6% Al_2O_3	61.01	14.8	15
Al. alloy 1350+12% SiC + 8% Al_2O_3	43.005	3.83	20
Al. alloy 1350+15% SiC + 10% Al_2O_3	49.091	5.78	25

The comparative results of UTS (MPa) for different percentage of particulate reinforcement is as shown in figure 7.

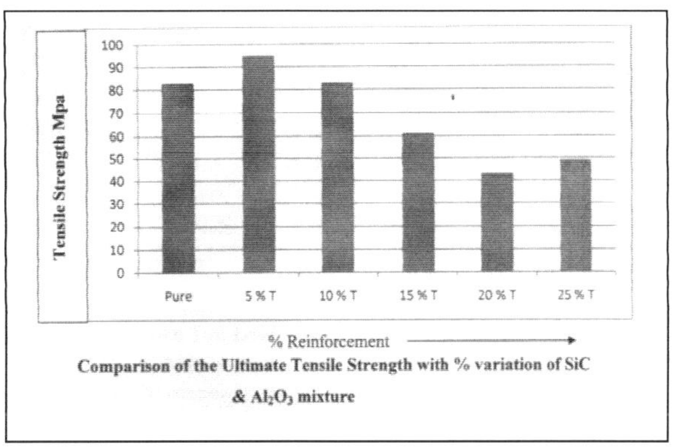

Figure 7. Comparison of Tensile Strength

IOP Conf. Series: Materials Science and Engineering 1189 (2021) 012009 doi:10.1088/1757-899X/1189/1/012009

6.2 Compression Test results

Experimentally obtained compression test for different particulate reinforcement percentage in AMMC is carried out and results are shown in table 3.

Table 3. Tensile Test Results for Different Composition

Al. alloy 1350	Load (KN)	Compressive Strength (N/mm^2)
Pure Al. alloy 1350	30.93	393.81
Al. alloy 1350 + 3% SiC + 2% Al_2O_3	29.18	371.53
Al. alloy 1350 + 6% SiC + 4% Al_2O_3	29.18	371.53
Al. alloy 1350 + 9% SiC + 6% Al_2O_3	30.64	390.12
Al. alloy 1350 + 12 % SiC + 8% Al_2O_3	30.32	386.04
Al. alloy 1350 + 15% SiC + 10% Al_2O_3	30.33	386.17

The comparative results of compression test for different percentage of particulate reinforcement is as shown in figure 8.

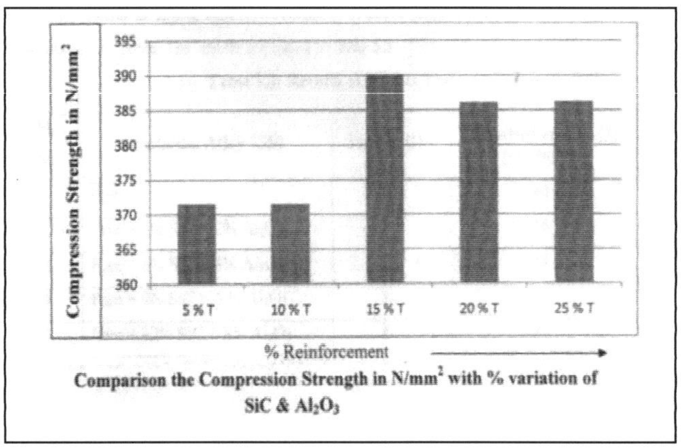

Figure 8. Comparison of Compression Strength

6.3 Impact Test results

Charpy impact test is a standardized high strain rate test which determines the amount of energy absorbed by the material during fracture. Experimentally obtained impact test for different particulate reinforcement percentage in AMMC is carried out and results are shown in table 4.

Table 4. Impact Test Results for Different Composition

Al. alloy 1350	Absorbed energy in Joules
Pure (Al. alloy 1350)	40
Al. alloy 1350 + 3 % SiC + 2% Al2O3	56
Al. alloy 1350 + 6 % SiC +4% Al2O3	38
Al. alloy 1350 + 9 % SiC +6% Al2O3	70
Al. alloy 1350 + 12 % SiC +8% Al2O3	46
Al. alloy 1350 + 15 % SiC +10% Al2O3	70

The comparative results of impact test for different percentage of particulate reinforcement is as shown in figure 9.

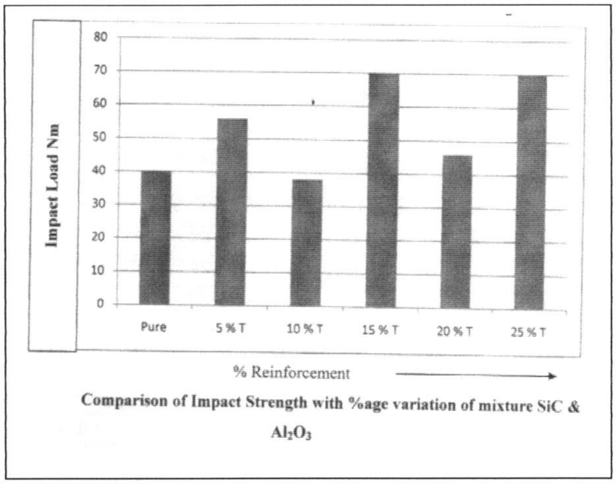

Figure 9. Comparison of Impact Strength

6.4. Hardness Test Results

Experimentally obtained Hardness test for different particulate reinforcement percentage in AMMC is carried out and results are shown in table 5.

Table 5. Hardness Test Results for Different Composition

Al. Alloy 1350	Rockwell Hardness Number (HRB)			Mean Hardness
	1	2	3	
Al. alloy 1350	37.1	37.1	36.8	37
Al. alloy 1350 + 3% SiC + 2% Al_2O_3	37.4	37.4	37.1	37.3
Al. alloy 1350 + 6 % SiC + 4% Al_2O_3	37.4	37.4	37.1	37.3
Al. alloy 1350 + 9 % SiC + 6% Al_2O_3	37.7	37.7	37.4	37.6
Al. alloy 1350 + 12% SiC + 8% Al_2O_3	37.1	37.1	36.8	37
Al. alloy 1350 + 15% SiC + 10 % Al_2O_3	37.4	37.4	37.1	37.3

The comparative results of hardness test for different percentage of particulate reinforcement is as shown in figure 10.

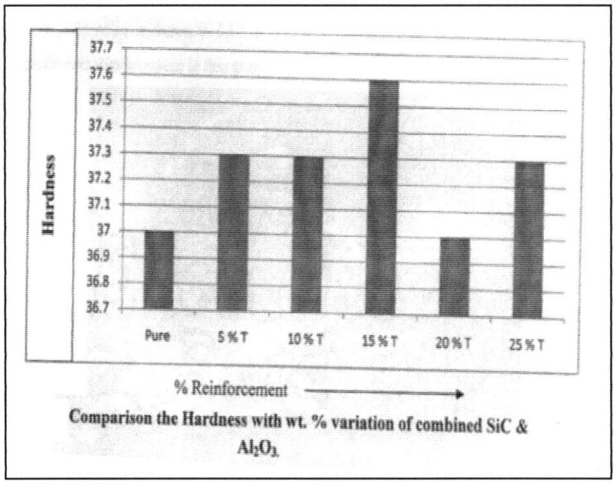

Figure 10. Comparison of Hardness Strength

6.5 SEM Analysis

SEM analysis is carried out for different particulate reinforcement percentage as Al. alloy 1350, Al. alloy 1350 + 3% SiC + 2% Al_2O_3, Al. alloy 1350 + 6% SiC + 4% Al_2O_3, Al. alloy 1350 + 9% SiC + 6% Al_2O_3, Al. alloy 1350 + 12% SiC + 8% Al_2O_3, and Al. alloy 1350 + 15% SiC + 10% Al_2O_3, and results are the obtained images are shown in figure 11, 12, 13, 14, 15 16 and 17.

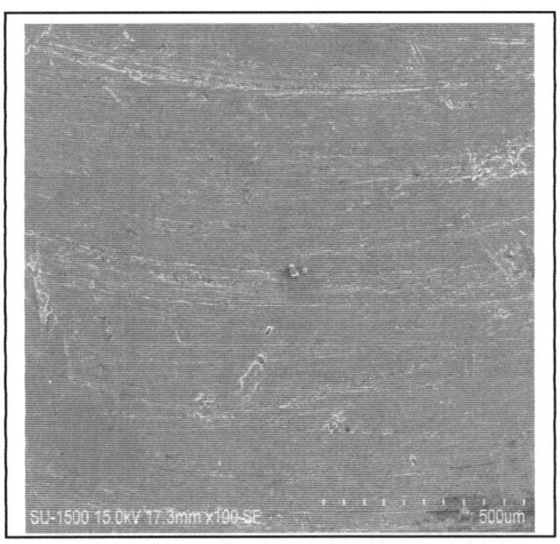

Figure 11. Pure Al. alloy 1350

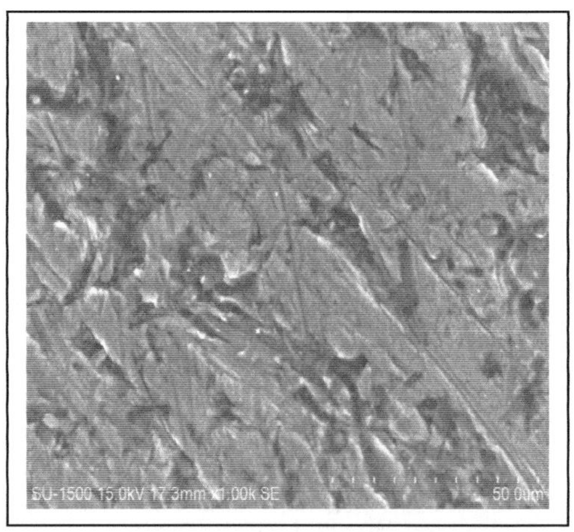

Figure 12. Al. alloy 1350+ 3% SiC + 2% Al_2O_3,

Figure 13 Al. alloy 1350 + 6% SiC + 4% Al_2O_3

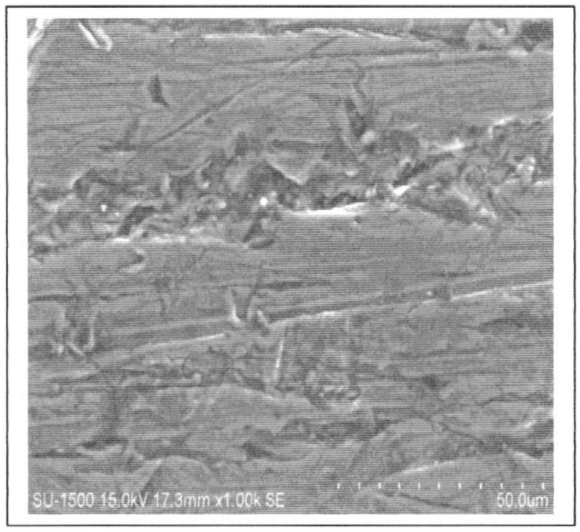

Figure 14 Al. alloy 1350 + 9% SiC + %6 Al_2O_3

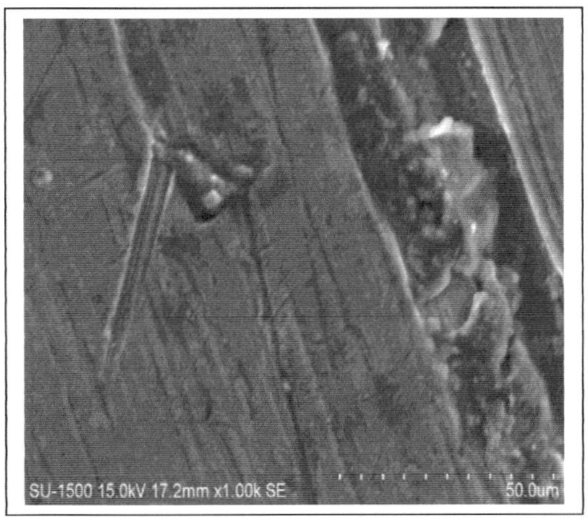

Figure 15 Al. alloy 1350 + 12% SiC + %8 Al_2O_3

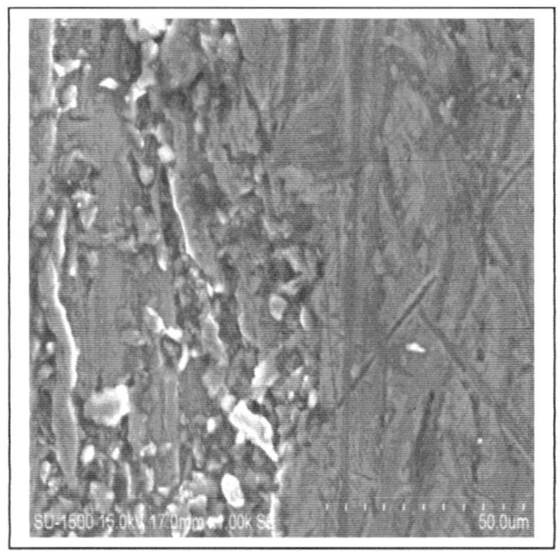

Figure 16 Al. alloy 1350 + 15% SiC + %10Al_2O_3

7. Conclusions

The mechanical characterization of aluminium metal matrix composites reinforced with varying weight fraction, particulate reinforcement of SiC & Al2O3 in the ration 3:2 and total particulate reinforcement percentage of 5%, 10%, 15%, 20% & 25% reveals the following conclusions.

a) The casting of Al. alloy 1350 with SiC & Al2O3 reinforcement composites has better tensile strength, compressive strength, impact strength as well as hardness.
b) Dispersion of SiC & Al2O3 particles in AMMC improves the hardness of the matrix material.

c) Optical micrographs of AMMC through SEM shows uniform distribution of SiC & Al2O3 particles, fabricated through stir casting technique.
d) The ultimate tensile strength & yield strength of AMMC tends to increase with increase in particulate reinforcement of SiC & Al2O3 in the matrix.
e) The hardness of AMMC increases with the addition of SiC & Al2O3 particles in the matrix.
f) The impact strength of AMMC increases with the increase in weight fraction of SiC & Al2O3 particles.
g) It is found that AMMC with 9% SiC & 6% Al2O3 particulate reinforcement shows better mechanical characteristics as compared to other percentages of particulate reinforcement and can be used in manufacturing automotive engine parts, due to improved strength to weight ratio.

8. References.

[1] Rajeshkumar Gangaram Bhandare, Parashuram, M. Sonavane, 2013 *Preparation of Aluminium Matrix composite by using stir casting method International Journal of Engineering and Technology* Vol 3, (2), pp 61-65.

[2] D. Sujan, M.E. Rahman, M. A. Malegue, C. K. Tan, 2012 *Physio-mechanical properties of Aluminium Metal Matrix composites with Al2O3 & SiC, world academy of science, International journal of chemical, Molecular, Nuclear, Materials & Metallurgical Engg.* Vol.6(8) pp 678-681

[3] Neelima Devi, C, Mahesh, V. Selvaraj N 2015, *Mechanical Characterisation of Al. SiC omposite, Internal Journal of Applied Engg. Research,* Vol. 1(4204) pp. 793-799.

[4] A.Venugopal, Mahesh N.Manoharan, 2015, *Evaluation of Mechanical properties of Al. MMC for Marine applications, ARPN Journal of Engg. & Applied Sciences,* Vol. 10(13), pp. 5557-5559.

[5] Arun L. R., Saddam Hussain B. Dr. Suneel Kumar N. Kulkarni, 2013, *Dynamic Behaviour of Hybrid Al. 6061 Metal Matrix reinforced with SiC & Fly ash particulates", Intl journal of Innovative research in science, Engg. & technology,* Vol. 2 (6), pp 2501 – 2506.

[6] B. Vijay Ramanath, C. Elanchezhian, M. Jai Vignesh, S. Rajesh, C. Parswajiram, A. Siddique Ahmed. Ghias 2014 *Evaluation of Mechanical properties of Al. Alloy Alumina – Boron carbide Metal Matrix composites Materials and Designs* Vol.58, pp. 332-338.

[7] K. K. Alaneme, B. O. Ademilna, M. O. Bodunrin, 2013, *Mechanical properties & corrosion behavior of Al. Hybrid composites reinforced with SiC and Bamboo leaf ash , Tribology in Industry,* Vol. 35 (1), pp 25-35.

[8] M. Vykuntarao, S Chiranjeevi Rao, Ch. Vinod Babu & M. V. Shekar Babu, 2015, *Influence of reinforcement particles on the mechanical properties of Al. based MMC, Chemical Science Review & letters,* Vol. 4(13), pp 335-341.

[9] Michal Oluwatosin Bodumrin Kenneth Kanayo. Alaneme, Lesley Heath chown, 2015, *Aluminium matrix hybrid composites: a review of reinforcement philosophies, mechanical, corrosion & tribological characteristics, Journal of materials Research & Technology,* Vol. 4(4), pp 434-445.

[10] Suryanarayanan, K. R. Praveen, S. Raghuraman, 2013, *Silicon Chabride reinforced Aluminium Metal Matrix composites for Aerospace applications", International Journal of Innovative Research in Science, Engg. & Technology,* Vol. 2, pp 6336-6344.

IOP Conf. Series: Materials Science and Engineering 1189 (2021) 012010 doi:10.1088/1757-899X/1189/1/012010

An overview of polymer nanocomposites: Understanding of mechanical and tribological behavior

Qurat Ul Ain[1], Rakesh Sehgal[1], M F Wani[1,3] and M K Singh[2,4]

[1] Tribology Laboratory, Department of Mechanical Engineering, NIT Srinagar, Hazratbal, Kashmir, 190006, India
[2]Department of Mechanical Engineering, IIT Kanpur, Kanpur 208016, Uttar Pradesh, India

mfwani@nitsri.ac.in[3], manjesh@iitk.ac.in[4]

Abstract. Polymer nanocomposites are polymeric materials reinforced with various types of nano-materials. Over last few decades, these materials have drawn a lot of attention particularly as a favorable alternative in extreme condition exposures. Polymer nanocomposites have potential applications in many fields such as automotive, high-barrier food packing films, extremely hard and wear resistant coatings/materials to eco-friendly composites. The main objective of this review article is to give current state-of-the-art of research activities on nanofiller-based-polymer-nanocomposites with particular focus on their application to improve tribological properties. Even an apparently simple object in reality is a technological marvel of nano-engineered materials. For example, a vehicle tyre is generally produced by mixing some specific polymers and nano-fillers. The tribological applications of polymer materials are likely to progress with the practice of polymer nanocomposites being applied to novel uses. The main motivation behind this review is to highlight the key mechanisms governing the effects of reinforcement on the mechanical and tribological behavior of polymer nanocomposites. It is observed that the improvement of one feature or characteristic may not interpret an enrichment of the other properties associated with the system. However, it is significant to comprehend the diverse elements or aspects accountable for performance of the material. To summarize, this overview presents basics of the complexities in the mechanical and tribological behaviour of polymer nanocomposites.

1. Introduction

Polymers are very common in our daily life. We are surrounded by things made of polymers such as paints, adhesives, packaging materials, cosmetics, fertilizers, tyres, fabrics etc. This is due to the fact that it's very easy to process polymers into complex shapes. Polymers are usually characterized by low energy density, large intramolecular entropy, nanoscopic length scale and k_BT energy scale (k_B is Boltzmann constant and T is temperature). There have been several attempts in recent times to use polymers in tribological applications to control friction and wear of surfaces [1-4]. Polymer composites are interesting for their low weight, high strength-to-weight, high stiffness-to-weight and low material cost [5]. These materials find applications in many fields including automobile body parts, tyres, cams, gears, brake materials, bearing cages, biomedical devices, artificial implants, food industries, energy storage and conversions and aerospace industries [6, 7]. Use of nanoparticle fillers in composites lowers their weight fractions in comparison to the use of microparticles. Additionally high surface-area-to-volume ratios of nanomaterials lead to improved adhesion. Nanomaterials are typically classified on the basis of dimensions ranging from 1 to 100 nm. The 0D-materials are particles with all the three dimensions below 100 nm such as carbon nanomaterials, aluminum oxide, calcium carbonate, copper oxide, silicon dioxide, silicon carbide, silicon nitride, titanium dioxide, zinc oxide and zirconium dioxide. The 1D-materials have two dimensions below 100 nm like carbon nanofillers and nanofibers and the 2D-materials have only one dimension in the nanometer scale comprising of layered sheet like constituents clays, graphene sheets, and molybdenum disulfide. Past studies suggest that nanoparticle based polymer composites typically termed as "nanocomposites" exhibit improved triblological performance [8–11].

Polymer nanocomposites are comparatively new variety of materials exhibiting great potential in various applications due to the fine balance of strength, hardness and toughness as well as excellent tribological behavior. Studies revealed that a vital path to the perfection of these characteristics is the uniform and homogenous dispersal of the nano-fillers, interfacial bonding and interaction between the particles and the polymer. The fillers are capable of improving the homogeneity of the transfer film and the adhesion to the counter surface. The mechanical/chemical reactions of nanoparticles during rolling/sliding, the capability of the dispersed nanoparticles in the polymer to support asperities in the counter-surface and the tribo-chemical reactions of nanofillers with the counter surface favor the above capabilities. The molecular weight, stereochemistry, degree of crosslinking, polymer blend composition and chemical affinity all are of fundamental importance in determining tribological performances. The polymer blend features are dependent on the dispersion phase in terms of size, degree of dispersion, and interfacial phase interaction between the components of the blend. These factors primarily need to be controlled to attain specific properties of the material [12-14]. Other critical parameters responsible for performance of polymer nanocomposites include contact area, load/stress (collective stress), sliding speed, topography of counter face, temperature and molecular relaxation.

The knowledge of complex structural changes occurring in a tribosystem at molecular scale is vital for designing the materials with desired characteristics. In spite of availability of large variants of nanomaterials, very few have been examined for tribological applications. Continuum theory has limitations in modeling polymer nanocomposites. Continuum modeling does not take into account the atomic-scale details and mechanisms like atomic stick-slip. Nanoscale analysis and modeling techniques that include the molecular structure of the contact is therefore necessary. Molecular simulations are generally used to overcome the limitations of continuum theory. Molecular dynamics simulations aimed at analyzing specific problems are possibly the most effective methods to elucidate complex mechanisms satisfactorily. The fundamental understanding of interacting molecular bonds and their nature is achieved with molecular simulation and modeling techniques [15]. These methods allow modeling of the intricate interfacial occurrences with extraordinary resolution in both length and time scales. These molecular simulations yield understandings into the molecular-scale energetics, structure, dynamics, thermodynamics, transport and rheological behavior under tribological loadings. The focus of this review article is to develop a comprehensive understanding of polymer nanocomposite's performances under nanoindentation.

2. Literature Review

The practice of using nanoparticles in polymer matrix for tuning tribological characteristics gained importance in mid-1990s. This particular field has potential for detail research as new nanoparticles are consistently and economically designed. The polymer based nanocomposites exhibit properties that depend on the area of interacting interface between the filler (nanoparticles or nanofibers) and the matrix (a polymer) for its improved mechanical and tribological behavior. Almost for all the polymer/nanoparticle systems, there is an optimal amount of the nanoparticles which can be incorporated in the system; otherwise it degrades the toughness as the stiffness and strength increase beyond the optimum limit. Most of the present theoretical systems are obtained from the continuum mechanics models. The continuum mechanics theories are not always valid at nano-scale; rather the discrete-particle based model is commonly used [16].

Experimental studies have revealed extraordinary and remarkable tribological performance of filler polymer composites and nano-filled polymer composites but have also predicted some contradictions. In several studies, addition of filler enhances wear resistance, whereas some studies have shown they deteriorated the performance of the nanocomposite [17]. It is evident from literature that the same filler exhibits contrary trends when the size of filler was varied. The chemical affinity of the fillers is changed with the reinforcement in polymer matrix. The previous studies have considered the influence of addition of the nanofillers on the formation of transfer film [18, 19]. It is also specified that the nanoparticles present on the interface render thin and firm transfer film formation [20]. Other works reveal that accurate dispersion of nanofillers is another vital factor to attain specific properties. The nanoparticles typically tend to agglomerate at higher filler contents thereby leading to shortcomings in efficiency of polymer matrix. The studies recognized that the increased wear resistance is due to rise in the thermal conductivity of nanocomposites. The increase in thermal conductivity enables the heat dissipation produced through friction which reduces the temperature of the working tribosystem [21]. The minimum amount of filler additions have depicted improvements in properties by reduced wear rates [9]. It is manifested by previous studies that the filler content lesser than 1.0 wt.% are sometimes enough for improving wear resistance by 1000 folds. The nanofillers have the capability to impart significant tribological and mechanical features for use in specific systems. There are numerous characteristics of these systems that must be considered to acquire a thorough understanding of the underlying mechanisms [19].

The metal, ceramic, or polymer "nanoparticles" present extensive types of 0D-fillers [22]. These nanoparticles may be either hard or soft, prepared from either distinct or multi-chemical components, exhibiting different shapes or forms [23]. Currently, numerous types of nanoparticles or nanofillers are reinforced into the polymer matrix to increase their tribological performances. Inorganic and hard nanoparticles are frequently used as they exhibit remarkable performance. The importance of these nanoparticles is correlated with their capability to increase elastic modulus and impact wear resistance of the base polymers. In order to reduce coefficient of friction, soft polymeric nanoparticles are also incorporated in the matrix [24, 25]. Existing technologies allow synthesizing nanoparticles with various kinds of forms and dimensions, either restrained or applied in the form of coating on diverse surface categories [26, 22]. 1D-nanofillers include nano-tubes, rods, fibers and wires. They are metal, ceramic, or polymer materials and their molecular structures could be amorphous or crystalline in nature. Studies reveal incorporation of nanofillers in a polymer matrix has improved performances by increasing load-bearing capability and toughness. The nanofillers facilitate the reduction in wear rate by forming a persistent and few nanometers thick transfer film on the counter face during interaction [27]. 2D-nanofillers include natural cationic clays, graphene sheets, metal chalcogenides, and various transition metal oxides. These fillers possess a layered like molecular structure having van der Waals bonds between the layers. The thin layer which has only few nanometers thickness exhibit extraordinary electronic, magnetic, thermal, mechanical and tribological characteristics. The usage of the nanomaterials in tribology field has drawn a lot of consideration over the recent years [28].

Simulations of nanoindentation on polymer nanocomposites are very rare. A recent report on nanoindentation simulation study for the quantitative nanomechanical characteristics of polyethylene (PE) and the polyhedral oligomeric silsesquioxane (POSS) composites (POSS-PE) is directed on computation of hardness and elastic modulus of polymers using different shapes of indenter. The different shapes greatly affected the hardness values and elastic modulus of flat indenter is greater than cube-corner and spherical shape indenter. With molecular dynamics (MD) simulation, several mechanical behaviors of polymer nanocomposites with various molecular interactions are studied [29, 30]. The nanoindentation study on graphene based polymer nanocomposites has been conducted using MD simulation. The effects of reinforcement of graphene sheets in PE polymer indentation are investigated. The nanoindentation exhibited that the material resistance is enhanced significantly with the presence of layered graphene sheets. The measured values of this study reveal that resistance to indentation of a single-layer graphene coated PE is around fourteen-fold than that of virgin PE. The alignment and the way graphene sheets disperse in the nanocomposites considerably affect the tribo-mechanical properties [31]. MD simulations have been used to study mechanical and tribological performance of polymer nanocomposite with wrinkled graphene layer placed in-between. PE and polymethylmethacrylate (PMMA) polymers are reinforced with graphene layer to improve interfacial mechanical and tribological properties. The effects of graphene wrinkles, nature of matrix, length of polymer chain and pulling-out velocity of graphene sheet (GR) are revealed. The results indicate that a flat GR in comparison to a wrinkled GR efficiently improved the counter face behavior. The pulled-out velocity of GR (wrinkled) influences the interacting counter face properties of both the polymer nanocomposites used. The effect of degree of polymerization of polymers on the interacting surface properties is insignificant in the graphene reinforced PE system, but very significant in the graphene reinforced PMMA system [32]. The current state-of-the-art understanding is focused on the enhancement of uniformity in dispersion by functionalization, chemical reformation and heat treatment or the supplement of stabilizers/compatibilizers among the polymer matrix and the nano-phases [33-39]. The ultimate challenge is that experimentally measured results of different properties cannot be simply applied and used from one system to another e.g., from a tribological test rig to a real application. The tribological performance is predictable based on theoretical modeling and computer simulation for any particular application only when the operating conditions of the application and test environment are identical. Polymers do not always line up with the rule of improved properties such as mechanical strength and surface hardness for decreased wear rate, thus the need of fine modification for tribological performance without compromising the bulk strength of these materials. Thus, there is possibility of a strong growth in the tribology filed of polymers in all present forms available as virgin/pristine, composites, nanocomposites and hybrids.

3. Conclusion

This overview presents some features of the existing state of progress of polymer-based nanocomposites with better mechanical and tribological performance, main focus on nanoindentation using molecular simulation and theoretical approaches. The evolving field of polymer nanocomposites is of immense potential. The vital benefit of nanocomposites over conventional composites is their capability towards reaching and enhancing strength and toughness properties equally. The ideal impact of nanomaterials on improving tribological performances of polymer nanocomposites is still an open area for research in future. The existing research studies in this field are limited. Therefore this might be another major area for polymer tribology research using experimental and theoretical approaches. The present research and development activities in the area of polymers are heading towards polymer nanocomposites and polymer surface engineering reforms and are developing rapidly. There is a need to explore distinctive processing methods for polymer nanocomposite's use in tribological applications. The tribological study on polymer nanocomposites is still in initial phase and we shall expect pioneering research work in this field.

4. References

[1] M K Singh, P Ilg, R M Espinosa-Marzal, M Kroger, N D Spencer, Effect of Crosslinking on the Microtribological Behavior of Model Polymer Brushes *Tribology Letters* **63(2)** 2016 p.17.

[2] M K Singh, P Ilg, R M Espinosa-Marzal, M Kroger, N D Spencer, Polymer Brushes under Shear: Molecular Dynamics Simulations Compared to Experiments *Langmuir* **31(16)** 2015 pp.4798-4805.

[3] M K Singh, C Kang, P Ilg, R Crockett, M Krooger, N D Spencer, Combined Experimental and Simulation Studies of Cross-linked Polymer Brushes under Shear *Macromolecules* **51(24)**, 2018 pp.10174-10183.

[4] M K Singh, Polymer Brush Based Tribology, in: J Katiyar, P Ramkumar, T Rao, J Davim (Eds) *Tribology in Materials and Applications* Springer, Cham, 2020 (pp. 15-32).

[5] P L Menezes, S V Kailas, and M R Lovell, Fundamentals of engineering surfaces. *Tribology for Scientists and Engineers* Springer, New York, 2013.

[6] K Friedrich, A K Schlarb, *Tribology of Polymeric Nanocomposites: Friction and Wear of Bulk Materials and Coatings* Elsevier, 2011

[7] L Guadagno, B D Vivo, A D Bartolomeo, P Lamberti, A Sorrentino, V Tucci, L Vertuccio, and L Vittoria, Effect of Functionalization on the Thermo-mechanical and Electrical Behavior of Multi-wall Carbon Nanotube/epoxy composites. *Carbon* **49(6)** 2011 pp.1919-1930

[8] C Baillie, *Green Composites: Polymer Composites and the Environment* CRC Press, 2005.

[9] D L Burris, B Boesl, G R Bourne, and W G Sawyer, Polymeric Nanocomposites for Tribological Applications *Macromol. Mater. Eng.* **292(4)** 2007 pp.387-402.

[10] A Sorrentino, M Tortora, V Vittoria, Diffusion Behavior in Polymer–clay Nanocomposites *J. Polym. Sci. Pol. Phys.*,**44(2)** 2006 pp.265-274.

[11] A Sorrentino, L Vertuccio, V Vittoria, Influence of Multi-walled Carbon Nanotubes on the β form Crystallization of Syndiotactic Polystyrene at Low Temperature. *Express Polym. Lett.* **4** 2010 pp.339-345.

[12] S Pavlidou, C D Papaspyrides, A Review on Polymer–layered Silicate Nanocomposites. *Prog. Polym. Sci.* **33(12)** 2008 pp.1119-1198.

[13] G Gorrasi, G Attanasio, L Izzo, A Sorrentino, Controlled Release Mechanisms of Sodium Benzoate from a Biodegradable Polymer and Halloysite Nanotube Composite. *Polym. Int.* **66(5)** 2017 pp.690-698.

[14] S Liparoti, G Landi, A Sorrentino, V Speranza, M Cakmak, H C Neitzert, Flexible Poly (Amide-Imide)-Carbon Black Based Microheater with High-Temperature Capability and an Extremely Low Temperature Coefficient. *Adv. Electron. Mater,* **2(6)** 2016 p.1600126.

[15] B Bhushan, *Nanotribology and nanomechanics: an introduction* Springer Science & Business Media 2008.

[16] A Gouldstone, K J Van Vliet, S Suresh, Simulation of Defect Nucleation in a Crystal. *Nature* **411(6838)** 2001 pp.656-656.

[17] B Bhushan, *Fundamentals of Tribology and Bridging the Gap between the Macro- and Micro/Nanoscales.* Springer, Netherlands, Dordrecht, 2001.

[18] B J Briscoe, S K Sinha, Tribological Applications of Polymers and their Composites: Past, Present and Future Prospects. *Tribology and Interface Engineering Series* (Vol. **55**, pp. 1-14) Elsevier, 2008.

[19] J P Davim, *Tribology of Nanocomposites* Springer Berlin Heidelberg, 2013.

[20] L Chang, Z Zhang, L Ye, K Friedrich, Tribological Properties of Epoxy Nanocomposites: III. Characteristics of Transfer Films. *Wear* **262(5-6)** 2007 pp.699-706.

[21] I Hutchings, P Shipway, Design and Selection of Materials for Tribological Applications. *Tribology Friction and Wear of Engineering Materials* Elsevier, Oxford, 2017.

[22] C F Higgs, M Marinack, J Mpagazehe, R Pudjoprawoto, Particle Tribology: Granular, Slurry, and Powder Tribosystems. *Tribology for Scientists and Engineers* Springer, New York, 2017

[23] G Schmid, The Nature of Nanotechnology. *Nanotechnology*. pp.3-39, 2010.

[24] G Gorrasi, A Sorrentino, Mechanical Milling as a Technology to Produce Structural and Functional Bio-nanocomposites. *Green Chem.,* **17(5)** 2015 pp.2610-2625.

[25] Q J Wang, Y CHUNG, *Encyclopedia of Tribology*, 2013.

[26] F Delogu, G Gorrasi, A Sorrentino, Fabrication of Polymer Nanocomposites via Ball Milling: Present Status and Future Perspectives. *Prog. Mater.* **86** 2017 pp.75-126.

[27] M B Elinski, Z Liu, J C Spear, J D Batteas, 2D or not 2D? The Impact of Nanoscale Roughness and Substrate Interactions on the Tribological Properties of Graphene and MoS_2. *J. Phys. D Appl. Phys.* **50(10)** 2017 p.103003.

[28] B Pan, J Tan, H Jia, J Chen, Y Tai, J Liu, Y Zhang, Q Niu, Tribological Behavior of Phenolic Nanocomposites Reinforced by 2D Atomic Crystal of Boron Nitride. *J. Polym.Mater.* **4** 2016.

[29] E Hu, Y Sun, F Zeng, J Qu, Nanoindentation Simulation of PE/POSS under Different Shapes of Indenters. *Acta Mech. Solida Sin.* **24(4)** 2011 pp.365-372.

[30] E Hu, Y Sun, F Zeng, The Enhancement Mechanism and Deformation Analysis of Polyethylene Incorporated with POSS by Nanoindentation Simulation. *Rev. Adv. Mater. Sci.* **33(1)** 2013 pp.85-91.

[31] A R Alian, M A N Dewapriya, S A Meguid, Molecular Dynamics Study of the Reinforcement Effect of Graphene in Multilayered Polymer Nanocomposites. *Mater. Design* **124** 2017 pp.47-57.

[32] F Liu, N Hu, H Ning, Y Liu, Y Li, L Wu, Molecular Dynamics Simulation on Interfacial Mechanical Properties of Polymer Nanocomposites with Wrinkled Graphene. *Comp. Mater. Sci.* **108** 2015 pp.160-167.

[33] X H Men, Z Z Zhang, H J Song, K Wang , W Jiang, Functionalization of carbon nanotubes to improve the tribological properties of poly (furfurylalcohol) composite coatings. *Compos Sci Technol* **68** 2008 p 1042–1049

[34] F J Carrión, C Espejo, J Sanes, M D Bermudez, Single-walled carbon nanotubes modified by ionic liquid as antiwear additives of thermoplastics. *Compos Sci Technol* **70** 2010 p 2160–2167

[35] F J Carrion, J Sanes, M D Bermudez, A Arribas, New single-walled carbon nanotubes-ionic liquid lubricant. Application to polycarbonate-stainless steel sliding contact. *Tribol Lett* **41** 2011 p 199–207

[36] O Coban, M O Bora, E Avcu, T Sinmazcelik, The Influence of Annealing on the Crystallization and Tribological Behavior of MWNT/PEEK Nanocomposites. *Polym Compos* **32** 2011 p 1766–1771

[37] M Pollanen, S Pirinen, M Suvanto, T T Pakkanen, Influence of carbon nanotube-polymeric compatibilizer masterbatches on morphological, thermal, mechanical, and tribological properties of polyethylene. *Compos Sci Technol* **71** 2011 p 1353–1360

[38] Y J Shi, L W Mu, X Feng, X H Lu, Tribological behavior of carbon nanotube and polytetrafluoroethylene filled polyimide composites under different lubricated conditions. *J Appl Polym Sci* **121** 2011 p 1574–1578

[39] X H Men, Z Z Zhang, J Yang, X T Zhu, K Wang , W Jiang, Effect of different functional carbon nanotubes on the tribological behaviors of Poly(Furfuryl Alcohol)-derived carbon nanocomposites. *Tribol Trans* **54** 2011 p 265–274

Preparation and Characterisation of Temperature Regulating Silk Cocoon Extract (Sericin)/ PVA Based Smart fabric

Vinutha P[1,2 a *], H N Divakar [3, b], K M Subbaiah[4,c] and Savitha M[5,d]

[1] PG student, Department of Industrial Production and Engineering, The National Institute of Engineering, Mysuru- 570008, India.

[2]Lecturer, Government Polytechnic Mirle, Department of collegiate and technical education, Bangalore-560001, India.

[3] Professor, Department of Industrial Production and Engineering, The National Institute of Engineering, Mysuru- 570008, India.

[4] Professor, Department of Industrial Production and Engineering, The National Institute of Engineering, Mysuru- 570008, India.

[5] Professor, Department of Industrial Production and Engineering, Sri Jayachamarajendra College of Engineering, JSS Science & Technology University, Mysuru- 570 006, India.

[a]vinuthap85@gmail.com, [b]hndivakar@nie.ac.in, [c]iphod@nie.ac.in [d]savitham@sjce.ac.in

Keywords: Sericin, PVA, Smart fabric

Abstract: Smart fabrics help to regulate body heat and provide thermal comfort to the body. Hence from many decades research on heat transfer analysis and temperature regulating wearable clothing takes important place. Hence, here we developed green, simplistic and economical sericin/PVA film using industrial waste sericin along with PVA. For this, the fabrication of sericin/PVA was carried out by employing polymer solution casting method and slow evaporation method by varying the sericin concentration by keeping PVA constant that is in the ratio of 2:1, 1:1. Characterization techniques like X ray diffraction, and UV visible spectroscopy were carried out and fluorescence effect were studied. This research showed that higher concentration sericin/PVA film can effectively absorb phonons of UV radiation and convert into fluorescence light thereby maintaining the temperature of the film/fabric. Hence this film/fabric can be used in industries where workers are constantly exposed to harmful UV radiations.

1. Introduction

Clothing which is considered as one of the basic needs of human being, plays an important role in maintaining thermo regulation of the human body [1]. It is also considered as fabric that wraps around the skin and acts as protective shielding for humans [2]. The end result of clothing on thermal adaption relies on type of material used and properties of the clothes. Consequently, the properties of cloth, depends on skin temperature and somantic sense, whereas thermal comfort of clothes depends on heat exchange properties of textiles [1, 3]. Hence generally thermally adaptive fabrics are used to overcome the confine application of conventional fabrics and to develop a fabric with enhanced mechanical properties like in tensile, compression and toughness and thermal properties like increase thermal stability, conductivity without affecting the fundamental properties of the fabrics.

IOP Conf. Series: Materials Science and Engineering 1189 (2021) 012011 doi:10.1088/1757-899X/1189/1/012011

Change in life style of very individual and improved civilization has made to discover an alternate for natural fibres which fulfils the limitations of conventional textiles [4]. Sericin a type of protein and gummy substance present at outer shell of the cocoon has considered as waste in silk production industries but it got unique properties like hydrophilic, antioxidant, anticancer and especially in protecting from UV radiation [5, 6]. Therefore responsible for photo stability of the cocoons. Sercin consists of various amino acids which are crossed link. Due to this ability of the sericin, it can be blend easily with other macromolecules particularly synthetic polymers enhances the physical properties of the materials [7-9]. Hence sericin can be used in natural and artificial fibres and improves the thermal stability of the blend [10]. This ability of the sericin opens an window for use of sericin in temperature regulating fabric research. Polymers which are macromolecules, formed by the joining the building blocks called monomers are widely used in textile industries to fulfil the requirements of the consumer. Few properties of polymers like light weight, high strength, resistant to chemicals and ability to process easily has made polymers to use in textiles [11-13].

2. Experimental

2.1 Materials

Multi voltine cocoons were kindly provided by central sericultural research and training institute, sriramapura, Mysore. Poly vinyl alcohol, distilled water, whatman filter paper and gloves were purchased from padmashri scientific, Mysuru. Distilled water is used throughout experiment.

2.2 Synthesis of silk extraction (Sericin)

Multi voltine cocoons were selected for sericin extraction and then pupae were removed by cutting the cocoons. Weight of the 5 cocoons was weighed using digital weighing machine and they weighed about 4gms. These weighed cut cocoons were immersed in 50ml distilled water and boiled for approximately half an hour. Then extracted solution was filtered using whatmann paper and collected in beaker. Hence by this process 8wt % of silk sericin was obtained as shown in figure.

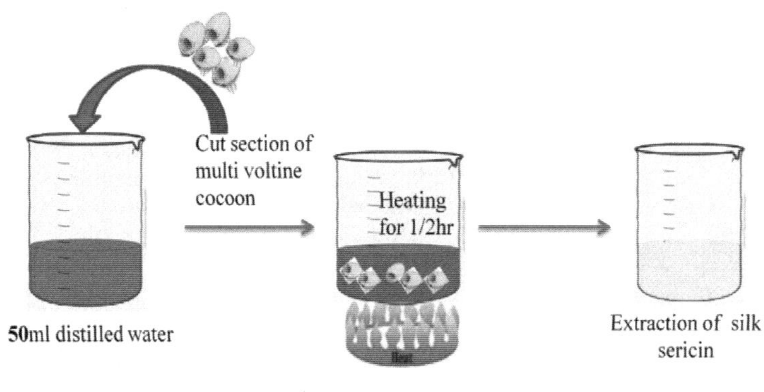

Figure 1: Schematic representation of extraction of silk sericin

2.3 Fabrication of silk cocoon extract (sericin)/ polymer based smart fabric

4gms of PVA granules were weighed using digital weighing machine and dissolved thoroughly in 50ml of boiling water by continuous stirring using magnetic stirrer. After it was completely dissolved, 8 wt % of PVA solution was obtained. Then prepared 20ml of PVA solution is measured and mixed thoroughly with different quantity of extraction of silk sericin at approximately 50^0C, so that homogeneously blended mixture is obtained w. For this different quantity of sericin like 40ml, 20ml was mixed continuously with 20ml of PVA that is by keeping PVA constant.

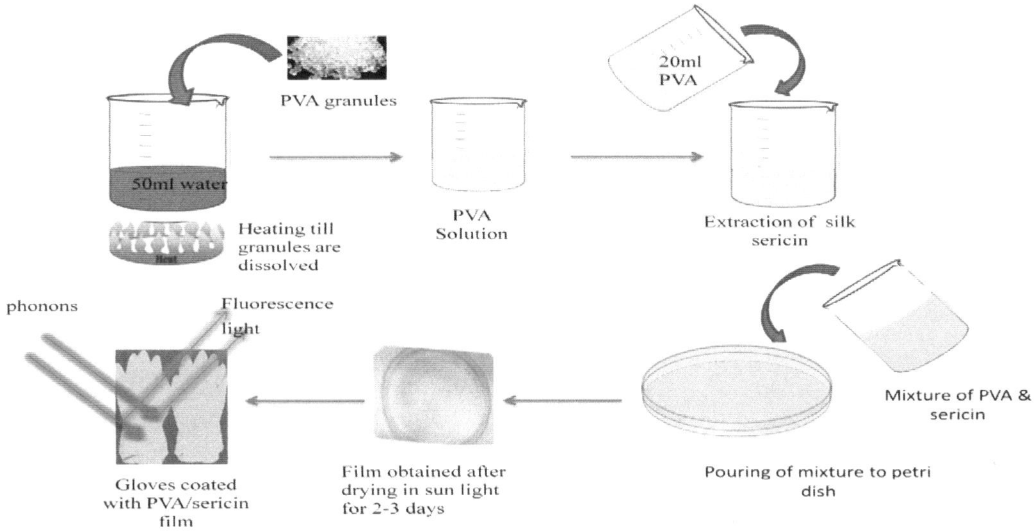

Figure 2: Schematic representation of fabrication of silk cocoon extract (sericin)/ Polymer based smart fabric.

2.4 Material characterization

The XRD patterns of sericin/PVA film were obtained from desktop X-ray powder diffraction instrumentation, Bruker, Germany and 2θ ranging from 14 to 70^0. Absorption spectrometry and transmittance spectrometry were carried out using UV Visible spectrometer from the range 100 to 700 nm.

2.5 Fluorescence effect

UV radiation absorption and fluorescence effect were determined using high resolution laser beam in dark chamber.

3 Results and Discussion

3.1 Ultraviolet- Visible Spectroscopy

In fig 3a, the PVA alone will not show any considerable absorption in any UV region. Absorption of higher concentration film (2:1 sericin/PVA film) has higher absorption rate. The maximum absorption starts from wavelength 200nm to 230nm, in UVC region. By the obtained plot it is clear that as the concentration of the silk extraction increases, the absorption rate also increases in UVC region.

Transmission of light/ radiation is more in case of PVA that indicates it alone is incapable of absorbing UV radiations especially in the range between 230 to 300nm. In the same region, if we observe higher concentration film (2:1 sericin/PVA film) transmission rate is considerably less that is almost zero at 230 nm. Hence higher concentration film has got an ability to absorb almost all incident radiations and hence places a very important role in absorbing UV radiation.

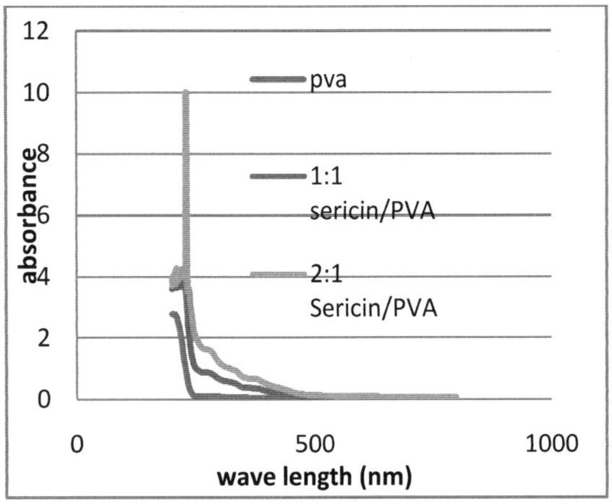

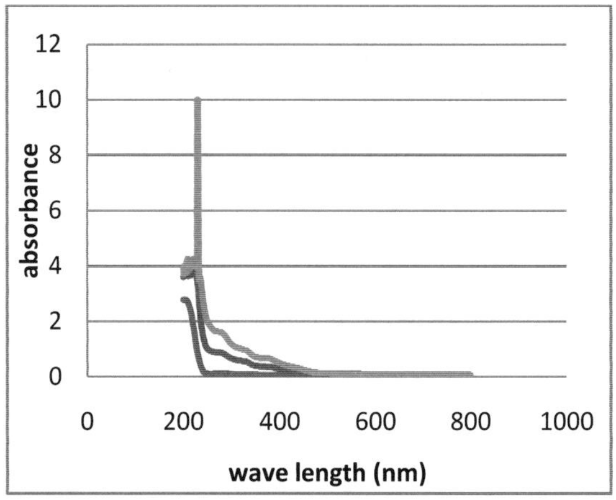

(a) (b)

Figure 3a & b: UV- visible absorption and Transmission spectra of sericin/PVA film

3.2 X-ray Powder Diffraction

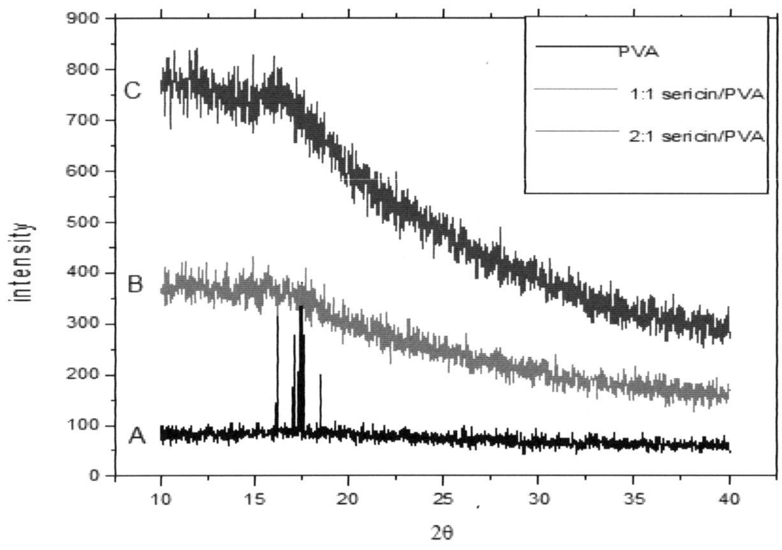

Figure 4: XRD of sericin/PVA films

Figure 4 shows different concentration of sericin in PVA. By monitoring the shape, position and intensity of the reflection or absorption type of crystallographic structure can be determined. Hence the below plotted graph obtained in the figure 4 indicates that crystallographic structure of various concentration of the sericin/PVA film. As shown in graph A, Pure PVA film has sharp peaks at 17 to 18.5^0 that indicate crystalline property. Further diffraction angle has no spark peaks indicating amorphous property. Hence PVA film is a semi crystalline material. As shown in graph B 1:1 sericin/PVA film has a broad valley without sharp edges, indicating amorphous property. Hence sericin/PVA is an amorphous material. Even graph C shows broad valley indicating amporphous property of the film. Therefore by observing the graph, it clearly indicates that as the concentration of the sericin in the developed film increases, amorphicity of the material also increases.

3.3 Fluorescence effect

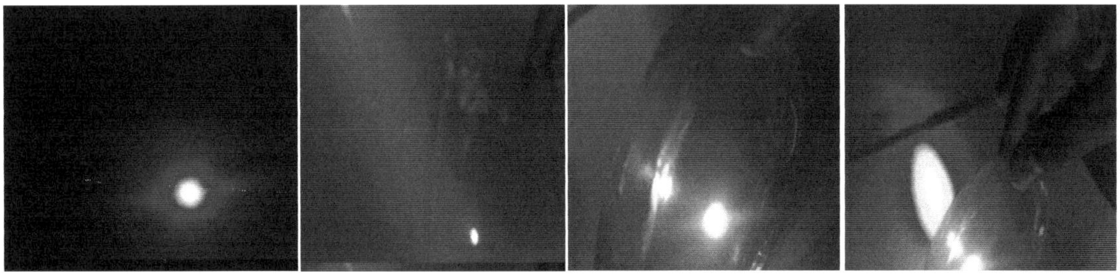

(a) (b) (c) (d)

Figure 5: Fluorescence effect of UV light on (a) plain paper, (b) 1:1 sericin/PVA film and (c) 2:1 sericin/PVA film

UV light/ radiation was made to fall on plain paper as shown in fig 5 a, it transmitted UV light (purple colour light) without absorbing. Then the same source of UV light was made to fall on PVA host film, again it transmitted hazardous UV light as shown in fig b. Later same source was made to incident on various concentration of sericin/PVA film, that absorbed phonons of UV light and transmitted non hazardous fluorescence light (green colour light) of various intensity as shown on fig c & d. Therefore by comparing various images of figure 5, higher concentration sericin/PVA film can effectively absorbed phonons of UV radiation and emit low intensity fluorescence light.

According to the principle of fluorescence, sericin//PVA film absorbs phonons (heat) of the UV source. This absorbed radiation excites the system and makes the electrons in the ground state to jump to higher energy level. After a lag time, approximately few nano seconds (life time of fluorescence), the excited electrons returns back to the ground state. While returning back the electron releases the stored energy in the form of photons (light). Hence according to Netwon's law of conservation of energy "phonons of the UV radiation are converted into photons". Therefore we can say that sericin/PVA laminated fabric absorbs UV radiation and releases a distinct green fluorescent colour as shown in figure

4. Conclusion

In this present research, sericin/PVA film has been developed a green, simplistic and economical film. For this, the fabrication of sericin/PVA was carried out by employing polymer solution casting method and slow evaporation method by varying the sericin concentration by keeping PVA constant that is in the ratio of 2:1, 1:1. The obtained results showed that increase in the concentration of the sericin in PVA increases the absorption capacity from 0.69 to 10 (Au) at 230nm wavelength. Transmission spectra of UV spectroscopy showed decrease in transmission from 19 to 0% at 230nm wavelength. The images of fluorescence effect of the developed film/fabric showed that presence of sericin in PVA has intensively reduced the adverse impact of UV phonons by converting phonons into fluorescence light, their by maintains the temperature of the developed film/fabric which is due to the strong conduction of phonons in the developed film. Therefore this developed amorphous sericin/PVA film is mechanically elastic and flexible, visibly transparent. Ability of developed film to absorb phonons of UV radiation and to convert harmful UV phonons into non hazardous fluorescence light and transmitting the heat. Hence, this film is the right choice to be used in gloves/fabrics especially in industries where workers are constantly exposed to harmful UV radiation and military persons who are directly exposed to sunlight so has to maintain comfortable temperature for the wearer.

Acknowledgments

This work was supported by NIE- Centre for research and development and provided financial assistant. The authors also thank sericultural research and training institute and faculty of The NIE, Mysuru.

References

[1] Eng CT. Breathable Garments with Thermo-Physiological Wear Comfort-A Review.
[2] Coyle S, Diamond D. Smart nanotextiles: materials and their application.
[3] Mäkinen H, Jussila K. Cold-protective clothing: types, design and standards. In Protective Clothing 2014 Jan 1 (pp. 3-38). Woodhead Publishing.
[4] Bibekananda basu, Polymers in Textiles
[5] Rajput SK, Singh MK. Sericin-a unique biomaterial. IOSR Journal of Polymer and Textile Engineering. 2015;2(3):29-35
[6] Kaur J, Rajkhowa R, Tsuzuki T, Millington K, Zhang J, Wang X. Photoprotection by silk cocoons. Biomacromolecules. 2013 Oct 14;14(10):3660-7.
[7] Sarovart S, Sudatis B, Meesilpa P, Grady BP, Magaraphan R. The use of sericin as an antioxidant and antimicrobial for polluted air treatment. Rev Adv Mater Sci. 2003 Dec 1;5(3):
[8] Barajas-Gamboa JA, Serpa-Guerra AM, Restrepo-Osorio A, Ã lvarez-López, Ingeniería y competitividad. Sericin applications: a globular silk protein 2016 Jul;18(2):193-20
[9] Pongsathorn Kumkun, Nantaprapa Tuancharoensri, GarethRoss, Sararat Mahasaranon, Jirapas Jongjitwimol, PaulD Tophamd and SukunyaRoss, Green fabrication route of robust, biodegradable silk sericinand poly(vinyl alcohol) nanofibrous scaffolds.
[10] Ramesan MT, Athira VK, Jayakrishnan P, Gopinathan C. Preparation, characterization, electrical and antibacterial properties of sericin/poly (vinyl alcohol)/poly (vinyl pyrrolidone) composites. Journal of Applied Polymer Science. 2016 Jun 20;133(24).
[11] plastics.americanchemistry.com/plastics/The-Basics
[12] Hu J, Lu J. Smart polymers for textile applications. InSmart Polymers and their Applications 2014 Jan 1 (pp. 437-475). Woodhead Publishing.
[13] Jocić D. Polymer-Based Smart Coatings for Comfort in Clothing. Tekstilec. 2016 Apr 1;59(2).
[14] Kundachira Subramani N. Opto-electrical characteristics of poly (vinyl alcohol)/cesium zincate nanodielectrics. The Journal of Physical Chemistry C. 2015 Sep 3;119(35):20244-55.

Microstructure and Wear Behavior of Microwave Treated WC-10Co-4Cr Composite Coating on AISI 4140 Alloy Steel

D G Pradeep* H S Nithin, B N Sharath, K S Madhu, C V Venkatesh

Department of Mechanical Engineering, Malnad College of Engineering, Hassan, affiliated to VTU, Belagavi, Karnataka, India

*Corresponding author e-mail address: pradeepdgmce@gmail.com

Abstract. WC-10Co-4Cr composite coatings have been coated with AISI 4140 low alloy steel substrate using high-speed oxygen fuel (HVOF) technology with a particle size from 0 to 50 microns. Microwave heating energy is used to enhance the mechanical and tribological properties of sprayed coatings as post-heat treatment technique. Microwave fused coatings exhibit greater surface roughness, porosity, micro hardness, and wear strength than the as sprayed deposition. The as coated and fused coatings have been characterized by x-ray Diffraction (XRD) and Scanning Electron Microscopy (SEM) procedure. Wear behaviors under normal loads of 5N, 10N and 15N have been investigated by a pin on a disc device at temperatures of 30 °C, 100 °C and 200 °C. The results revealed that microwave-treated coatings are highly recommended to protect the sliding surface, such as a high-speed spindle, under different environmental conditions.

Keywords: HVOF; Wear; WC-10Co-4Cr; Microwave Treated.

1. Introduction

In many applications, for example, automotive, aerospace, power plants and various industrial sectors, wear, corrosion, and thermal stability are anticipated phenomena. Therefore, in such applications, improved wear and dynamic resistance are crucial. These can be improved by hard thermal spray coatings in an effective and reasonable manner. These coatings may be formulated in a defensive atmosphere or in the air, preserving fair deposition efficiency at moderately low costs [1]. Thermal spray technique has been used for a long time to produce wear-resistant materials on the surface [2-3]. Accordingly, several studies have shown that, the wear phenomena is influenced by various possessions such as geometry, normal load, relative surface motion, rubbing surface texture, sliding speed, device rigidity, material type, lubrication and vibration [4-5].

Consequently, several research issues relating to friction and wear by sliding fields were created to address Wear Mechanism [6]. In many applications steel alloy AISI 4140 is commonly used as a result of high mechanical strength at high temperatures [7-9]. Carbide-based coatings are most suitable for wear applications. The HVOF technique is most preferable for carbide coatings, since it exhibits low temperature flame with high velocity compared to other thermal spray processes. Therefore, WC decarburization is significantly reduced during spraying [10-11]. WC-10Co-4Cr coating composition is widely used composite coatings for sliding and abrasive wear and slurry erosion because of higher hardness and fracture toughness [12-14]. Furnace heating is most often and conventional post treatment technique used to for thermal spray coatings to homogenies and strength the coating structure [15]. Microwave heating is the modern post treatment technique reported as cost effective solution for enhancing the thermal spray coatings [16-18]. The researcher reported that, Microwave fused Tribaloy-T400 and Cr3C2 reinforced Tribaloy-T400 composite coatings exhibited better mechanical and wear properties than as-sprayed coatings [19]. Hence, investigation on the effect of microwave fusing on carbide coatings is of high interest. In the present study, WC-12Co-4Cr is coated using HVOF technique over AISI4140 steel alloy. The as-sprayed coating is post treated using a domestic microwave oven. The sliding wear behavior of both as-sprayed and microwave fused WC-12Co-4Cr coatings are investigated at room temperature, 100 °C and 200 ° C. The phase, microstructure analysis and wear mechanism is examined using XRD, SEM and EDS techniques.

2. Experimentation

2.1 Materials and development of coating

In the present work, AISI 4140 low alloy steel was used as a substrate. The procured substrate was in the form of a rolled plate and plate was sectioned for the dimension of 75 mm x 75 mm x 25 mm in size. The base metal was blown at a pressure of 0.5 MPa with 16 # grit before spraying to achieve surface roughness, Ra 8~10μm for stronger bonding strength between the surface of the substrate as well as the coating. The WC-10Co-4Cr coating powder of 0 to 50 μm was deposited over the substrate by HVOF method and the parameters are shown in table 1. The composite feedstock powder was developed through agglomeration and sintering and had spherical grains.

Table 1. Spray parameters for HVOF

Powder	WC-10Co-4Cr
Hydrogen (lpm)	600
Oxygen (lpm)	300
Air (lpm)	200
Nitrogen (lpm)	Carrier gas 40
Distance (mm)	300
Traverse velocity (m/min)	80
Angle between gun to job	90°
Coating thickness (μm)	200

2.2 Microwave energy post treatment

Use of Microwave energy for post-treatment is designated as microwave fusing in the present investigation. Microwave fusing focused on the development of homogenized surface coating characteristics for a various application. As-coated WC-10Co-4Cr samples were cut into 10mm×10mm×5 mm form for the domestic microwave fuse. Figure 1 represents the as-sprayed coated samples were properly positioned on the refractory brick. Graphite sheet (extractor) of 1 mm thick with 99% purity was placed above the coated sample to produce carbon.

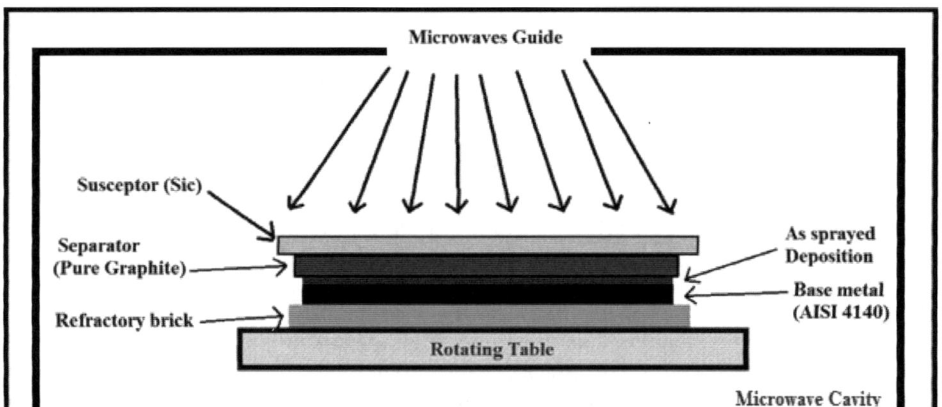

Figure 1. Microwave processing.

Silicon carbide used as a susceptor and placed above the graphite sheet and heats the coated sample through the absorption of microwaves. As soon as the setup is ready, the power 900W and frequency of 2.45 GHz has been set to start the fusion of as sprayed coating about 10 min. The fused coated samples were slowly cooled under atmospheric conditions. Scanning electron microscope (SEM) and x-ray diffractometer (XRD) was used to observe the morphology and phase analysis for both as-coated and fused coatings.

2.3 Wear studies

Sliding wear test was carried out using pin on disc tribometer (TR-20LE-PHM 400-Ducom) at different temperatures under dry conditions as per ASTM G99-05. The test temperatures are RT, 200°C and 400°C. Wear test was conducted with normal loads of 5N, 10N and 15N. A computerized interpretation system is utilized for continuous record of both height loss and frictional force. Wear rate is expressed in mm^3/m by converting the height loss into volume loss using the pin cross section area. Alumina disc is used as the counter body. Wear mechanism was observed by witnessing the worn surfaces and wear debris through SEM. The elemental composition of wear debris is determined using EDS (JSM 6380LA, JEOL, JAPAN). Dry sliding wear study parameters as shown in Table 2.

Table 2. The dry sliding wear parameters

Wear test method	weight loss method
Speed of disc	420 rpm
Normal load applied in steps	5N, 10N, 15N
Sliding temperature	Room temp, 100°c, 200°c
Duration considered	6 min
Sliding distance	700 m
Track diameter	90 mm

3. Results and Discussions

3.1 Microstructural Characterization

Figure 2 (a) and (b) illustrate the microstructure of as sprayed and microwave fused coatings respectively. The as sprayed deposition reveals a usually heterogeneous microstructure with several micro pores, microcracks and unmelted/partial melted particles as shown in the figure 2(a).

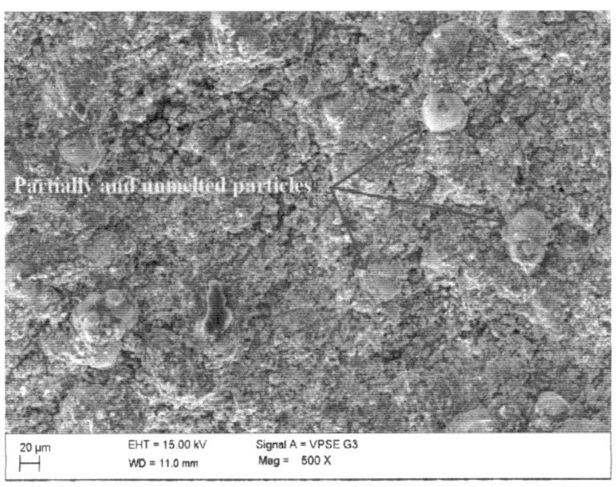

Figure 2a. Surface morphology of as-coated sample

Figure 2b. Microwave fused coating sample

Compared to the as sprayed coated samples, the microwave fused coating showed reduced pores and dense microstructures (Figure. 2b). A progressive effect on the characteristics of the HVOF coating was noticed via the microwave fusing. The Unmelted and partially melted particles leads to melting and flowed due to constant microwave exposure, likely filling up the pores. As a result of fusing a relative compact, uniform microstructure is observed in the post processed samples. The cross-sectioned morphology as-sprayed and microwave fused WC-10Co-4Cr coating is shown in Figure 3 (a) and (b) respectively. The porosity and surface roughness of the as-sprayed coating was 1.2 ± 0.20% and 6.52 µm respectively.

Figure 3a. As sprayed coating at the interface

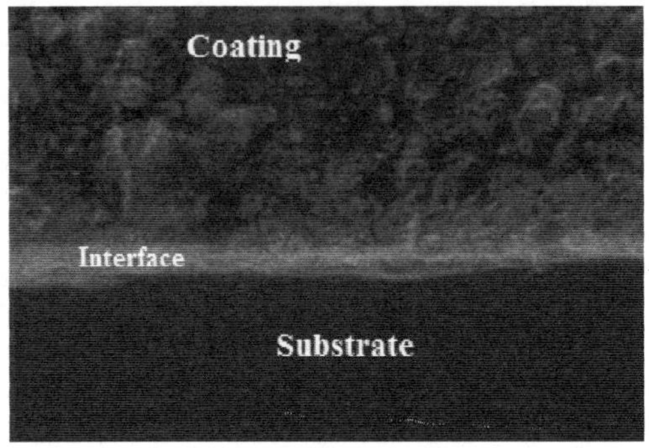

Figure 3b. Microwave treated coating at the interface

Throughout the sprayed coating, the microwave reels the unmelted and partially melted particles for a period of 10 min. Microwave fusing removes the micro cracks, pores and voids which are present in the as-sprayed coating. Fusing leads to the growth of diffusion layer at the interface of the substrate-coating which is observed Figure 3(b). Metallurgical bonding has been formed between the coating and the substrate, thereby improving the adhesion to the substrate. Microwave fused coating shows the porosity of 0.70 ± 0.3 and surface roughness Ra 4.21 µm. The rapid cooling processes of the microwave resulted in finer coating structure than those of conventional post-heat processes [20].

3.2 XRD analysis

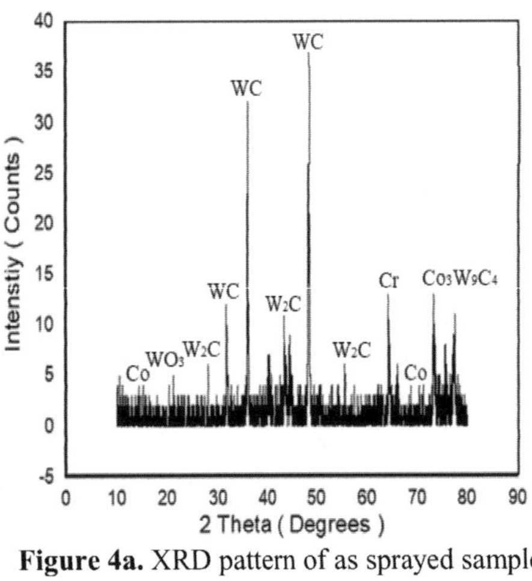

Figure 4a. XRD pattern of as sprayed sample

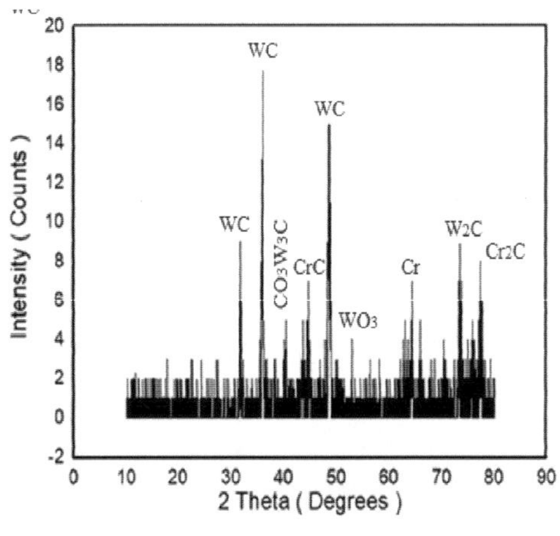

Figure 4b. XRD pattern of Microwave fused coated sample

Figure 4(a) and (b) represents the XRD patterns of as-sprayed and fused WC-10Co-4Cr coating on AISI 4140 steel alloy. In coated samples, the major presence of the WC phase was detected [21]. The various phases such as W_2C, WO_3 and $Co_3W_9C_4$ corresponds to minor peaks. Occurrence of WC oxidation and decomposition during inflight time coating process leads to above mentioned phases. The chemical reaction described below is more likely to occur when the WC to W_2C oxidants relative to other potential reactions [22]. A cobalt and chromium binding phase of the amorphous is attributed to the curve between 20 ° and 50 ° angles in 2θ (Figure 4.a). The amorphous phase of this coating is mostly found due to its rapid cooling effect [23]. In microwave fused coatings, no significant change in constituents is observed in as that of the as-sprayed coatings. The XRD pattern of fused coating indicate that effect of oxidation and decarburization are negligible. While tungsten is oxidized into metallic oxide during microwave fusing and also carbon entry occurs from graphite seperator [24].

3.3 Microhardness studies

The micro hardness test was carried out across the thickness of both as sprayed and microwave fused coatings. The microhardness report for the different loads of both coatings is shown in the figure 5. For 10N load, the microhardness of both as sprayed and fused coatings were 645 ± 20 HV1 and 738.7 + 15 HV1 respectively. Similarly, for the load of 20N, 621.3 ± 15 HV1 and 715.0 ± 15 HV1 were detected for both as sprayed and fused coatings respectively. The bare metal microhardness test was also performed and 199 ± 15 HV1 and 194.3 ± 10 HV1 values were found for both 10N and 20N loads respectively. For both the loads the microwave fused coating exhibited better micro hardness than the as-sprayed coating. Effective metallurgical bonding and uniform homogenized coating structure are the potential descriptions for an improvement in microhardness of the fused coating.

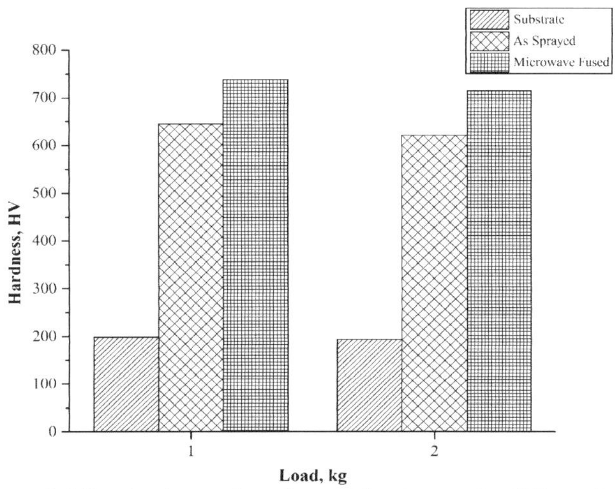

Figure 5. Microhardness of coatings with substrate for different loads

3.4 Dry sliding wear behavior

Figure. 6(a, b and c) demonstrates the friction coefficients of AISI 4140 steel, WC-10Co-4Cr coating and Fused coating at RT, 100 and 200 °C with 5N, 10N and 15N normal loads respectively. Tribological properties of as-sprayed and fused coatings are compared. Wear rate was determined considering the weight loss for all the loads and wear is represented in mm^3/m.

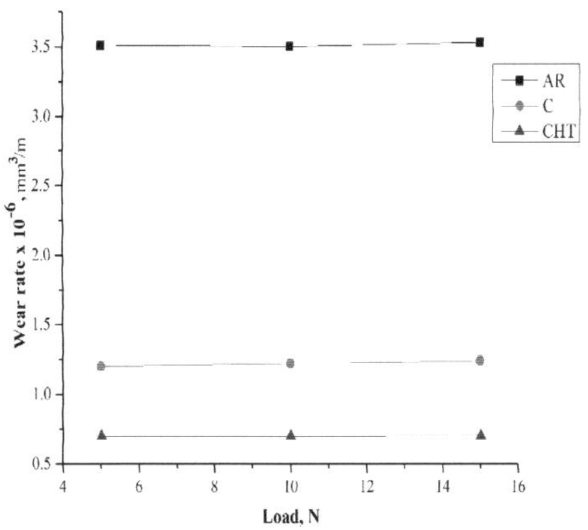

Figure 6a. Wear rate of room temperature for various loads

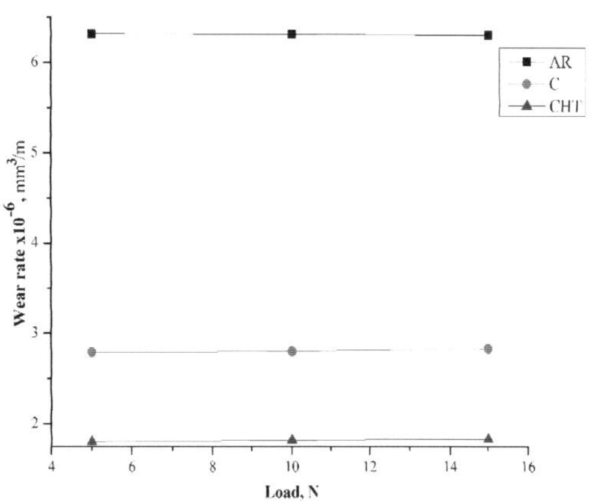

Figure 6b. Wear rate of 100 ^{0}C for various loads

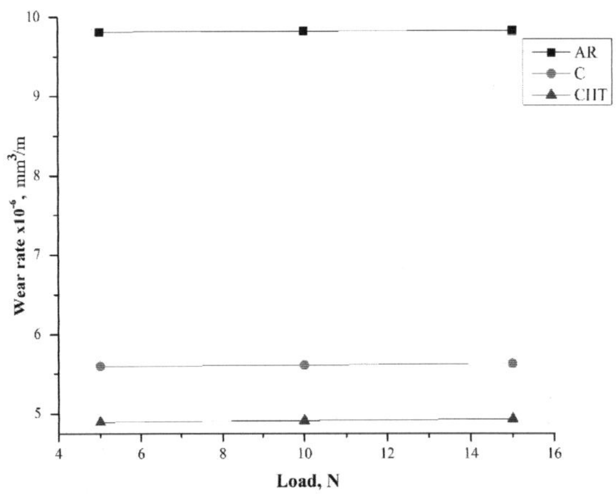

Figure 6c. Wear rate of 200 ^{0}C for various loads

From the figure 7 indicates that, fused coatings shows better wear resistance comparatively in all the test temperatures and loads. Fused coating shows the highest wear resistance at RT condition. Also the wear graphs indicates that significant wear rate change is not observed with respect to load. Similar observation have been made in all the test temperature. The flow of metallic binder during fusing resulted in strengthening the coating structure by adhering the carbide particles [25].

This was due to the high stiffness caused by the metallurgical bonding during the microwave fusing through diffusion of elements from the substrate to the coating. However, carbides like WC, W_2C and Cr2C give high temperature strength and prevent more material loss during sliding [26]. Substrate, as coated and microwave fused wear rate values are listed in Table 3.

Table 3. Wear rate values for normal load and temperature

Samples	Room temperature			100 ^{0}C			200 ^{0}C		
	5 N	10 N	15 N	5 N	10 N	15 N	5 N	10 N	15 N
Substrate	3.51* 10^{-6}	3.50* 10^{-6}	3.53* 10^{-6}	6.32* 10^{-6}	6.31* 10^{-6}	6.30* 10^{-6}	9.81* 10^{-6}	9.82* 10^{-6}	9.83* 10^{-6}
As coated	1.20* 10^{-6}	1.22* 10^{-6}	1.24* 10^{-6}	2.79* 10^{-6}	2.80* 10^{-6}	2.83* 10^{-6}	5.60* 10^{-6}	5.61* 10^{-6}	5.63* 10^{-6}
Microwave fused	7.017* 10^{-7}	7.015 *10^{-7}	7.020* 10^{-7}	1.80* 10^{-6}	1.82* 10^{-6}	1.83* 10^{-6}	4.89* 10^{-6}	4.31* 10^{-6}	4.33* 10^{-6}

Figure 7a and 7b represents the SEM images of the worn surface of as-sprayed and fused WC-10Co-4Cr coatings under 20 N normal load at 200 °C temperature respectively. It is observed from the worn surface that the coatings are characterized by groove, delamination and flaking pit for both as sprayed and fused condition. Whereas the fused coating shows the smoother wear surface as compared to as-sprayed coating. Micro cracks are observed for both worn surfaces, as a result of fracture of metal oxide formed due to oxidation of coating elements during high temperature wear exposure. Flaking pits are more frequently observed in as-sprayed worn surface which leads to severe material loss. Fused coating shows the smoother worn surface and experiencing the minimal delamination and micro crack areas. The microwave fusing of the coating leads to better adherence between the coating splats, this prevents the coating from flaking and delamination [27]. Also, the dispersion of carbon form the graphite separator enhances the wear property of the fused coating.

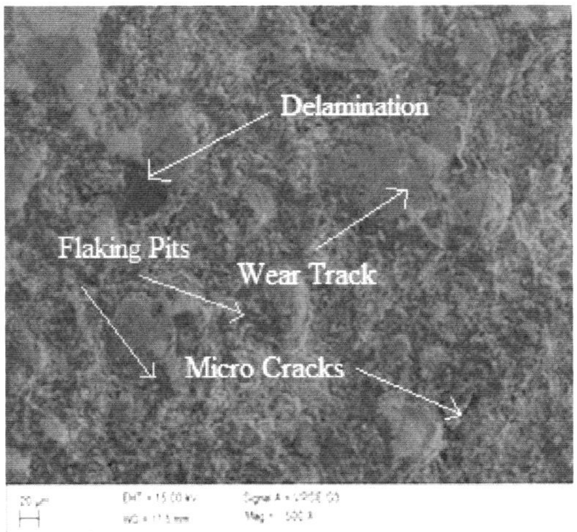

Figure 7a. Surface morphology of as sprayed coating worn surface

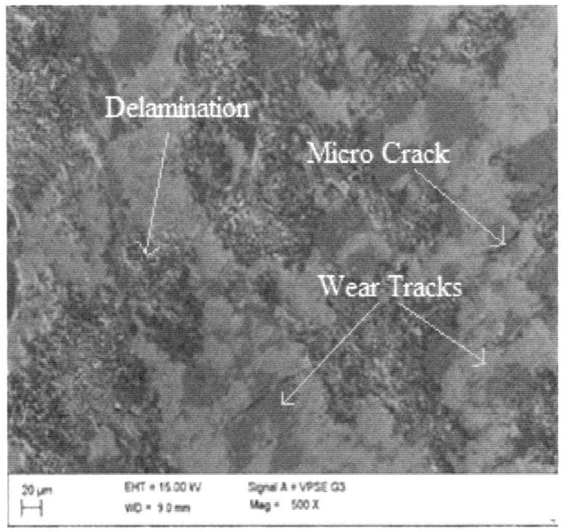

Figure 7b. Microwave fused coating worn surface

4. Conclusions

1.WC-10Co-4Cr feedstock was successfully coated over AISI4140 low steel alloy using HVOF spray technique with a particle size of 0 to 50 μm and an average coating thickness of 250 μm.
2.The micro hardness for 10N was 645 ± 20 HV1 and 738,7 + 15 HV1 for both as sprayed and microwave fused coatings. Furthermore, both as sprayed and fused coatings were observed for loads of 20N, 621.3 ± 15 HV1 and 715.0 ± 15 HV1 respectively.
3.In comparison, the microwave-treated coating showed a lower rate of wear than the substrate and spray coatings for specific temperature-related to applied loads.
4.Microwave fusing resulted in the development of metallurgical bonding between coating-substrate interfaces and fusing leads to better cohesion between the splats.
5.The dispersion of carbon to the coating from the graphite separator sheet during the fusing process will be beneficial for wear resistance

References

[1] L. Pawlowski (Ed.), *The Science and Engineering of Thermal Spray Coatings*, Wiley, Chichester, 1995.
[2] Rajinikanth V, Venkateswarlu K. *An investigation of sliding wear behavior of WC–Co coating*. Tribol Int 2011; 44:1711–9.
[3] Nithin HS, Desai V, Ramesh MR. *Elevated temperature solid particle erosion behaviour of carbide reinforced CoCrAlY composite coatings*. Materials Research Express. 2018 Jun 20;5(6):066529.
[4] M.B. Peterson and W.O. Winer, *Wear Theory and Mechanisms, Wear Control Handbook*. ASME, New York, 1980, pp. 35-80.
[5] M.A. Chowdhury and D.M. Nuruzzaman, *'Experimental Investigation on Friction and Wear Properties of Different Steel Materials'*, Tribology in Industry, 2013, vol. 35, no. 1, pp. 42-50.
[6] O. Barrau, C. Boher, Gras R, Rezai-Aria F. *Analysis of the friction and wear behaviour of hot work tool steel for forging'*, Wear, 2003, vol. 255, no. 7, pp.1444-1454.
[7] Choo SH, Lee S, Golkovski MG. *Effects of accelerated electron beam irradiation on surface hardening and fatigue properties in an AISI 4140 steel used for automotive crankshaft*. Materials Science and Engineering: A. 2000 Nov 30;293(1-2):56-70.
[8] Campbell Jr FC. *Manufacturing technology for aerospace structural materials*. Elsevier; 2011 Aug 31.
[9] Medina-Flores A, Oseguera J, Santiago P, Ascencio JA. *Structural analysis of AISI-SAE 4140 steel nitrided by post-discharge microwave.* Surface and Coatings Technology. 2004 Nov 1; 188:7-12.
[10] Y. Qiao, T.E. Fischer, A. Dent, *Surf. Coat.* Technol. 2003, 172 (1) 24.

[11] H. Zhao, Z. Ding, Y. Zhang, Q. Wang, Proc. Int. *Therm. Spray Conf.* 2007, 884.

[12] S. Hong, Y. Wu, B. Wang, Y. Zheng, W. Gao, and G. Li, *High- Velocity Oxygen-Fuel Spray Parameter Optimization of Nanostructured WC–10Co–4Cr Coatings and Sliding Wear Behavior of the Optimized Coating,* Mater. Des., 2014, 55, p 286- 291.

[13] S. Bhandari, H. Singh, H. Kumar, and V. Rastogi, *Slurry Erosion Performance Study of Detonation Gun-Sprayed WC-10Co-4Cr Coatings on CF8M Steel Under Hydro-Accelerated Conditions,* J.Therm. Spray Technol., 2012, 21, p 1054-1064.

[14] Shetty DK, Wright IG, Mincer PN, Clauer AH. *Indentation fracture of WC-Co cermets.* Journal of Materials Science. 1985, 20(5):1873-82.

[15] Thostenson, E.T., Chou, and T.W. *Microwave processing: fundamentals and applications in composites*. Composites Part A: Applied science and Manufacturing, 1999, 30, 1055–1071

[16] Rajkumar K, Rajan P, Charles JM. *Microwave heat treatment on aluminium 6061 alloy-boron carbide composites.* Procedia Eng. 2014 Jan 1;86:34-41.

[17] Thostenson ET, Chou TW. *Microwave processing: fundamentals and applications.* Composites Part A: Applied Science and Manufacturing. 1999 Sep 1;30(9):1055-71.

[18] Rao SR, Padmanabhan G. *Fabrication and mechanical properties of aluminium-boron carbide composites.* International journal of materials and biomaterials applications. 2012 Aug;2(3):15-8.

[19] Prasad CD, Joladarashi S, Ramesh MR, Srinath MS, Channabasappa BH. *Influence of microwave hybrid heating on the sliding wear behaviour of HVOF sprayed CoMoCrSi coating.* Materials Research Express. 2018 Jul 26;5(8):086519.

[20] C. Durga Prasad, Sharnappa Joladarashi, *Effect of microwave heating on microstructure and elevated temperature adhesive wear behavior of HVOF deposited CoMoCrSi-Cr_3C_2 coating,* Surface & Coatings Technology Vol. 374 (2019) 291–304.

[21] Guilemany JM, Espallargas N, Suegama PH, Benedetti AV. *Comparative study of Cr3C2-NiCr coatings obtained by HVOF and hard chromium coatings.* Corrosion Science. 2006;48(10):2998-3013.

[22] C. Borgnakke, R.E. Sonntag, *Fundamentals of Thermodynamics*, seventh ed., Don Fowley, 2009.

[23] Kang AS, Grewal JS, Cheema GS. *Effect of thermal spray coatings on wear behavior of high tensile steel applicable for tiller blades.* Materials Today Proceedings. 2017;4(2 Pt A):95-103.

[24] Kun-ying Ding, Peng-cheng Lu, *Nanoindentation Behaviors of HVOF WC-10Co-4Cr Coating after Heat Treatment,* Advanced Materials Research 2013, Vol. 629 pp 14-18.

[25] Zhu Y C, Yukimura K, Ding C X, and Zhang P Y, *Thin Solid Films* 2001, 388, 277.

[26] Zoei MS, Sadeghi MH, Salehi M. *Effect of grinding parameters on the wear resistance and residual stress of HVOF-deposited WC-10Co-4Cr coating*. Surface and Coatings Technology. 2016;307(Pt A):886-891.

[27] Doddamani M, Mathapati M, Ramesh MR. *Microstructure and tribological behavior of plasma sprayed NiCrAlY/WC-Co/cenosphere/solid lubricants composite coatings*. Surface and Coatings Technology. 2018 Nov 25;354:92-100.

Computational study of effect of Ceramic coatings and piston bowl depth on heat fluxes of Re-entrant type piston

B Suresha[1*], Anantha Padmanabha[1], Amith R[1], Ishwara Prasanna S[1] and I M Janamdar[1]

[1]Centre for Composite Materials Research, Department of Mechanical Engineering, The National Institute of Engineering, Mysuru – 570008, Karnataka, India

[*]Corresponding author: sureshab@nie.ac.in

Abstract. The current research article investigates the variation of heat flux with geometric modification on the re-entrant type piston bowl by varying bowl depth by computational domain. The contribution of ceramic coating on heat flux is also investigated in this research report. Computational study was carried out by using ANSYS 19.2 software in order to obtain the augmented results. The variation of piston bowl depth of Re-entrant type piston is accomplished as there is a novel need to study the variation in heat flux of piston at different bowl depths. The impact of the thickness of the ceramic coating is also evaluated. The focus of this article is to study the effects of combustion on the piston of a compression ignition engine. It was noticed that with the decrease of bowl depth and addition of ceramic coating there is a decrease in heat flux. Around 27.75% depreciation in the heat flux was noticed for the uncoated piston with a bowl depth of 15mm in comparison with 27 mm bowl depth and a 25% decrement with the addition of a ceramic coating of 0.4 mm and also a 55% decrement in the heat flux was observed with 1mm coating to the piston with 15 mm bowl depth in reference to 27 mm bowl depth of respective coating thickness. Disparity of maximum heat flux and heat losses with increasing bowl depth is also analyzed in the inquiry.

Keywords: Re-entrant type piston, Thermal barrier coating, Heat flux, Piston bowl depth

1. Introduction

Enhanced torque production and high load carrying capability is the noteworthy benefit of diesel engines. The shape of the combustion chambers play a vital role in the comprehensive mixing of air and fuel in such engines. The variation of the piston bowl profile is the renowned factor in order to obtain optimal engine performance and emissions [1]. Squish, swirl and Turbulence are highly influenced by piston crown geometry and the intake system [2]. Piston bowl geometry has a greater influence on intensified squish near Top Dead Centre. Thus, the attainment of a thorough combustion process is effected by piston bowl shape [3]. To acquire complete and uniform mixing, there is always a potential need to change the shape of the piston bowl contour [4]. Donateo et.al also showed that the process of changing the shape of the piston is cost-effective [5].

The present work involves the usage of re-entrant type piston crown geometry with chamfer. Comparatively lower emission rates are the conspicuous advantage of Re-entrant type piston [6]. Narrow throat of re-entrant type piston imparts potent squish and reverse squish flow thereby augment the air-fuel mixing [7]. Foremost reason for lesser emissions is that there is a remarkable increment in squish, swirl and Turbulence Kinetic Energy during compression stroke as a result of symmetric vortices of Re-entrant piston bowl [8-10]. The optimum re-entrant angle is prominent in having decremental HC, CO and smoke emissions [11, 12]. Mamilla et al. concluded that there is an enhanced brake thermal efficiency and peak pressure heat release rate with re-entrant piston [13]. Chamfering is basically done to split the fuel into two different portions. One of the portions goes to the depth of the bowl and works similarly as that of the re-entrant bowl and fuel in the chamfered region utilizes surface air for comprehensive combustion. Rapid air-fuel mixing is the utmost motto in the utilization of a chamfered bowl [14]. The prime limitation of re-entrant piston is that the increase in the re-entrant angle and complexity of crown geometry results in diminished thermal stability of the entire structure. The inquiry of the after-effects of combustion and thermal stability of re-entrant piston is the prime focus of the current investigation. Therefore, to inflate the thermal stability there is a potential need to study the consequence of coatings on the piston crown.

Thermal barrier coatings (TBCs) are generally coated on any material to degrade heat rejection [15]. Enhanced power density, fuel efficiency increment and better combustion characteristics leading to multi-fuel capacity is the pronounced supremacy of TBCs in diesel engines [16, 17]. Ahmanemi et al. noticed that TBCs improved the engine power by 8% and there was a notable decrement in specific fuel consumption by 15-20% [18]. The current work puts forward the study of the effect of ceramic D8 coatings on the piston crown. Jalaludin et al. in one of their research articles concluded that lower conductivity of ceramic coating helped in a 98% reduction in heat flux when compared to uncoated piston [19]. Elevated thermal efficiency and declined emissions are observed by Sethuraman et al. by coating creaming coatings on piston crown [20].
In this research article computational study was carried out by using ANSYS 19.2 commercially available software. The entire practical conditions are incarnated in the computational domain. The chief reason for opting for a computational study is to clearly study the scenario of piston bowl depth variation and to obtain the augmented results.

2. Experimentation

Fig. 1 depicts the entire workflow of the current research investigation. Initially, the model was created as the per the dimensions. Later the entire simulations were carried out using ANSYS 19.2 general purpose software.

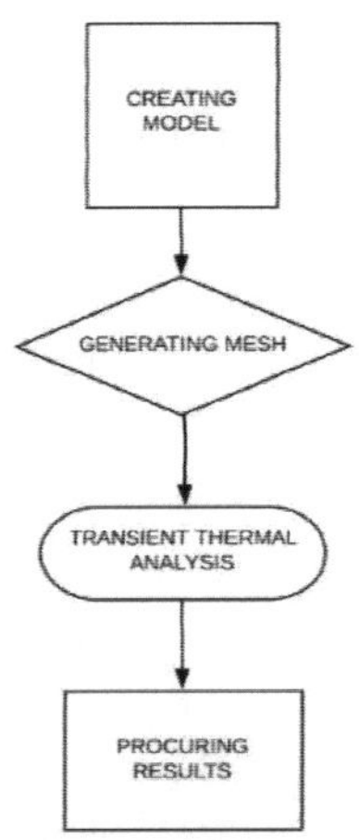

Figure 1. Work flow chart

IOP Conf. Series: Materials Science and Engineering 1189 (2021) 012013 doi:10.1088/1757-899X/1189/1/012013

2.1 Creating model

General purpose package software CATIA V5 has been employed for modelling chamfered re-entrant type piston. The designed piston can be utilized in a diesel engine. The dimensions of the piston inculcated in the present inquiry have been adopted as mentioned by Muhammet Cerit and Mehmet Coban [21]. The initial 3D model of the designed piston is as portrayed in Fig. 2.

Figure 2. 3D model of the Piston

The main focus of this analysis is to illustrate the effect of variation of piston bowl depth on the thermal stability of the piston. The initial depth of the piston bowl was 27 mm. The depth variations have been done in three iterations namely 25 mm, 20 mm and 15 mm. Chamfer radius of 5 mm and the length of 160 mm were kept constant in each of these cases as represented in Fig. 3.

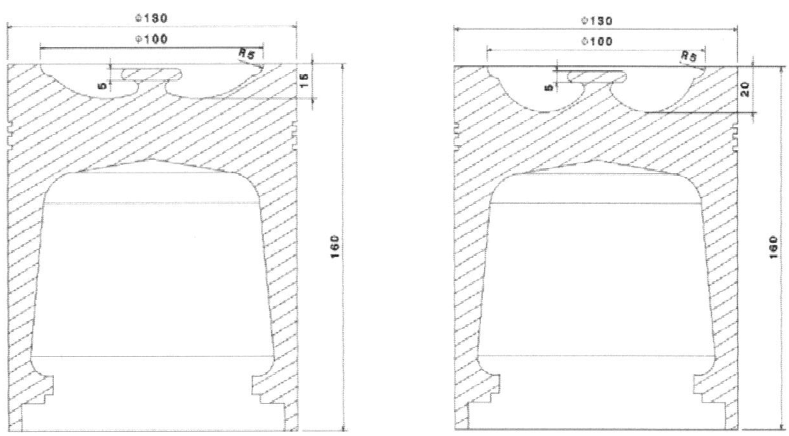

a) 15 mm b) 20 mm

IOP Conf. Series: Materials Science and Engineering 1189 (2021) 012013 doi:10.1088/1757-899X/1189/1/012013

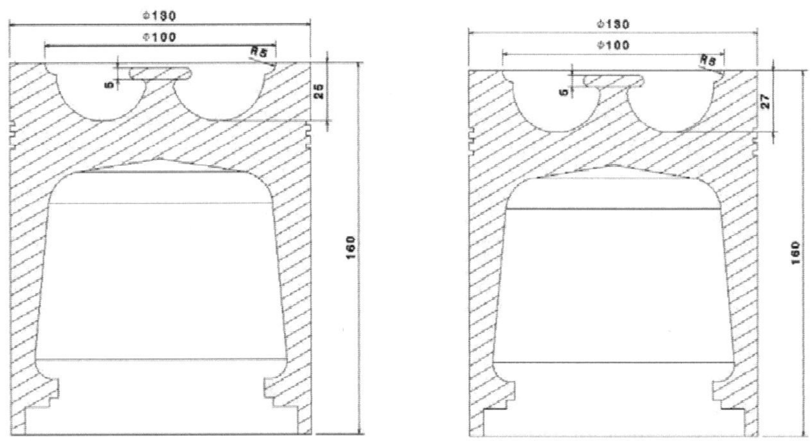

c) 25 mm d) 27 mm

Figure 3. Cross-sectional view of pistons with different bowl depths (all dimensions are in mm)

2.2 Generating mesh

Meshing plays a vital role in obtaining accurate results. The entire piston has been discretized into nodes and elements. Table 1 shows the mesh details of coated and uncoated Re-entrant type piston with 27 mm bowl depth.

Table 1. Meshing data of uncoated and coated piston with 27 mm bowl depth

Mesh details	Uncoated piston	Coated piston
Number of elements	165057	181262
Number of nodes	265130	298376
Average element size (mm)	1	1

Bob et al. scrutinized the intensive effect of the shape of nodes in a mesh in obtaining augmented results [22]. There are typically two kinds of meshes namely structured and unstructured meshes. Unstructured triangular meshes are currently a part of this study. Massive parallel computations can be done comprehensively by such kind of meshes [23]. Unconditionally well adaptable way of unstructured mesh, when compared with structured mesh with complex geometry, makes it more predominant in computational fluid dynamics [24]. Less computational time and during the process of adaptive mesh refinement, no moderation is required in terms of the basic algorithm, the mesh always remains as unstructured triangular mesh.

Stable and accurate results are also one of the major criteria for implementing triangular meshes. Also, second-order accuracy can be obtained more effortlessly [25, 26]. Fig. 4 renders the completely meshed uncoated chamfered piston.

The mesh details of the coated piston crown is also shown. A coating of 1mm thickness and 0.4 mm thickness has been created with ANSYS 19.2 software. As expected with the coating, the number of cells

IOP Conf. Series: Materials Science and Engineering 1189 (2021) 012013 doi:10.1088/1757-899X/1189/1/012013

increased. Fig. 4b shows this trait. Table 1 clearly portrays that there is a significant increment with the number of cells.

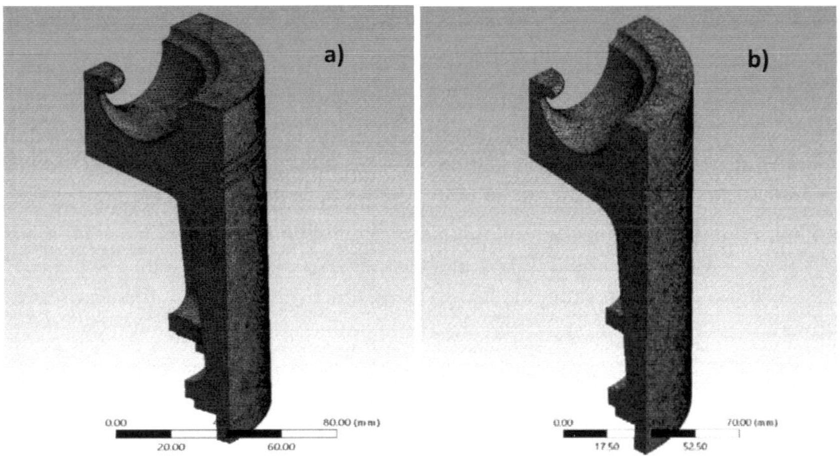

a) Coated b) Uncoated

Figure 4. Mesh for coated and uncoated pistons

2.2.1 Mesh independent test

Initially, mesh independent test was done on uncoated re-entrant type piston with 27 mm bowl depth with the maximum temperature being 1200 K. To account for the independency of the mesh, two distinct mesh sizes namely medium mesh and fine mesh were utilized that was available in Ansys 19.2 software which is depicted in Fig. 5. From the outcomes, it can be concluded that there is no significant variation in the obtained results. The computational time was very high in solving fine mesh grids when compared to a medium-mesh. Drastic abatement in computational time in getting results of medium meshes is the prime reason for their usage. The procedure utilized by Yang et al. is followed in this study [27].

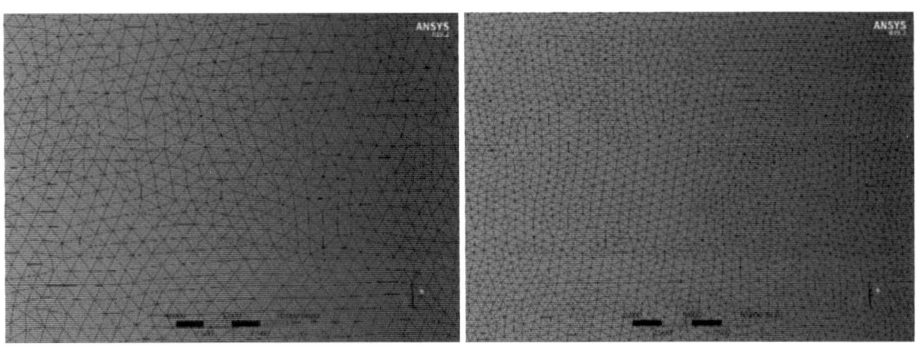

a) Medium mesh b) Fine mesh

Figure 5. Meshes used to conduct mesh independent test

2.2.2 Material properties

The current investigation utilizes a cast alloy as a piston material in the computational domain, having a density of 2680 kg/m^3 and Young's modulus being 73 GPa. It possesses a thermal conductivity of 152 W/m^2 °C and has a specific heat of 963 J/ kg °C. Ceramic 8D coating is used on the piston crown region with 0.4mm and 1mm thickness respectively. Basically, ceramic coatings have high Knoop hardness. The coating used in the present work is encompassed with a density of 4900 kg/m^3 and has an isotropic thermal conductivity of 4.5 W/m C.

2.2.3 Boundary conditions

The boundary conditions on various parts of the piston are adapted from the work done by Kumarasekaran and Safdari [28]. Fig. 6 depicts the various boundary conditions. The maximum temperature was 1200 K and it is found on the piston crown region. quadrant of the piston design is used to portray the entire boundary conditions.

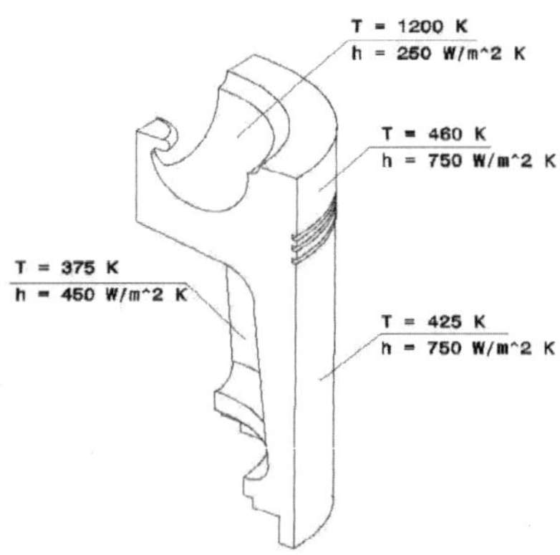

Figure 6. various boundary conditions on the piston

2.3 Transient thermal analysis

Transient thermal Analysis of piston was showcased in the present work by general-purpose ANSYS 19.2 software. The main motive for employing transient thermal analysis is to obtain the changes in temperature and heat flux profiles with reference to time. As it was noticed that temperature changes drastically with time in an IC engine, it was imperative to use transient analysis. In any practical application, the amount of heat entering the piston is not always equal to the amount of heat convected and conducted through the piston rings to air or any other coolant.

In order to realize the distribution of temperature and variation of heat flow rate across the domain, solution for fundamental equations like heat conduction, contact equation and convection equation has to be delineated. The equations for transient thermal analysis is as represented by Subodh et al. [29].

The general heat conduction equation is given by,

$$K\nabla^2 T + q_e - \rho C\frac{\partial T}{\partial t} = 0 \quad (1)$$

where K stands for thermal conductivity, q_e represents internal heat generation, ρ represents the density of the material. C stands for specific heat.

Conductive boundary representation by variational formulation is shown by

$$\frac{\partial \chi_k^{(e)}}{\partial\{t\}^{(e)}} = \left[\begin{matrix} k_{11} & k_{12} & k_{13} \\ k_{21} & k_{22} & k_{23} \\ k_{31} & k_{32} & k_{33} \end{matrix}\right] \begin{Bmatrix} t_i \\ t_j \\ t_k \end{Bmatrix} = [k]^{[e]} \{t\}^{(e)} \tag{2}$$

$[k]^{[e]}$= Global stiffness matrix
The global matrix is obtained as mentioned by Samria et al [30]

The governing equation for the contact boundary can be portrayed as

$$q_c = h_c\,(T^e - T^p) \tag{3}$$

The variational formulation for contact boundary between two elements is noted as

$$\chi_{bcont.} = \frac{h_c}{2} \int_{s_i}^{s_j} [\{t\}^e - \{t\}^p]^2\, 2\pi r ds \tag{4}$$

Upon differentiation of the equation (4) with respect to surface yield Temperature, a linear equation is obtained as rendered in equation (5) which is the part global set of equations

$$\frac{(\partial \chi_{bcont.})^e}{\partial\{t_s\}^e} = \frac{2\pi h_c r_{mr_{ij}}}{6\cos\theta} \begin{bmatrix} 2-\frac{\varepsilon}{2} & 1 \\ 1 & 2+\frac{\varepsilon}{2} \end{bmatrix} \cdot \begin{Bmatrix} \{\{t_s\}^e - \{t_s\}^p\}_1 \\ \{\{t_s\}_e - \{t_s\}_p\}_2 \end{Bmatrix} \tag{5}$$

The general heat convection equation is given by

$$-K \left[\frac{\partial T}{\partial n}\right] = h\,[T - T_\infty] \tag{6}$$

Variation formulation is represented by,

$$\frac{\partial \chi_{bconv}}{\partial\{t_s\}} = \frac{2\pi h r_{mr_{ij}}}{6\cos\theta} \begin{bmatrix} 2-\frac{\varepsilon}{2} & 1 \\ 1 & 2+\frac{\varepsilon}{2} \end{bmatrix} \begin{Bmatrix} t_{si} \\ t_{sj} \end{Bmatrix} - \begin{Bmatrix} (ht_\infty)_1 \\ (ht_\infty)_2 \end{Bmatrix} \tag{7}$$

$$\text{Here, } r_m = \frac{r_i + r_j}{2},\; r_{ij} = r_j - r_i,\; r_j = r_m - \frac{r_{ij}}{2},\; \varepsilon = \frac{r_{ij}}{r_m} \tag{8}$$

$$\frac{\partial \chi_{bconv}}{\partial\{t_s\}} = \{H_s\}\{t_s\} - \{t_\infty\} \tag{9}$$

Thus, a combination of variation integrals of the heat transfer equation is

$$([K] - [H])\{T\} = \{h_f\} \tag{10}$$

where $[K]$= conduction matrix, $[H]$ =convection matrix, $\{T\}$ =unknown temperature at each nodal point, h_f = column vector with a known quantity. The global matrices introduced above are of the order of (N×N) where N stands for the number of nodes.
Equation (10) is solved at every nodal point to attain the desired solution.

3. Results and discussion

3.1 Validation

The present work has been numerically validated. The general heat conduction equation has been solved to obtain augmented values of heat flux. The values obtained for heat flux numerically is compared with the results delivered from the computational domain. The volumetric expansion accounting for the volumetric deformation is also solved numerically. Equation (11) depicts the co-relation for obtaining volumetric expansion.

a) General heat conduction equation

$$K\nabla^2 T + q_e - \rho C\frac{\partial T}{\partial t} = 0 \quad (11)$$

b) Equation of volumetric thermal expansion

$$\Delta V = 3\alpha V_0 \Delta T \quad (12)$$

where α is the coefficient of thermal expansion and V_0 is the initial volume.

The results obtained in the computational domain are in good agreement with those obtained mathematically.

3.2 Impact of heat flux on piston crown without coating

The variation of heat flux is more pronounced in the piston bowl region as portrayed by the Fig. 7. So the crown region is more intensely studied in this section. From these figures, it can be noticed that the red region i.e. the region with maximum heat flux occupies the majority of surface area within the piston crown. Main cause for the present characteristic is that the crown region is exposed to higher combustion temperature as the entire process of combustion is incarnated within the crown region. It can also be perceived that the maximum temperature of the entire combustion cycle is taken up by the piston crown. In the case of a 15mm depth piston bowl, the heat flux value inflated to around 13395000 W/m^2. There is an unpredictable variation of maximum heat flux with the increment in the bowl depth. At 27 mm there is a comprehensive enhancement of heat flux to 14880000 W/m^2. With the bowl depth augmentation, there is an accrual of maximum heat flux. From the images, it can be observed that as the depth increases there is a conspicuous narrowing of the region where increased heat flux is found. It can be stated that as there is a narrowing of the red region, a considerable decrement in its area can be observed. This can be the prime reason for the escalation of maximal heat flux with bowl depth. It can be discerned that both increases in maximum heat flux and narrowing of the area is very less as outlined by Fig. 6. Maximum heat flux variation with respect to depth is as illustrated in Table 2.

Table 2. Variation of maximum heat flux with respect to bowl depth

Bowl depth (mm)	Max heat flux (W/m^2)
15	13395000
20	14254000
25	14801000
27	14880000

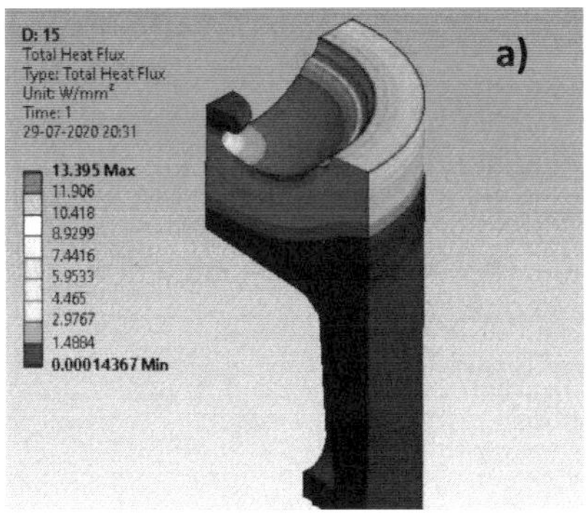

a)15 mm bowl depth with larger red region

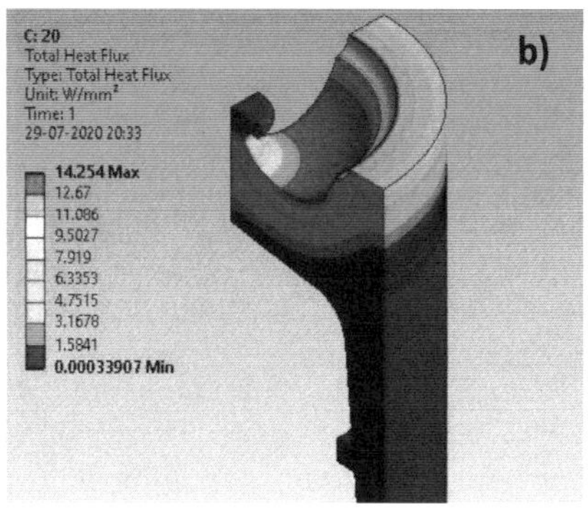

b) 20 mm bowl depth with slightly narrowed region

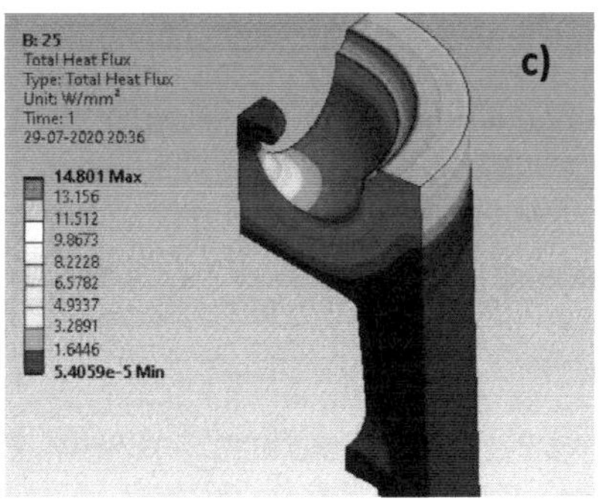

c) 25 mm bowl depth with narrowed red region

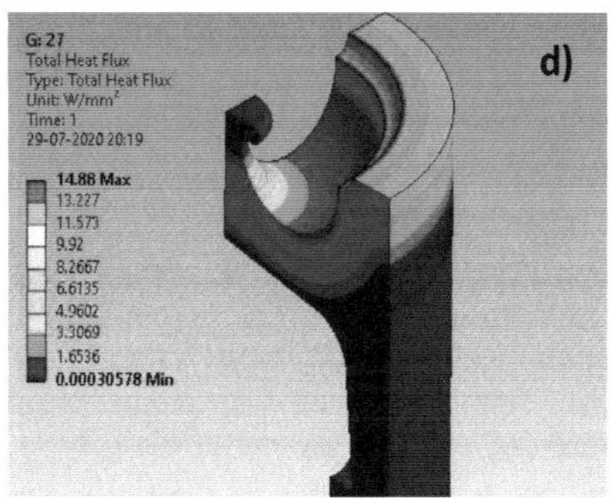

d) 27 mm bowl depth with narrowed red region

Figure 7. Variation of heat flux in different bowl depths

3.2.1 Expansion behaviour of piston with varying depth

The presence of heat is the supreme motive for the expansion of the piston. Analysis of volumetric deformation in reference to piston bowl depth is encompassed in the Fig. 7. The outcome renders that with the increase the bowl depth, volume expansion is decreased. With the inflated bowl depth, there is an enhanced surface area. Hence, the area available for heat flow is increased. This might be the probable reason for diminished volumetric expansion with bowl depth increment. Fig. 7 clearly depicts the trend of decrement. These data show that there is a negligible volumetric deformation. This depicts the fact that irrespective of bowl depth, geometry as a whole has a stable configuration and has the capability to withstand the fluctuating temperature.

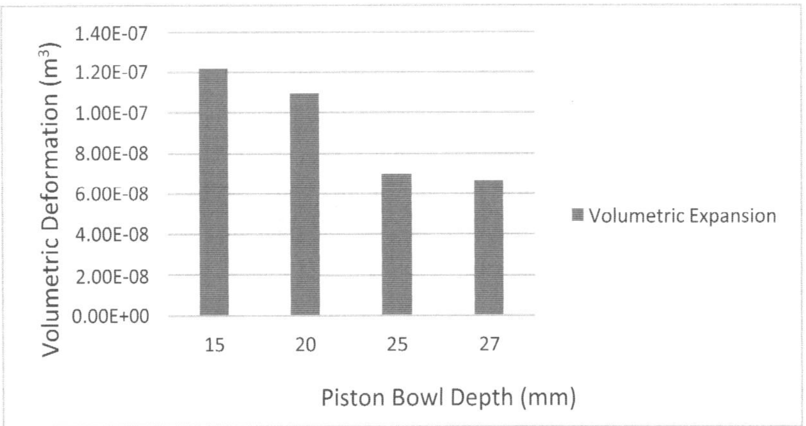

Figure 8. Variation of volumetric deformation with bowl depth

3.3 Effect of ceramic coating on variation of heat flux

Ceramic coatings are applied on the piston crown in 2 iterations of 0.4 mm and 1mm. A significant decrement in heat flux is the major insight with coating the piston crown region. Substantial reduction of heat flux around 20- 25% with 0.4 mm coating thickness and around 50- 55% with 1 mm coating thickness is realized. These can be accounted for the fact that most of the heat on the crown region has been absorbed by the ceramic coating. It is a known fact that heat flux is a product of heat transfer coefficient and the difference of temperature between the coating and the modelling surface. There is a drop in the heat flux with inflating surface temperature. This mechanism of ceramic coatings play a vital role in its insulation behaviour. It is evident that the ceramic coatings have effectively acted as a thermal barrier. Fig. 8 shows the heat flux feature on 27 mm bowl depth with 0mm, 0.4mm and 1mm coating. It clearly depicts that the awful diminishing of heat flux is the function of the thickness of the ceramic coating. With no coating, a larger variation of heat flux is encompassed in the crown region. This fluctuation is chiefly due to the more complex geometry of the piston crown. With the coating, extremely high heat is resisted due to the high insulating capability of the coating. This in turn, would enhance the heat flow rate within the combustion chamber thereby raise the combustion efficiency and fuel efficiency of an engine.

3.3.1 Thickness effect on heat flux

Comprehensive heat flux decrement with inflating thickness can be accounted for from the Fig. 9. There is a clear trend of increasing average heat flux with bowl depth in the absence of thermal barrier coating. With the incremental coating thickness, there is a substantial decrement in heat flux. More the thickness of the coating, more is the heat being retained in the combustion chamber and better the entire process of combustion. 0.4 mm thickness shows the trend of increment of heat flux with increasing bowl depth. The trend remains unchanged even for 1 mm coating thickness. It can be noticed at 0.4mm and 1mm thickness with respect to uncoated piston has a decrement of heat flux. There is a percentage depreciation of 27.29%, 21.60%, 20.87% and 20.30% for 15 mm, 20 mm, 25 mm and 27 mm bowl depth respectively. The percentage depletion of 56.22%, 56.30%, 55.84% and 55.53% can be seen for 15 mm, 20 mm, 25 mm and 27 mm bowl depth respectively in reference to the uncoated piston. Noteworthy outcome is that for 1mm coating thickness there is a substantial decrement of heat flux in comparison of 0.4 mm thickness. Enhanced Knoop hardness with incremental thickness is the major reason for its stability [31]. The entire thermal load, as well as the cooling load on the piston, is reduced by the increased TBC thickness [32].

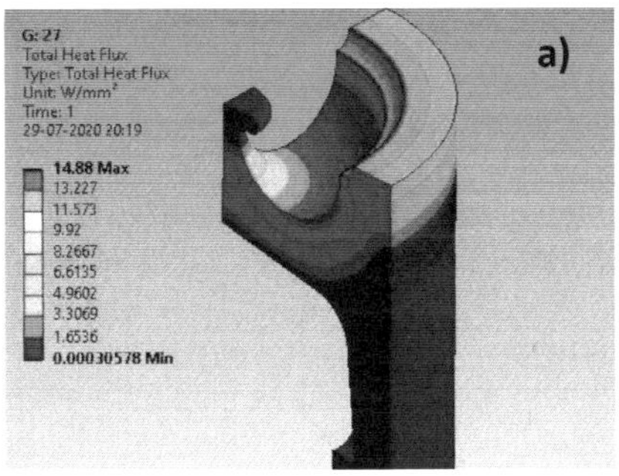

a) 0 mm thickness

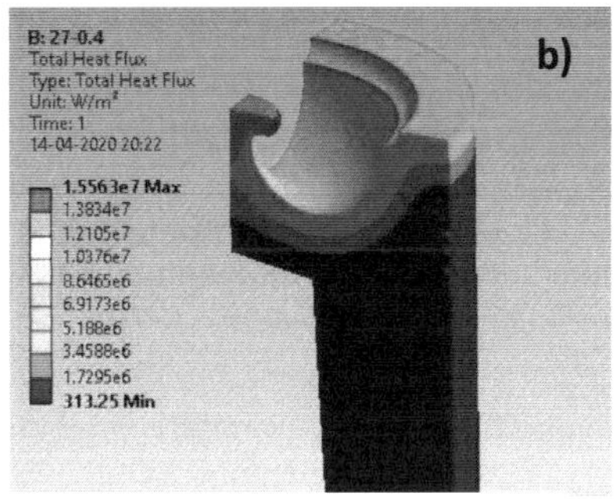

b) 0.4 mm thickness

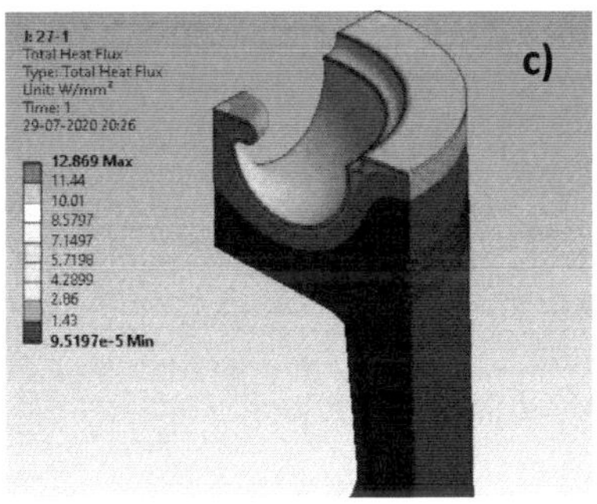

c) 1mm thickness

Figure 9. Variation of heat flux on 27 mm bowl depth with varied thickness of ceramic coating

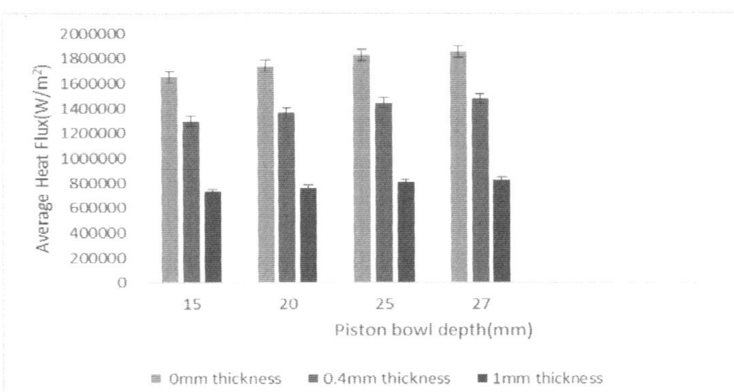

Figure 10. Behaviour of heat flux with piston bowl depth

For longer life of an engine usage of agreemental thickness of TBC is recommended. Lesser the thickness, a larger thermal gradient would be created between the coating and crown which might reduce the life of the piston [33]. Thus, a significant decrement in heat flux with coating thickness would uplift the thermal stability of an engine.

3.4 Surface to volume ratio effect

Surface to volume ratio effect on heat flux has been studied briefly in this work. The values of surface to volume in reference to piston bowl depth for various coating thickness are as shown in Table 3.

Table 3. Variation of surface to volume ratio with respect to piston bowl depth and coating thickness

Coating Thickness (mm)	Piston bowl depth (mm)	Surface to volume ratio (m^{-1})
0.00	15	153.94
0.00	20	157.61
0.00	25	162.26
0.00	27	164.98
0.40	15	182.59
0.40	20	189.08
0.40	25	197.53
0.40	27	201.50
1.00	15	181.80
1.00	20	188.14
1.00	25	196.40
1.00	27	200.28

There is an increment in surface to volume ratio with the increasing bowl depth as expected. This trend remains unchanged with increasing coating thickness. The trait of heat flux variation with the surface to volume ratio is portrayed by the Figs. 11, 12, and 13. The plots clearly show that with the increment of the surface to volume ratio there is an inflating heat flux and the raise in heat flux occurs linearly. The average heat flux is enhanced by 5.26%, 10.56% and 12.28% in an uncoated piston crown. At 0.4mm coating the inflation is 5.04%, 11.36%, 13.90%. Increment of 5.05%, 11.51% and 14.05% is seen with 1mm coating thickness. The significant heat flux increase shows that heat losses are also increased with increased surface to volume ratio. These results are in good agreement with the outcomes of Helgi et al. in one of their research articles [34]. Amir et al. also noticed that with a drastic enhancement of the heat flux with the surface to volume ratio, there is a pronounced inflation of heat losses [35]. These heat losses are not at all permissible in an engine which in turn may reduce the engine efficiency. Thus, it can be stated that the lesser the surface to volume ratio lesser is the heat losses. Thermal stability is higher with lesser surface to volume ratio. These results also show lesser the surface to volume ratio lesser is the bowl depth. This can be correlated to the fact that lesser bowl depth has low heat flux and in turn significantly lesser heat losses. The piston bowl with lesser depth has supreme thermal stability when compared to pistons with increased bowl depth.

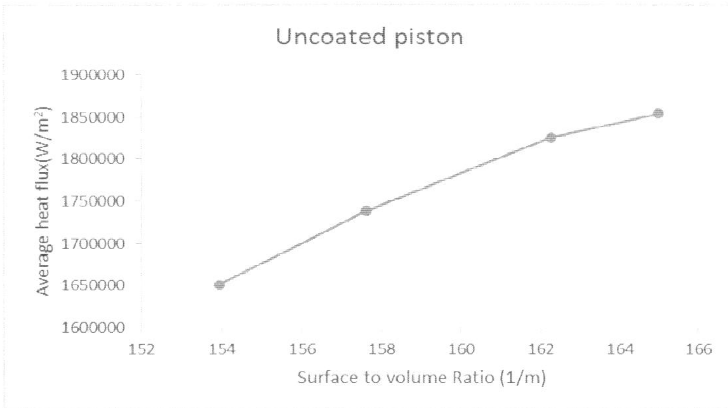

Figure 11. Average heat flux vs surface to volume ratio (without coating)

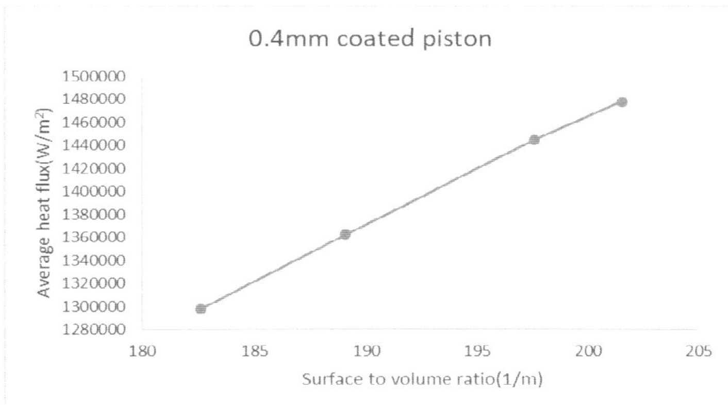

Figure 12. Average heat flux vs surface to volume ratio (With 0.4mm coating thickness)

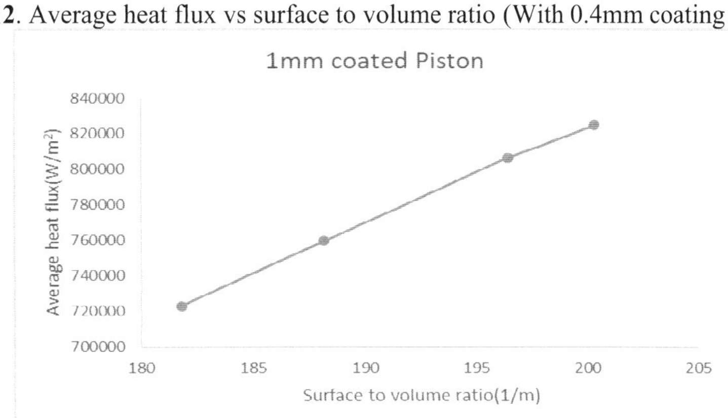

Figure 13. Average heat flux vs surface to volume ratio (With 1mm coating thickness)

IOP Conf. Series: Materials Science and Engineering 1189 (2021) 012013 doi:10.1088/1757-899X/1189/1/012013

4. Conclusions

- The narrowing of the region with maximum heat flux can be accounted for the inflated maximum heat flux with increasing bowl depth.
- Principal heat flux variation is majorly seen on the piston crown region, as this is more prone to the combustion process. Further, the decrement in the piston bowl depth, the average heat flux is decreased.
- Around 27.75% decrement of heat flux is seen for 15 mm bowl depth when compared to 27 mm bowl depth.
- A comprehensive decrement of heat flux can be portrayed in the presence of coating. Further, with the enhanced coating thickness, there is a conspicuous decrement in the heat flux.
- Coating thickness of 0.4 mm showed a pronounced diminishing of heat flux by 20-25%. More than 50-55% heat flux decrement can be observed in the presence of 1mm thickness coating.
- With a lesser thickness, when subjected to longer duration of work, a larger temperature gradient may be created which would result in the depreciation of engine life.
- Surface to volume ratio increases with the increase in the bowl depth and linear increment of heat flux can be seen with increasing bowl depth.
- Lesser is the surface to volume ratio lesser the heat losses. Greater the bowl depth more is the heat flux and more is the heat losses. Thus, the piston with a lesser surface to volume ratio and increased ceramic coating (15 mm bowl depth with 1mm coating) can be potentially stable in transient thermal conditions.

References

[1] Donateo T, De Risi A, Laforgia D. On the Computer-Aided Conversion of a Diesel Engine to CNG-Dedicated or Dual Fuel Combustion Regime. InASME 2012 Internal Combustion Engine Division Spring Technical Conference 2012 May 6 (pp. 933-943). American Society of Mechanical Engineers Digital Collection.

[2] Fansler TD. Turbulence production and relaxation in bowl-in-piston engines. SAE Transactions. 1993 Jan 1:616-28.

[3] Song J, Yao C, Liu Y, Jiang Z. Investigation on flow field in simplified piston bowls for DI diesel engine. Engineering Applications of Computational Fluid Mechanics. 2008 Jan 1;2(3):354-65.

[4] Channappagoudra M, Ramesh K, Manavendra G. Comparative study of standard engine and modified engine with different piston bowl geometries operated with B20 fuel blend. Renewable energy. 2019 Apr 1;133:216-32.

[5] Donateo T, Tornese F, Laforgia D. Computer-aided conversion of an engine from diesel to methane. Applied energy. 2013 Aug 1;108:8-23.

[6] Williams TJ, Tindal MJ. Gas flow studies in direct injection diesel engines with re-entrant combustion chambers. SAE Transactions. 1980 Jan 1:120-31.

[7] Kurtz EM, Styron J. An assessment of two piston bowl concepts in a medium-duty diesel engine. SAE International Journal of Engines. 2012 May 1;5(2):344-52.

[8] Payri F, Benajes J, Margot X, Gil A. CFD modeling of the in-cylinder flow in direct-injection Diesel engines. Computers & fluids. 2004 Sep 1;33(8):995-1021.

[9] Prasad BV, Sharma CS, Anand TN, Ravikrishna RV. High swirl-inducing piston bowls in small diesel engines for emission reduction. Applied Energy. 2011 Jul 1;88(7):2355-67.

[10] Song J, Yao C, Liu Y, Jiang Z. Investigation on flow field in simplified piston bowls for DI diesel engine. Engineering Applications of Computational Fluid Mechanics. 2008 Jan 1;2(3):354-65.
[11] Soni DK, Gupta R. Numerical analysis of flow dynamics for two piston bowl designs at different spray angles. Journal of Cleaner Production. 2017 Apr 15;149:723-34.
[12] Mamilla VR, Mallikarjun MV, Rao GL. Effect of combustion chamber design on a DI diesel engine fuelled with jatropha methyl esters blends with diesel. Procedia Engineering. 2013 Jan 1;64:479-90.
[13] Jaichandar S, Annamalai K. Effects of open combustion chamber geometries on the performance of pongamia biodiesel in a DI diesel engine. Fuel. 2012 Aug 1;98:272-9.
[14] Styron J, Baldwin B, Fulton B, Ives D, Ramanathan S. Ford 2011 6.7 L power stroke® diesel engine combustion system development. SAE Technical Paper; 2011 Apr 12.
[15] Pierz PM. Thermal barrier coating development for diesel engine aluminum pistons. Surface and Coatings Technology. 1993 Dec 3;61(1-3):60-6.
[16] Ramaswamy P, Seetharamu S, Verma KB, Raman N, Rao KJ. Thermomechanical fatigue characterization of zirconia (8% Y2O3-ZrO2) and mullite thermal barrier coatings on diesel engine components: Effect of coatings on engine performance. Proceedings of the Institution of Mechanical Engineers, Part C: Journal of Mechanical Engineering Science. 2000 May 1;214(5):729-42.
[17] Zhu D, Miller RA. Thermal barrier coatings for advanced gas turbine and diesel engines.
[18] Ahmaniemi S, Tuominen J, Vippola M, Vuoristo P, Mäntylä T, Cernuschi F, Gualco C, Bonadei A, Di Maggio R. Characterization of modified thick thermal barrier coatings. Journal of Thermal Spray Technology. 2004 Sep 1;13(3):361-9.
[19] Jalaludina HA, Abdullahb S, Ghazalib MJ, Abdullahc B, Abdullahc NR. Experimental study of ceramic coated piston crown for compressed natural gas direct injection engines. Procedia Engineering. 2013 Jan 1;68:505-11.
[20] Sathyamoorthi S, Prabhakaran M, Abraar SM. Numerical investigation of ceramic coating on piston crown using Finite Element Analysis. International Journal of Scientific Engineering and Applied Science (IJSEAS). 2016;4:258-63.
[21] Cerit M, Coban M. Temperature and thermal stress analyses of a ceramic-coated aluminum alloy piston used in a diesel engine. International Journal of Thermal Sciences. 2014 Mar 1;77:11-8.
[22] Weathers B, Sauls J, Powell G. Transient thermal analysis of screw compressors, part II: Transient thermal analysis of a screw compressor to determine rotor-to-housing clearances.
[23] Jones MT, Plassmann PE. Computational results for parallel unstructured mesh computations. Computing Systems in Engineering. 1994 Aug 1;5(4-6):297-309.
[24] Lyra PR, Almeida RC. A preliminary study on the performance of stabilized finite element CFD methods on triangular, quadrilateral and mixed unstructured meshes. Communications in numerical methods in engineering. 2002 Jan;18(1):53-61.
[25] Lindquist DR, Gilest MB. A comparison of numerical schemes on triangular and quadrilateral meshes. In11th International Conference on Numerical Methods in Fluid Dynamics 1989 (pp. 369-373). Springer, Berlin, Heidelberg.
[26] Pan D, Cheng JC. A second-order upwind finite-volume method for the Euler solution on unstructured triangular meshes. International journal for numerical methods in fluids. 1993 Jun 30;16(12):1079-98.
[27] Li J, Yang WM, An H, Maghbouli A, Chou SK. Effects of piston bowl geometry on combustion and emission characteristics of biodiesel fuelled diesel engines. Fuel. 2014 Mar 15;120:66-73.

[28] Kumarasekaran K, Safdari YB. Thermal stress analysis of a novel design air-gap insulated piston. SAE Technical Paper; 1994 Apr 1.

[29] Sharma SK, Saini PK, Samria NK. Experimental thermal analysis of diesel engine piston and cylinder wall. Journal of Engineering. 2015;2015.

[30] Sharma SK, Saini PK, Samria NK. Modelling and analysis of radial thermal stresses and temperature field in diesel engine valves with and without air cavity. International Journal of Engineering, Science and Technology. 2013;5(3):111-23.

[31] Zhu D, Miller RA. Thermal conductivity and elastic modulus evolution of thermal barrier coatings under high heat flux conditions. Journal of Thermal Spray Technology. 2000 Jun 1;9(2):175-80.

[32] Buyukkaya E, Cerit M. Thermal analysis of a ceramic coating diesel engine piston using 3-D finite element method. Surface and Coatings Technology. 2007 Nov 25;202(2):398-402.

[33] Ramaswamy P, Shankar V, Reghu VR, Mathew N, Kumar SM. A model to predict the influence of inconsistencies in thermal barrier coating (TBC) thicknesses in pistons of IC engines. Materials Today: Proceedings. 2018 Jan 1;5(5):12623-31.

[34] Fridriksson HS, Tuner M, Andersson O, Sunden B, Persson H, Ljungqvist M. Effect of piston bowl shape and swirl ratio on engine heat transfer in a light-duty diesel engine. SAE Technical Paper; 2014 Apr 1.

[35] Kakaee AH, Nasiri-Toosi A, Partovi B, Paykani A. Effects of piston bowl geometry on combustion and emissions characteristics of a natural gas/diesel RCCI engine. Applied Thermal Engineering. 2016 June 5;102:1462-72.

Study on Performance and Emission Characteristics of Four stroke Gasoline engine under Formulated Neem oil as base lubricant

Girish A R[1], Shashidhara Y M [2],
[1]Research Scholar, Malnad College of Engineering-573202 (INDIA)
[2] Professor, Malnad College of Engineering - 573202 (INDIA)
girishar.4691@gmail.com[1], shashi.yms@gmail.com[2]

Abstract

An attempt is made in this paper to use the neem oil as a lubricant to operate a gasoline engine. Basically, the raw oil is modified twice through esterification method and made into Neem Trimethylolpropane ester (NTMPE). A four stroke single cylinder petrol engine is utilized to conduct experiments. Engine is lubricated under various blends of NTMPE with mineral oil. The performance and emission study is made under these blends of lubrication. The obtained results are compared with the outcome under petroleum oil mode of lubrication. The outcomes show that 10 % drop in fuel consumption and marginal improvement mechanical efficiency is noticed when the engine is operated under NTMPE20-MO80 mode of lubrication. About 9 % increase in thermal efficiency is observed under NTMPE20-MO80 blend. Emission results shows that 8 % drop in unburnt HC and 10 % lower CO, under various blends of biolubricants and petroleum oil. Particularly, when the engine is operated under NTMPE20 – MO80 mode of lubrication, lowest CO and unburnt HC are observed.

Keywords: Vegetable oil, Neem oil, Trimethylolpropane, Transesterification, Biolubricant

1. Introduction

Vegetable based lubricants are recognized as substitute to petroleum oils for the reason that they possess certain advantageous technical properties and it is eco-friendly. Plant based lubricants usually show high lubrication property, high VI, high flash point, low toxicity and low evaporative losses [1, 2]. But, it also has certain drawbacks namely low oxidative and thermal stability [3, 4]. To replace usual petroleum oil-based lubricating oils and synthetic oils, high oleic varieties of plant based oils are considered to be potential candidates because of their superior oxidative stability [5, 6]. Many literatures shows that, to overcome the low temperature fluidity and low thermal and oxidative stability of vegetable oils, chemical modification of vegetable oil, genetic modification of seeds and adding additives were recommended [7, 8].

A review is made on the latest developments in production of lubricants from plant based oils through chemical formulation methods namely epoxidation of vegetable oils, estolide formation and esterification/transesterification. Above mentioned methods shows that there is considerably increase in the physical properties of the base plant oils, and efficient to produce plant based lubricants [9]. Biolubricant production of Ester Trimethylolpropane (ETMP) was carried out through esterification of Fatty acid of Jatropha oil with Trimethylolpropane. It is reported that, the product was successfully synthesized to nearly 55 %. Flash point was greater than 300 °C, pour point was observed as low as –30°C, and viscosity is 79 centipoises [10].

Experiment is carried out for the production of ester through esterification of Fatty Acid Rubber seed oil (FARSO) with TMP. It is reported that there is increase in the flash point, high Viscosity Indices (VI) and lower pour point with the esterification process. RSOTMP ester shows comparative physicochemical properties which can be utilize as efficient biolubricant base stock [11]. Through transesterification of Sesame Methyl Ester (SME) with trimethylolpropane, sesame seed oil lubricant was produced. Result shows that the viscosity index is 193, pour point is -21°C and flash point is 196°C. It as high potential as base stock for Sesame biolubricant formulation [12]. Coconut and palm oil is used as lubricants in four stroke engine were evaluated and performance was compared with petroleum based engine oil. It is reported that, with vegetable oil based lubricant there is increased fuel efficiency and reduction in concentration of oxides of carbon and nitrogen in the engine emission [13]. Experiment is carried out on pure petroleum oil, different blends of petroleum based oil and pongamia oil and pure pongamia oil as lubricants for the CI engine run with pongamia oil as biodiesel. Results show that there is drop in fuel consumption and highest thermal efficiency during medium and high load conditions is observed under the use of pure pongamia oil as a lubricant [14]. The alternative for mineral oil based lubricant, rapeseed oil based bio-lubricant in a VCR diesel engine is investigated. Results reveal that there is reduction in Cu, Fe, Al wear, ash and soot content in combined use of biofuel/biolubricant of rapeseed oil when compared to biofuel/mineral oil combination [15]. Study is conducted using alkyl-rapeseed oil methyl ester and mineral oil blend in definite proportions as two stroke crankcase lubricants. It is reported that alkyl-rapeseed oil methyl ester generates lower emissions which reduces green house gases, improves engine life due to higher lubricity. It also enhance oxidative stability of gasoline [16].

This paper presents the utilisation of Neem oil as engine lubricant for lubricating a four stroke S.I. engine. Through experiment, performance characteristics and emission study of the engine are determined under the blends of modified neem oil and mineral oil lubrication. The outcome of experiment are compared with those attain under petroleum oil lubrication.

2. Structural modification of Neem oil

In the first phase of oil formulation, Neem oil is transesterified with methanol to obtain their ester forms such as Methyl ester of Neem oil (MENO). Further, in the second phase, ester is again transesterified with Trimethylolpropane (TMP) to obtain Neem Trimethylolpropane ester (NTMPE) and it is referred as modified neem oil for the experimental study.

3. Properties of NTMPE oil

The physico-chemical properties of modified neem oil is shown in the Table 1.

Table 1. Properties of NTMPE oil

Properties	NTMPE	Mineral Oil
Flash point (^{0}C)	124	200
Fire point(^{0}C)	126	208
Viscosity Index	240.73	123

Viscosity, 40°C (mm^2/s)	12.34	19.26
Viscosity, 100°C (mm^2/s)	4.469	1.28

4. Experimental setup

A single cylinder air cooled four stroke petrol engine developing 1.3kW at 4200 rpm is used to conduct experiments. The engine is attached to an electric generator for determining the performance. Figure 1 depicts the experimental setup to carry out the experiments

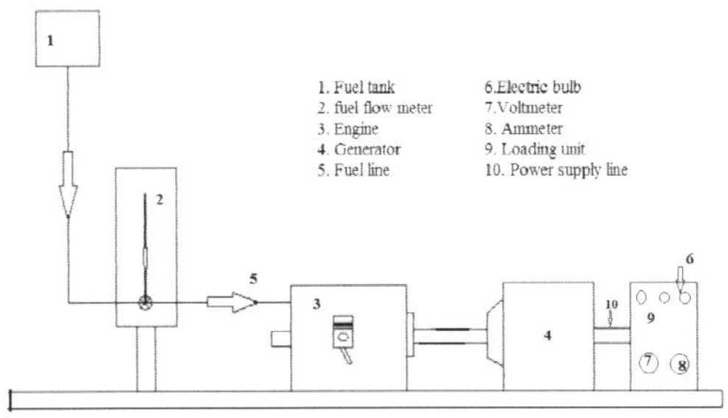

Figure 1. Experimental Set Up

Methodology

Basically, NTMPE and mineral oil are blended in volume for different ratios such as 10 % NTMPE and 90 % mineral oil (NTMPE10-MO90), NTMPE20-MO80, NTMPE30-MO70 NTMPE40-MO60 and NTMPE50-MO50. The engine characteristics like SFC, brake thermal efficiency, mechanical efficiency are calculated. Further, emission components such as CO and HC are measured using flue gas analyzer.

5. Results and Discussion

5.1. Efficiency

5.1.1. Fuel Consumption

Figure 2 shows difference in Specific Fuel Consumption (SFC) with respect to Brake Power for various blends of NTMPE and mineral oil. The results revealed that, at higher load operations, the lowest fuel consumption is seen under the blend NTMPE 20 – MO80. About 10% drop in fuel consumption under this blend of lubrication is observed, compared to petroleum oil mode of lubrication. Though, engine consume more fuel to produce a kW power under the blends, NTMPE30-MO70, NTMPE40-MO 60 and NTMPE50-MO50 modes of lubrication.

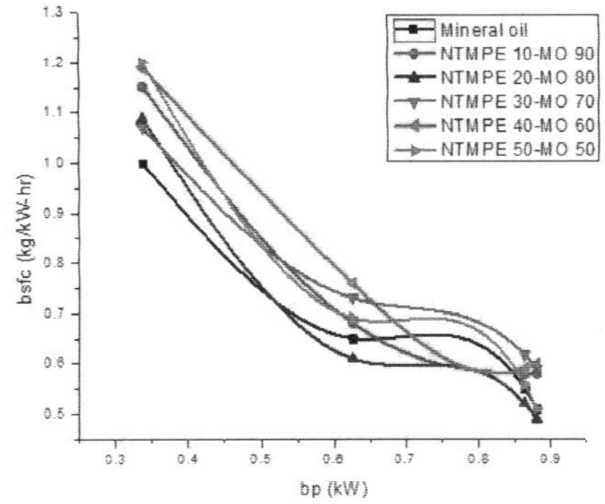

Figure 2. Difference in BSFC with respect to brake power for blends of NTMPE and mineral oil

5.1.2 Brake thermal efficiency

Figure 3 represent difference in brake thermal efficiency with respect to brake power for various proportions of NTMPE and mineral oil. The result shows that, at higher load operations, there is about 10 % increase in brake thermal efficiency under the blend, NTMPE 20- MO80 and it may be due to increase in mechanical efficiency under the same blend. However, the engine efficiency is marginally dropped under the remaining blends of operation. The lowest thermal efficiency is observed under NTMPE 50 – MO50 mode of lubrication.

.

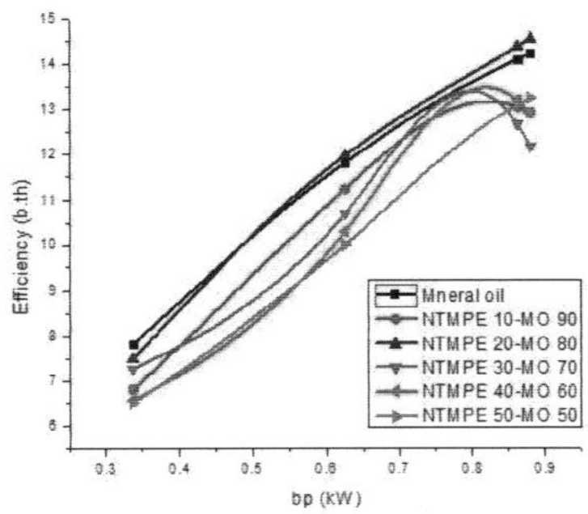

Figure 3. Plot of thermal efficiency with respect to brake power for blends of NTMPE and mineral oil

5.1.3 Mechanical efficiency

Figure 4 depicts difference in mechanical efficiency with respect to Brake Power for various blends of NTMPE and mineral oil. At higher load operation it is seen that the mechanical efficiency under NTMPE10-MO 90 and NTMPE20-MO80 mode of lubrications is marginally improved compared to mineral oil lubrication. However, mechanical efficiency under the blends, NTMPE30-MO70, NTMPE40-MO 60 and NTMPE50-MO 50 is lower than the mineral oil lubrication and the lowest efficiency is observed under NTMPE 50 – MO 50 mode of lubrication. The decrease in mechanical efficiency under the blends NTMPE 30 % and above may be due to the effect of oleic acid quantity present in the fatty acid composition.

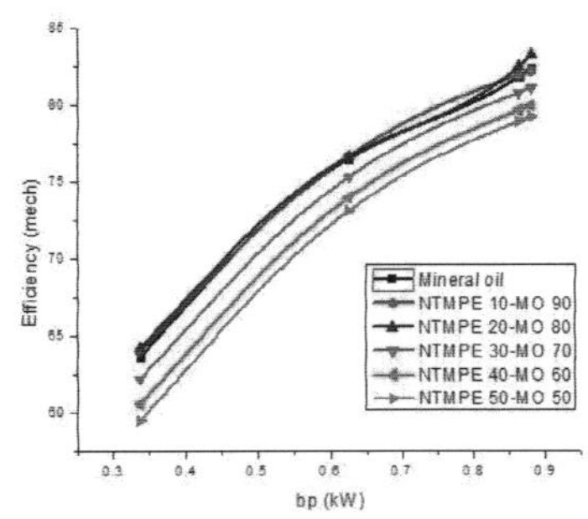

Figure 4. Difference in mechanical efficiency with brake power for blends of NTMPE and mineral oil

5.2 Emission Study

5.2.1 Carbon Monoxide (CO)

Figure 5 shows difference in CO emission with respect to brake power for optimum blends of bio lubricant and mineral oil. Engine emits about 10% lower carbon monoxide values under NTMPE 20- MO80 compare to petroleum oil mode of lubrication because of higher flash point and lower volatility.

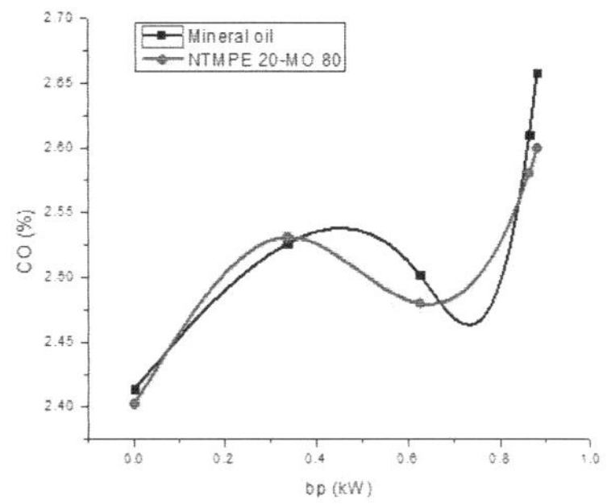

Figure 5. CO emission with respect to brake power for blends of bio lubricant and mineral oil

5.2.2 Hydrocarbon (HC)

Figure 6 shows the difference in unburnt hydrocarbon with respect to brake power for optimum blends of bio lubricant and mineral oil. The results revealed that, at full load operation, there is about 8 % decrease in unburnt hydrocarbon emission under NTMPE 20-MO 80 mode of lubrication compared to petroleum oil mode of lubrication.

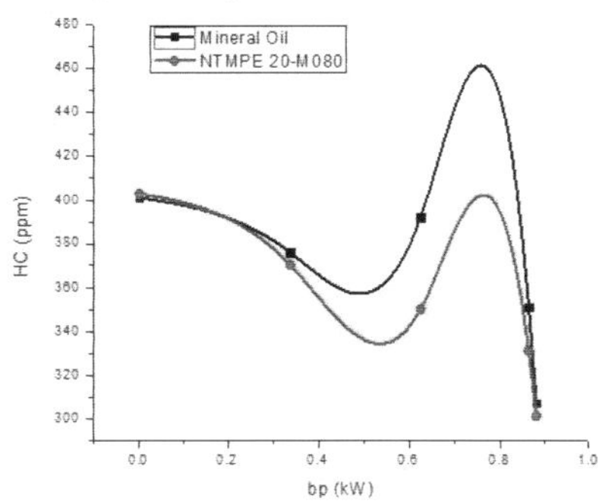

Figure 6. Hydrocarbon emission with respect to brake power for blends of bio lubricant and mineral oil

The decrease in CO and unburnt hydrocarbon may be owing to the high viscosity compared to mineral oil. The high viscosity results in lower the engine friction. Lower the engine friction increases the volumetric efficiency of the engine, due to this quantity of air intake to the engine is more. Therefore, there is sufficient amount of oxygen is in the chamber to burn the fuel completely. It may be the reason why the emission is decreases under bio lubricant compared to petroleum oil mode of lubrication.

6. Conclusions

Lubrication under the modified Neem oil, blended with mineral oil is observed to be better than the pure mineral oil lubrication. The blend, NTMPE20-MO80, found to be the better lubricant in terms engine performance and emissions compared to the petroleum oil lubrication in the engine. The mechanical efficiency of the engine is marginally improved under NTMPE20-MO80 blend type of lubrication compared to petroleum oil. Engine consumes about 10 % less fuel under NTMPE20-MO80 blend of lubrication compared to mineral oil. The thermal efficiency of the engine is improved by about 9 % under this mode of lubrication. Further, emission characteristics under NTMPE20-MO80 blend shows there is about 10 % drop in Carbon monoxide emission and unburnt hydrocarbon emission is decreased by 8 % compared to mineral oil mode of lubrication.

7. References

[1] H.M. Mobarak, E.NizaMohamad, H.H.Masjuki, M.A.Kalam, K.A.H.AlMahmud, M. Habibullah, A.M.Ashraful, "The prospects of bio-lubricants as alternatives in automotive applications", Renewable and Sustainable Energy Reviews 33(2014), PP 34–43

[2] Nagendramma P, Kaul S. "Development of ecofriendly/biodegradable lubricants: an overview". Renew Sustain Energy Rev 2012;16:764–74.

[3] Ebtisam K. Heikal, M.S. Elmelawy , Salah A. Khalil, N.M. Elbasuny, "Manufacturing of environment friendly biolubricants from vegetable oils", Egyptian Journal of Petroleum, 2016.

[4] AnjanaSrivastava, PreetiSahai"Vegetable oils as lube basestocks: A review", ISSN 1684–5315, African Journal of Biotechnology, Vol. 12(9), 27 February, 2013, PP 880-891.

[5] Rani.S, Tarun M S, M L Joy, K Prabhakaran Nair "Vegetable oils as lubricant base oil: A Review", International Journal of Scientific & Engineering Research, Volume 5, Issue 7, July-2014, ISSN 2229-5518, PP 708-710.

[6] Amit Suhane, A.Rehman, H.K.Khaira, "Potential of Non Edible Vegetable Oils as an Alternative Lubricants in Automotive Applications" International Journal of Engineering Research and Applications (IJERA), ISSN: 2248-9622, Vol. 2, Issue 5, September- October 2012, PP.1330-1335.

[7] M.K. Nizam,Hayder A Abdul Bari, "The use of Vegetable Oil in Lubricant as Base Oil: A Review",National Conference on Postgraduate Research (NCON-PGR) 2009, PP 123-127.

[8] Sevim Z. Erhan, SvajusAsadauskas, "Lubricant basestocks from vegetable oils", Industrial Crops and Products II (2000), PP 277-282.

[9] Josiah McNutt, Quan (Sophia) He, "Development of biolubricants from vegetable oils via chemical modification", Journal of Industrial and Engineering Chemistry 36 (2016), PP 1–12.

[10] Noor HafizahArbain and JumatSalimon, Synthesis and Characterization of Ester Trimethylolpropane Based JatrophaCurcas Oil as Biolubricant Base Stocks, Journal of Science and Technology, pp 47-58.

[11] AinaatulAsmaa' Ishak, JumatSalimon, Synthesis Of Rubber Seed Oil And Trimethylolpropane Based Biolubricant Base Stocks, Malaysian Journal of Analytical Sciences, Vol 17, No 3 (2013),PP 414 – 421.

[12] Matthew ChukwudiMenkitiab, OchejeOcholib, KechinyereBibianOyohc&OkechukwuDominic Onukwulib, Synthesis and Tribological Evaluation of Sesame Oil-Based Trimethylolpropane Ester, Journal of the Chinese Advanced Materials Society, http://dx.doi.org/10.1080/22243682.2015.1013154.

[13] Jagadeesh K Mannekote, Satish V Kailas, "Experimental Investigation of coconut and Palm oils as lubricants in Four stroke engine", ISSN 1881-2198, Tribology Online, 6, 1(2011), PP 76-82.

[14] Bekal S, Bhat N R "Bio-lubricant as an alternative to mineral oil for a CI engine— an experimental investigation with pongamia oil as a lubricant. Energy Sources 2012;34:1016–26.

[15] S. Arumugam, G. Sriram 1, R. Ellappan, Bio-lubricant-biodiesel combination of rapeseed oil: An experimentalinvestigation on engine oil tribology, performance, and emissions of variable compression engine, Energy 72 (2014), PP 618-627.

[16] C.Venkatesan, K.Vignesh, P.Kannadasen, G.Ramasivam," Rapeseed Oil based Lubricant reduces smoke emission in Two-Stroke Petrol engines", International Journal of Engineering Technology, Management and Applied Sciences, September 2014, Volume 2 Issue 4, ISSN 2349-4476, pp 168-179.

A Review on Ceramic Coatings for Low Carbon Steel Methods Materials and Applications

Umashanker L[1], T P Bharathesh[2], Roopashree C[3] and Saravanan R[4]

[1]Department of Mechanical Engineering, AMC Engineering College, Bengaluru, Karnataka 560083, India
[2]Department of Mechanical Engineering, Akshaya Institute of Technology, Tumakuru, Karnataka 572106, India
[3]Department of Mechanical Engineering, University Visvesvaraya College of Engineering, Bengaluru, Karnataka 560056, India
[4]Department of Mechanical Engineering, AMC Engineering College, Bengaluru, Karnataka 560083, India

Abstract. Ever since the Second World War many Researcher have significantly contributed to the research in ceramics. The aesthetics of components, durability, and wear resistance properties of components are increased by ceramic coatings. Surface coatings of ceramics to metals and to certain non-metals are the hot cake of research in ceramics now. Ceramic coatings are robust in structure and possess a higher level of lubricity, the property of oxidation limit their use to temperatures under 12000°C making it much favourable for hot forging dies. Ceramic coatings on metals have proved to enhance the metal functional properties; they make the material free from corrosion, enable it to behave as a non-conductor under certain circumstances, give a better level of abrasion resistance and expose the capability of maintaining their integrity even at elevated temperatures of about 45000° Fahrenheit. This property of ceramics has increased its popularity in the field of automotive industries and forging industries. Ceramic coating increases the working lifespan of the dies enabling them to produce many components before they are worn out. In the current investigation various research works conducted by different researchers across the world in the field of ceramic coatings on metal surface is being investigated. The gap in the research findings is used for further experimental investigations. Keywords: Ceramic coatings, low carbon steel, thermal barrier coatings,

1. Introduction

The Ceramic coating on metals has emerged as one of the most adoptable methods of preparing metals to get prevented from rusting become wear resistant and to make it harder. However, the word surface coating on metals sounds very simple but in actual practice the process is not that simpler. The process of applying ceramic coating onto a substrate is a multi-stage process. It involves processes like surface preparation, abrasive blasting, cleaning, painting, curing and final equipment cleaning. Metal corrosion is a basic destructive process that brings down the life of components that are exposed to corrosive atmosphere resulting in greater economic losses. Thus, it does not lead one to any surprises, that the research on corrosion and protection of materials from corrosion is being concentrated [1]. The starting point of corrosion of any material is the material surface. Ceramic coatings help us to produce a material surface that is free from corrosion. The success of coating largely depends on mechanical, chemical, and physical bonds that are responsible for the coating adherence and ultimate strength of ceramic layer.

2. Literature Survey

Ever since 1970, there has been a significant improvement in the field of ceramic coating on metal surfaces. Different environments in the modern world demand different metal and their alloys that can perform at elevated conditions. From the year 1980 onwards, ceramics has found its successful

IOP Conf. Series: Materials Science and Engineering 1189 (2021) 012015 doi:10.1088/1757-899X/1189/1/012015

application in the automobile industry [2]. The first successful application in automobile industry was found in gas turbine wings later the piston, piston crown, cylinder lining and valves have succeeded gas turbine wings [3]. Different metal ceramic combinations have shown different properties. When hardness and wear resistance are of consideration carbides like the silicon carbide and the tungsten carbide are used in the form of dispersoids. Whereas resistance towards oxidation even at high temperature is acquired by coating layers of oxides like TiO_2, SiO_2, Al_2O_3, etc., these coatings also impart hardness and wear resistance to the base material. These deposits are better than that of nickel deposits [4]. Owing to the extremely good quality and excellent properties in the past two decades, ceramic coatings find extensive applications in the surface modifications. Base alloys when must be protected from corrosion, oxidation, and minimization of damages due to wear ceramic coatings are the best suit. Another important function of ceramic coating is that they minimize the thermal barrier coatings [6]. Ceramic coatings in thin layers on the surface of metals have always helped in enhancing mechanical performance of metallic substrates. By coating ceramics on the metal surfaces, heat resistance, wear resistance, corrosion resistance electrical insulation and many other properties can be improved. [7]. there are many ceramic coating methods for protective coatings on mild steel. The methods differ in terms of coating quality, complexity of process, investment costs and deposition efficiency.

2.1. Ceramic Surface coating: Low Carbon Steel's Resistance towards corrosion: its effect

Greater is the carbon content in steel, greater will be its corrosion rate this is an imperative factor that determines the corrosion resistance of steel [7]. Providing a protective coating on the surface is one of the most suitable ways to improve the anti-oxidation ability of carbon steels. A large number of experimental investigations have been made on the process of coating different materials on the base metal and the results were different from each other several such investigations include zinc composite coating, where in zinc metal along with ceramic, metal oxides and polymers is coated. Composite coatings provide different properties such as resistance to corrosion and sometimes superior electrochemical properties. Ghaziof Soroor [8] investigated to study the coating properties; he made a comparative analysis using cyclic voltammetry, Tafel test, XRD and ESEM. He arrived at the conclusion that corrosion resistance property of Zn-Ni coatings saw an impressive improvement due to the addition of alumina nano particles as compared to Zn-Ni-Al_2O_3. While coatings of Zn-Ni-Al_2O_3, displayed a uniform and compact deposits as compared against Zn-Ni. Popoola A. P. I [9] found that with Zn-ZnO-Yttria coatings on mild steel, an increase in the corrosion resistance, hardness and the wear resistance saw a significant improvement. He claims that Yttria can be added to increase the wear resistance surface hardness and corrosion resistance of mild steel. Also, by adding Yttria, finer microstructure can be obtained, and this is the root cause for increasing the hardness value, corrosion resistance value and wear resistance value of Mild steel.

The ideal emerging materials that find suitability for wider engineering applications are the advanced ceramics. These materials because of their superior properties are being widely used in some biomedical applications, cutting tools, IC engines, gas turbine blades and spacecraft. Advanced ceramics differ from that of ceramics in their composition, processing techniques and its microstructure. To make them fit to be used in other engineering applications; extensive research has to be carried out in terms of mechanical, optical biomedical and electrical properties [10].

2.2. Ceramics in nuclear applications

Advanced Ceramics have found its application related to nuclear fission and fusion reactors. These are used in fission reaction in nuclear fuel cycles and in confinements of nuclear wastes. In case of fusion reaction these are used to sustain the fusion nuclear fuel cycle. In the nuclear reactors the material failure is due to swelling. These swellings on material surface are due to the knock of atoms by the neutron's atoms possessing high energy. the gaseous come together and forma a bubble in crystals or along the boundary. In addition to this various defect such as the vacancies, dislocations voids, loops and introduction of interstitial spaces are created. Crystalline defects in materials affect various other properties such as the thermal diffusivity, electrical conductivity, and chemical stability [11]. The swellings of Al_2O_3 and the AlN ceramics is larger than that of SiC and Si_3N_4 ceramics. The degradation of base material is due to the micro crack or void formation on the material surface. This can be identified by the occurrence of linear swelling by more than 1%. As compared to compounds with ionic bonding covalent bonded SiC and Si_3N_4 exhibit exceptional tolerance.

In rotary engine applications, advanced ceramic materials find a greater potential. Since it is not possible to manufacture a totally compact rotary engine, thus planning and production of metal ceramic bonds exhibit a greater demand. Couple in Constant of Thermal Enlargement (CTE) existing between the metals and ceramics becomes one of the major hindrances for the use of ceramic components in rotary engines. In practice the thermal application may produce excessive strain at the interface between metal and ceramics. The strain produced will be sufficient to create a catastrophic failure of ceramic material. Thus, it becomes a necessity to investigate different strategies for change in integrity of ceramic materials

2.3. *Thermal and Environmental Barrier Coatings*

With the introduction of advanced ceramics to the production of gas turbine components, enhances the operational temperature of the turbine, which results in the higher turbine efficiency and reduces the impact on the environment. Sometimes the ceramic components used in the gas turbines are subjected to hot corrosion and recession due to the action of hot corrosive gases and vapors that are moving at greater velocities. The effect of these corrosive hot gases on ceramics makes it very much essential to use Environmental Barrier Coatings (EBC) which have a greater resistance to the destructive combustion atmosphere and possess the capability to act as diffusion barriers which anticipate with the gaseous matter from reaching the ceramic coating. With the increased operational temperature, the use of Thermal Barrier Coatings (TBC) will become more essential. These TBC's are poor thermal conductors and allow a significant fall in temperature across the coatings, this enables the basic metal substrate to perform at temperatures lesser than that of combustion temperature. For the determination of CTE match, the deposition, and other properties of EBC's on ceramics with silicon as base and also EBC/TBC are considered for CTE match. While absence of harmful chemical interface interactions, actual TBC thermal conduction, hot corrosion resistance and EBC recession resistance are most essential for a CTE match [12].

2.4. *Surface coating applications*

Under this heading the applications of surface coatings in the fields like food industry, electronics industry, sports and technology, aeronautics and transportation departments, chemical and crude oil industries are explained. In the modern world, surface coatings have found its extensive use in some specialized areas such fields are sports equipment (e.g. horse hooves, swimming equipment, clothing, and golfing equipment), biomedical/orthopedics, art and bronze industries [13]. By adopting surface coatings on a base material, a wide range of properties can be attained. The commonly use base materials are Diamond Like Carbon (DLC), nitrides, metals and alloys TBC and decorative coatings.

A majority of cutting applications are made proficient by spraying a thin layer of resistant coating. The coatings on the base materials vary in thickness in microns. This coating is responsible for improving the cutting-edge wear resistance and is responsible for reducing the friction and diffusion. This includes high speed cutting, machining of hard materials and other nonferrous material that cannot be cut otherwise. There is a significant contribution of surface engineering to the field of transport industry. Nearly to an approximate of 6% of the total cost involved in manufacturing an engine and the transmission system are involved in the coating technologies. The three major categories that find the application of surface coating are the Power units, fixed permanent structures and majorly the vehicle components. Coatings are used widely in the field of power generating units and power transmission systems. These surface coatings prevent the power units from excessive wear and corrosion. To improve the wear resistance property and increase the life of brakes and suspension of an automobile, they are treated with thermally sprayed coatings.

The aerospace industry has been in practice of surface coatings over a period of 50 years from now. The surface coating on the gas turbine engines inhibits properties such as the high temperature strength, load bearing properties and the corrosion resistance. To protect the parts against corrosion they are thermally sprayed with polymer coatings. The gears and the bearings used in the aircrafts are coated with MOS_2 using PVD magnetron sputtering thus lowering the heat generation in the transmission system and prevents the meshing gear temperature rise. Surface modification can improve a components performance. The best example for this is the landing gear of the light combat aircraft. This was initially designed from maraging steel and thus requires a surface modification to enhance its wear resistance. The age-old practice of chromium plating exposes machining problems. Thus, by nitriding was developed and this process does not require any post machining operations also it does not reduce the fatigue life of the material unlike the hard chromium plating.

2.5. Methods used in surface coatings

In the current section we will be discussing entirely on the recent methods of surface coatings. Plasma spraying is the most popular technique used for the preparation of a large range of coatings. Plasma sprayed coatings find their applications in chemical industries, biomedical field, automobile, and aerospace industries. To improvise certain properties like the wear and corrosion resistance along with biocompatibility, plasma spraying is performed at an elevated temperature of 10000° K in consideration with higher specific energy densities and higher cooling rates. Oxidation problem is reduced due to the use of argon, hydrogen, or nitrogen. Despite of its high cost involved in the production of plasma coatings this process is much preferred because of the advantage that even metals possessing higher melting points can also be plasma sprayed. Before plasma coating is applied one should ensure that the surface on which it will be coated is free from metal oxides, oil, and any other powders on the surface. The coating thickness obtained by plasma spraying ranges from 2.5-2500 μm.

Sol – gel method this method has become the most popular method of preparing the ceramic coating as this is much simpler, excellent adhesion of the coating to the substrate material, low processing temperature. For employing the Sol-Gel thin films on base metal surface, different techniques are being used such as the dip coating, spray coating and the spin coating. Surface modification works greatly depend on the laser process as compared to the age-old heat treatment process as it is very economical and advantageous as compared to other processes. Few of the advantages of the laser treatment process are it has a short processing time, faster heating and cooling rate, lesser energy consumption, fine coating properties, accurate process control local heating and is an environmentally friendly process. Using the laser technology only the superficial layer of the metal substrate can be modified without altering the subsiding layers. Pulsed laser deposition which popularly known as the PLD is a much simpler technology that permits the growth of very thin films over a wider material range. This technique offers flexibility to produce various microstructure coatings, from dense and compact to columnar and a porous structure.

2.6. Developments in the field of ceramic coating

Earlier to 1990's, the application of ceramic coating was limited to protect certain metals components such as the Ni Super alloys, Ti –alloys and the Co-Cr alloy components. In the later years, the applications were extended to a wider range of materials. The corresponding section highlights some of the newly emerged applications.

2.7. Heat resistant Ceramic Coatings used in Gas Turbines

The late 1990's, gave rise to power plants utilizing highly efficient turbines that ran on both the gas and steam. The operating temperature of these turbines was relatively high. To sustain such a high temperature, the turbine parts were manufactured from high heat resistant super alloy. The combined advances made in cooling technology and heat resistant coating technology have liberally contributed to the development of these highly efficient turbines. Gas turbines that run at a temperature of 15000° C make use of blades made from a single crystal Ni-based super alloy. In such extreme conditions the use of TBC becomes very much essential in addition to this steam cooled technology also must be employed.

2.8. High temperature Corrosion Resistant Ceramic Coatings used for Waste to Energy Boilers

The modern-day boilers promote the energy conservation and reduction of the harmful gases like CO_2 and oxides of various harmful gases. A higher performance is expected from these plants that are facilitating the demands of the society such as the economy globalisation. The high efficiency waste to energy boilers runs utilising the steam under a temperature of 4000°C and a pressure of 3.9 MPa. The fluidized bed boilers which consume the industrial waste such as the biofuels, RPF, RDF, etc., are usually operated at temperature range of 4500°C to 5400°C and pressure varies from 5.9 to 9.8 MPa. The combusted gases of waste to energy boilers create a corrosive environment. The amount of HCL, Sox, present in the flue gases coming from WTE is quite larger than that of flue gases coming from the fossil fuels boilers. There is a larger amount of deposition on high temperature boiler components. These deposits are of low melting point ashes that contain highly concentrated chlorides. To prevent these boilers from corrosive is becomes mandatory to use corrosion materials and corrosive resistant coatings to enhance the life and reduce the maintenance activity. In addition to this, it also becomes essential to make use of highly durable ceramics available at a lesser cost in combination with suitable metals in the production of boilers and its equipment's.

The stable operation of the boiler for a longer period is made possible using advanced refractory and many metallic materials such as the ceramic tiles, metal spray coatings and welded joining of Ni based alloys. Thus, the refractory materials are made of high purity silicon carbides and alumina finds its use. High temperature furnace damages and slugging of boiler walls are prevented using silicon carbide refractories. Use of Silicon carbide tiles for water cooled furnace has portrayed better performance, lesser maintenance, and greater durability. The boiler super heaters work under extreme environmental conditions of erosion and corrosion caused by the soot blowers at a temperature above 4000°C. To enhance the working life period of super heaters, cermet and ceramic spray coatings are applied. This application can even be extended to boilers that operate under 5000°C temperature and 9.8 MPa steam conditions.

Experimental investigations have shown that dual layer coatings of TiO_2 625 alloy cermet coating using HVOF spray processes and Cr_3C_2-NiCr alloy cermet and ZrO_2 (Yttria stabilised zirconia) or Ni based alloys using plasma jet spray process on exposing them in super heaters that were maintained at a temperature of around 5000°C over a period of about 1.3 to 2 years. This pronounced a confirmation of its operating life span of more than three or more years. This life span could not have been obtained by any other metal spray coatings.

A detailed investigation of the causes for deterioration was made. The investigations revealed that the deterioration of coatings was mainly due to the corrosive gas penetration. This penetration onto the metal coating interface has caused swelling of the coating layer. The swelling progresses further as the adhesive strength between the metal and the coating reduces and results in the breaking down of the layer. From this, a dense coating is essential for increasing the lifetime. Some of the important material factors responsible for governing the durability are the resistance towards corrosion, porosity exhibited by the coating material, adhesive strength of coating material with the base material and thermal properties of base material and the coating. In case of Ni based alloys or YSZ, the superficial layer forms the diffusion barrier for corrosive gas penetration.

2.9. Wear-Resistant Ceramic Coatings for Bio implants

There is an increasing demand for the improved performance and increased life of the joints. This is an urgent socioeconomic need that has to be fulfilled. This can be facilitated using ceramics. By coating a ceramic film, the wear resistance and the life of the joints are improved. Though the concept sounds simpler, establishing a coating on a joint that is successful clinically and commercially successful has prone a challenge. The best found alternate to this is the adhesion of the coating to the substrate. The limitation of this technique is that if there is a failure the ceramic film forms a third body which increases the wear of the surfaces. In orthopedic application superficial films on joints formed by the vapor deposition or the chemical vapour deposition, zirconium oxide vapours and the DLC are extensively used.

2.10. High temperature protective ceramic coatings for stainless steel

Perfect adherence of ceramic coatings to the substrate is a must for high temperature shielding coatings. It is seen that the oxide films over the alloy's free surface grow uncontrollably on heating during hot rolling or any other hot working conditions carried out at a temperature of 12000ºC in the presence of oxygen for a period of certain couple of hours. The formation of multi-layered oxide layer on the surface of stainless steel causes a deterioration of metal layer varying from 1% to 2% this affects the quality of the stainless steel. Thus, greater care is taken to eliminate or minimize the process of high temperature oxidation. Thus, ceramic coatings have fascinated the protection of metal surface from high temperature oxidation.

2.11. Ceramic coatings extended opportunities Solid oxide fuel cell (SOFC)

In planar solid oxide fuel cells, the use of ceramic coatings is being investigated for extending the operational time of Stainless steel. SOFC 's is most attractive due to their higher efficiency cleaner emission as related to the combustion and have a higher potential for both stationary application and portable application. A fuel stack is formed by interconnecting the fuel cells. By interconnecting individual fuel cells mechanical strength and compartmentalized fuel flow and air to individual electrode is attained. The interconnections should exhibit electrical conduction along the stack at higher operating temperatures. Amongst SOFC's the ferritic stainless steel is considered the most promising as the stacks are operated at a temperature below 800^0C.

3. Conclusions

From the current literature survey, it is seen that, the air plasma spraying technique is widely used for thermal barrier coatings owing to its increased thermal and surface properties. Due to its high wear resistance and low thermal conductivity surface modification techniques are in use. Advanced ceramic materials such as the yttria and the zirconia find its applications in gas turbines.

4. References

[1] Krastev Dimitar. "Improvement of corrosion resistance of steels by surface modification." *Int. J. Corrosion resistance. InTech,* (2012): 296-316.

[2] Srinivasulu K, Manisha Vidyavathy, "A review of research, methods, materials, applications and opportunities", *Int. J. of Advanced Engineering Technology.*

[3] Serdar salman, Ramazan Kose, Levent Urtekin, Fehim Findik, "An investigation of different ceramic coatings thermal properties", *Materials and Design* **27** (2006) 585-590.

[4] Y. Adraider, Y. X. Pang, F. Nabhani, S. N. Hodgson, M. C. Sharp, A. Al-Waidh, "Laser-induced deposition of alumina ceramic coating on stainless steel from dry thin films for surface modification", *Ceramic International* **40** (2014) 6151-6156.

[5] Pleshivtsev V. G., Filippov G. A., Yu A. Pak, and Livanova O. V. "Effect of carbon content and stressed state on the corrosion rate of pipe steel in heating systems." *Metallurgist* **53 (7)** (2009): 502-505.

[6] Ercenk Ediz, Ugur Sen, and Senol Yilmaz. "Structural characterization of plasma sprayed basalt–SiC glass– ceramic coatings." *Ceramics International* **37(3)** (2011): 883-889.

[7] Sainz M. A., Miranzo P., and Osendi M. I. "Sintering behaviour and properties of YAlSiO and YAlSiON glass-ceramics." *Ceramics International* **37, 5** (2011): 1485-1492.

[8] Ghaziof Soroor, and Wei Gao. "Mechanical and chemical properties of Zn-Ni-Al2O3 nanocomposite coatings." *World Academy of Science, Engineering and Technology, Int. J Chemical, Molecular, Nuclear, Materials and Metallurgical Engineering* **9 (8)** (2015): 997-1001.
[9] Popoola A. P. I., Aigbodion V. S., and Fayomi O. S. I. "Anti-corrosion coating of mild steel using ternary Zn- ZnO-Y2O3 electrodeposition." *Surface and Coatings Technology* **306** (2016): 448-454.
[10] Ruhi G., Modi O. P., and Dhawan S. K. "Chitosan- polypyrrole-SiO_2 composite coatings with advanced anticorrosive properties." *Synthetic Metals* **200** (2015): 24-39

Acknowledgments

Umashanker L received his Bachelor of Engineering in Mechanical Engineering from Karnataka University, Dharwad, Post-graduation in Production Engineering from Basaveshwara Engineering College Bagalkot, VTU, Presently working as Associate Professor in the Department of Mechanical Engineering, AMC Engineering College. Bangalore.
T P Bharathesh received his Bachelor of Engineering in Mechanical Engineering from Bangalore University, Post- graduation in Production Engineering from Shivaji University, Maharashtra, India and PhD in Metal Matrix Composites from the Sri Krishnadevaraya University, Ananthapur, Andhra Pradesh, India. He is currently working as a Professor, in the Department of Mechanical Engineering, Akshaya institute of Technology, Tumakuru, India. His areas of interest are Metal Matrix Composites, Thermal Spray Coatings, Additive Manufacturing and Non- Traditional Machining. He has published over 15 research papers.
Roopashree C, M.Tech in Product Design and Manufacturing (2015- 2017) from SSIT, Tumakuru. MBA in Financial Management (Pursuing Project Work) Under IGNOU, Bangalore. Diploma in Finance Management From IGNOU, Bangalore. PG Diploma in Finance Management From IGNOU, Bangalore. BE in Industrial Engineering Management (2007-2011) from SSIT, Tumakuru. Presently working as Assistant Professor (Guest Faculty) at UVCE, Bangalore.
Saravanan. R received his Bachelor of Engineering in Mechanical Engineering from Anna University, Chennai, India and Post-graduation in Production Engineering and System Technology from the Davangere University, Davangere, India. He is currently working as an Assistant Professor in the Department of Mechanical Engineering, AMC Engineering College, Bengaluru, India. His areas of interest are Metal Matrix Composites, Machine Tools and Operations, Metal Forming and Non-Traditional Machining. He published more than 8 research papers

Influence of ageing time on microstructure, hardness and corrosion of UNS Zeron 100 super duplex stainless steel

D Disha[1], M B Davanageri[2], Ritesh Kumar [3]

[1,2,3]Department of Mechanical Engineering, Sahyadri College of Engineering and Management, Karnataka,India.

.

Abstract. The present study is to investigate the influence of ageing time on the microstructure, hardness and corrosion properties of UNS Zeron 100 super duplex stainless steel. The material was solution treated at 1050°C and water quenched, further the ageing has been carried out at 850°C for 30, 60 and 90 minutes. Hardness test and corrosion test were carried out for both solution treated as well as aged specimens for different ageing period. Results of hardness test indicate an increase in hardness as the ageing period increases. Corrosion test results showed good resistance to corrosion as the ageing time increases. Changes in mechanical properties were observed due to sigma phase formation. Scanning electron microscope (SEM) analysis showed the sigma phase precipitation in duplex stainless steel. EDX analysis shows the consumption of ferrite and formation of chromium which indicates the formation of sigma phase. Optical image analysis showed the increment of sigma phase as the ageing time is increased.

1. Introduction

Duplex stainless steels (DSS) are called "duplex" as they have assemblage of two phases of ferrite and austenite. Zeron 100 (DSS) has found wide industrial application in pharmaceutical, chemical, desalination, pulp, paper industries, power generation and mining industries. The effect of ageing temperature and ageing time effects on a 25Cr-6.7Ni-0.32N-3.0Mo-2.5W (DSS)[1].The ageing temperature of 600-900°C and ageing time of 1 to 240 minutes which resulted in formation of chi phase and sigma. Ageing significantly influenced on the mechanical properties due to precipitation of inter-metallic phases. The sigma phase effects on Z3CN20M cast (DSS). At temperature of 600-900°C, formation of sigma phase occurs at ferrite phase boundaries[2]. Further, authors have observed that for temperature range less than 700°C and aged for less than 4 hours yield strength and tensile strength was less when compared to aged specimens. Study carried out on SAF2507 SDSS at the temperature range of 200°C-700°C there was no significant change in mechanical properties. And at 800°C formation of sigma phase starts in SDSS[3]. The effects of solution treatments (1020°C 1080°C 1200°C) and different continuous cooling rates (1, 0.5, 0.25 and 0.1°C/second) on 2205 DSS and reported that precipitation of sigma phase was obtained in lower cooling rates and at lower solution treated temperature[4]. It was observed that the sigma phase formation occurs at 850°C significantly high[5]. DSS was heat treated at 700°C and 750°C which resulted the inter-metallic phases which were formed in USN S31803 DSS. The formation of chi and sigma phase was noticed. It was found that chi phase precipitation was initiated at ferrite/ferrite grain boundaries and chi phase precipitation occurs before sigma phase precipitation[6]. From the above literature it was observed that there is less focus on Zeron 100 SDSS. Hence the present study will give the insight to the influence of ageing on microstructure and mechanical properties of Zeron 100 SDSS.

2. Experimentation

The commercially available material Zeron 100 duplex stainless steel was selected to carry out the experiments to study effect of ageing on Microstructure, Mechanical and Corrosion properties. The prepared specimens for these tests were polished with metallographic techniques[7]. The material was initially solution treated at 1050°C for 120 minutes and quenched in water to obtain the balanced microstructure of austenite and ferrite. The aging was carried out at 850°C for 30, 60 and 90 minutes and all the specimens were quenched in water. Scanning electron microscope (SEM) was used to

IOP Conf. Series: Materials Science and Engineering 1189 (2021) 012016 doi:10.1088/1757-899X/1189/1/012016

study the microstructure and the corresponding elemental analysis. The hardness test was conducted using Rockwell hardness test rig. Optical image analyzer as shown in Figure 1 was used to study the volume fraction of the phases observed in the material after the ageing. To determine the corrosion resistance of the specimen Salt spray or fog test is conducted in corrosion analyzer as shown in Figure 2.

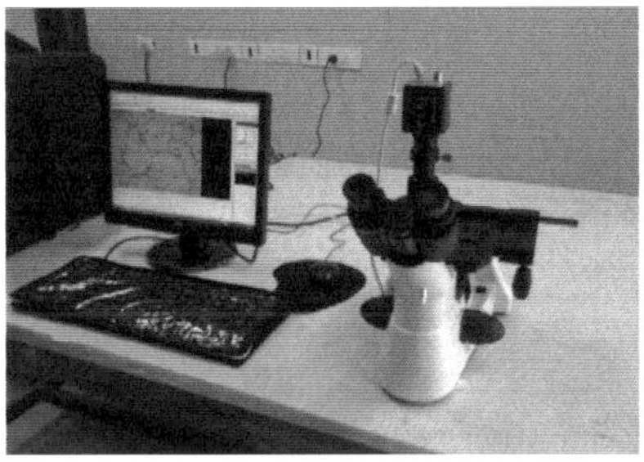

Fig.1 Optical image analyser

Fig.2 Corrosion Analyzer

3. Result and Discussion

The following sections describes results of solution treatment, heat treatment

3.1 Solution Treatment

The solution treatment was carried at 1050°C for 120 minutes. The Figure 3 and Figure 4 are the SEM and optical micrographs of solution treated specimens. It was observed that all the precipitates which exist in the Zeron 100 DSS are dissolved and only equal phase microstructure of ferrite and austenite was obtained. The chemical composition after the solution treatment is presented in the Table 1.The phase volume fraction analysis resulted in 57% of ferrite and 43 % austenite for solution treatment.

Table 1 Chemical Composition of UNS Zeron 100 SDSS

C	Mn	Si	S	P	Cr	Ni	Mo	Cu	W	Fe
0.30	0.476	0.494	0.009	0.027	25.378	7.153	3.437	0.598	0.620	61.778

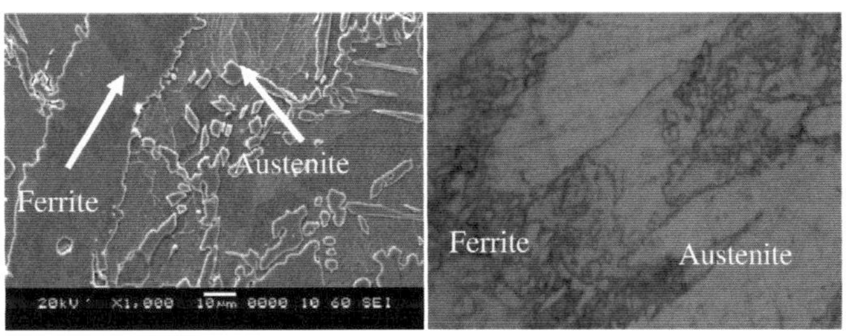

Fig 3 and 4. SEM and optical image of solution treated specimens.

3.2 Heat Treatment

The heat treatment was done at 850°C for 30, 60 and 90 minutes and quenched in water. The precipitation of secondary inter-metallic phases was observed. When AISI ZERON 100 SDSS was heated to the formation of sigma phase precipitates could be observed in the microstructure of SDSS.

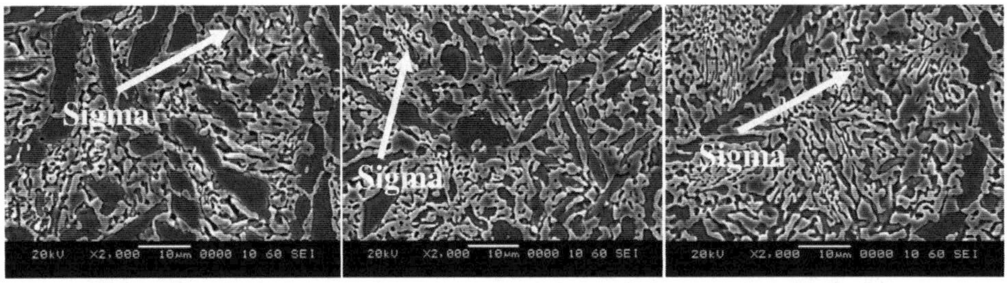

Fig 5. SEM images of heat treated specimens at 850°C (a) 30 minutes (b) 60 minutes (c) 90 minutes

It precipitates clearly at the austenite and ferrite boundary region and as the aging time increases the ferrite phase of DSS is consumed and sigma phase precipitation increases as observed in the Figure 5 a, b and c for 30, 60 and 90 minutes of ageing time respectively.
Figure 6 a, b and c are the phase volume fraction of Zeron 100 duplex stainless steel at ageing time for 30, 60 and 90 minutes respectively. The heat treatment process introduces microstructural changes promoting inter-metallic changes. As the ageing time of heat treatment increases the phase volume of sigma phase increases as observed in the Figure 6 a, b and c. The phase volume fraction of all the three phases for different ageing time is given in the table 2.

Table 2 Phase volume fraction of Zeron 100 SDSS at different ageing time

Heat Treatment Temperature 850° C	Phase Volume Fraction (%)		
Ageing Time(min)	Ferrite	Austenite	Sigma
0	56.62	43.38	0
30	53.84	40.85	5.31
60	46.65	45.48	7.87
90	43.07	42.40	14.53

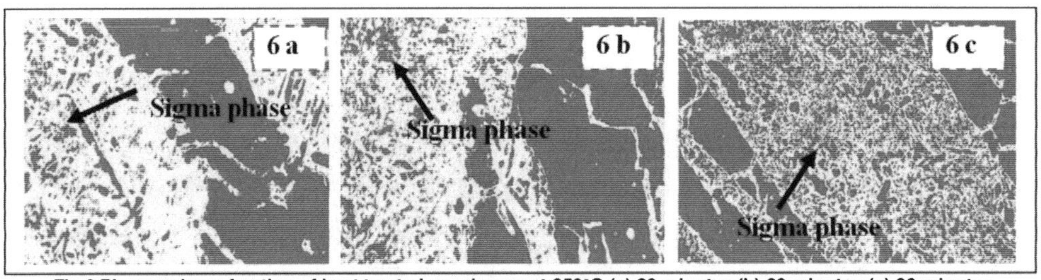

Fig 6 Phase volume fraction of heat treated specimens at 850°C (a) 30 minutes (b) 60 minutes (c) 90 minutes

3.3 EDX Analysis results

indicated that there is increase of chromium as the ageing time is increased. Although the increment observed was not much but there was a decrement in ferrite. This is because the sigma phase nucleates at the ferrite phase by consuming chromium and molybdenum. The table 3 shows the chemical composition for different ageing times in terms of weight and atomic mass percentage. It clearly indicates the increase in chromium as the ageing time increases. Result obtained after Optical Image Analysis concludes that as the aging time increases the diffusion rate goes on increasing and hence we can see more ferrite promoting elements gets concentrated and hence sigma phase goes on increasing.

3.4 Hardness

Hardness value has a direct relationship with the sigma phase formation. Specimens were tested for Rockwell hardness value at 850°C for different aging time. Due to the precipitation of inter metallic phase called sigma phase, the Rockwell hardness value is found to be increasing as the aging time increases.

Table 3 Mass percentage of different elements at different ageing time

Ageing Time(min)	Solution treated specimens		30		60		90	
Element	Weight%	Atomic%	Weight%	Atomic%	Weight%	Atomic%	Weight%	Atomic%
Molybdenum	3.65	1.99	3.80	2.01	3.79	1.99	3.92	2.09
Chromium	28.18	28.36	28.35	27.70	28.62	27.81	29.12	28.59
Fe	60.06	56.28	58.94	53.63	58.05	52.53	58.63	53.59

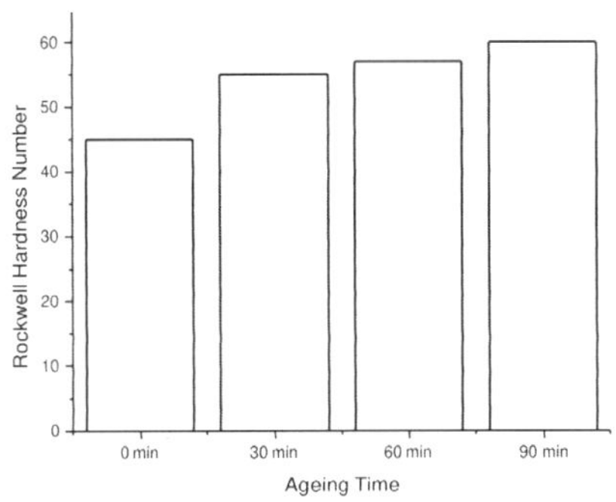

Fig.7 Rockwell hardness value for different aging time at 850° C

There is no much difference in the Rockwell hardness number between the aging times, but it is found to be increasing. Figure 7 shows us the Rockwell hardness value for different aging time at 850° C.

3.5 Corrosion

The examinations of corrosion resistance have revealed that the ageing process after super saturation do not cause significant changes in the anti-corrosive properties. The σ phase precipitates occurring along the boundaries of primary solidification grains, according to the reaction δ ferrite→phase+γ' secondary austenite. The prolonged annealing time results in the ferrite decomposition hence improving the corrosion resistance. The test carried out here showed no formation of white and red rust after 90 min of test as observed in Figure 8 and 9

Fig.8 Before corrosion treatment Fig.9 After corrosion treatment

4. Conclusion

- Solution treated Zeron 100 only showed austenite and ferrite phases, however after heat treatment bright sigma phases rich in chromium were formed at the interface of ferrite and austenite phases.
- SEM analysis showed the formation of sigma phase at the interface of ferrite and austenite phases. Optical image analysis revealed the increase in the formation of sigma phase as the ageing time of heat treatment was increased.
- EDX analysis showed the formation of chromium-molybdenum rich phase.
- Due to the formation of sigma phase the hardness value was found to be increasing in the heat treated specimens as compared to the solution treated specimens, this also indicates the reduction in toughness and the brittle nature of Zeron 100.
- The chromium rich phase improved the localized corrosion resistance in heat treated Zeron 100.

5. References

[1] Akisanya A R, and Renton N C (2012), Effect of ageing on phase evolution and mechanical properties of a high tungsten super-duplex stainless steel, Materials Science and Engineering: A, Vol.535, pp.281- 289.
[2] Wang,Y Q et al(2013),Effect of sigma phase precipitation on the mechanical and wear properties of Z3CN20. 09M cast duplex stainless steel. Nuclear Engineering and Design, Vol.259, pp.1-7.
[3] Villalobos,Detal(2009),Microstructural changes in SAF2507 super duplex stainless steel produced by thermal cycle,Matéria (Rio de Janeiro), Vol.14(3), pp.1061-1069.
[4] Chen T H, and Yang J R (2002), Microstructural characterization of simulated heat affected zone in a nitrogen-containing 2205 duplex stainless steel, Materials Science and Engineering: A, Vol.338(1- 2),pp.166- 181.
[5] Magnabosco, R (2009), Kinetics of sigma phase formation in a duplex stainless steel,Materials Research, Vol.12(3), pp.321-327.
[6] Escriba, D M et al (2009), Chi-phase precipitation in a duplex stainless steel,Materials Characterization, Vol.60(11), pp.1214-1219.
[7] Davanageri, M et al (2016). Dry Sliding Wear Behavior of Super Duplex Stainless Steel AISI 2507: A Statistical Approach. Archives of Foundry Engineering, Vol 16 (4), pp 47-56.

IOP Conf. Series: Materials Science and Engineering 1189 (2021) 012017 doi:10.1088/1757-899X/1189/1/012017

Experimental and numerical modal analysis of short carbon fiber reinforced polyethersulfone composites

Harshavardhan B[1,2*], R Ravishankar[1], B Suresha[2], and S Srinivas[3]

[1]Department of Mechanical Engineering, Sri Jayachamarajendra College of Engineering, Mysuru-570 006, India

[2]Department of Mechanical Engineering, The National Institute of Engineering, Mysuru-570 008, India

[3]Product Development Department, Brakes India Private Limited, Nanjangud-571 302, India

Abstract. Damping is the resistance to the vibratory motion. Viscous dampers are generally used to resist the repetitive motion and dissipate the energy out of the vibrating system. However, material of the vibrating body itself will have ability to dissipate the vibratory energy which is often referred as material damping. Even though, material damping is very small compared to viscous damping, in many practical situations, this may also play significant role. This work was focused on the evaluation of material damping of short carbon fiber (SCF) reinforced polyethersulfone (PES) composites by adopting free vibration test methodology. In addition, Young's modulus of the PES/SCF composites was determined through their natural frequencies. Investigation was made by reinforcing the PES with 10, 20 and 30 wt.% of SCF using twin screw extruder and injection moulding machine. Free vibration response of the clamped-free composite beam was captured through an accelerometer sensor. Free vibration test revealed that PES composites showed reduction in damping with increase in SCF content due to stiffening of the polymer. Natural frequencies of PES composites increased with increase in SCF loading. Tensile modulus enhanced by 267% with respect to neat polymer for 30 wt. % of SCF reinforcement. Also, numerical frequency analysis was carried out using Abaqus finite element package and simulated results closely fitted with the experimental data.

1. Introduction

In aerospace, marine and automobile industries, components are generally made by polymer composites reinforced with carbon fibers[1]–[3]. Carbon fiber reinforced plastics will have high strength to weight ratio [4], [5]. Damping is one of the most important property of the material due to which energy is dissipated from the vibrating system in the form heat [6]. Pure polymers will have better damping properties due to their visco-elastic nature. Generally, addition of fibers to polymers enhances the stiffness of composite which may result in reduction in their damping properties. Modal analysis is a tool to determine the vibration parameters such as natural frequencies and their corresponding mode shapes[7],[8]. Vibration techniques can be used as non-destructive tests to detect and locate the flaws in fiber reinforced composites[9]. Modal analysis is also used to optimize the dynamic response of the structures. Present modal investigation was carried out in order to understand the effect of short carbon fibers on damping behaviour of polyethersulfone composites.

2. Materials and Methods

2.1. Materials

The polyethersulfone (PES) was procured from the company Solvay (USA) and was used as matrix material. The short carbon fiber (SCF) based on polyacrylonitrile (PAN) was purchased from the company Teijin (Japan) and was used as reinforcement material. Materials formulation adopted in the present work is referenced in table 1. The properties of PES and SCF are stated in the previous work [10].

Table 1. Material formulations and designations [10].

Composites	Matrix wt. % PES	Fiber wt. % SCF	Density (g/cm^3)	Elastic modulus (GPa)
C_{00}PES	100	00	1.364	2.760
C_{10}PES	90	10	1.395	5.886
C_{20}PES	80	20	1.415	8.798
C_{30}PES	70	30	1.423	9.191

2.2. Fabrication methods

PES and SCF were mixed with different fiber weight fraction of 0, 10, 20 and 30% in ZE 40 twin screw extruder (KraussMaffei Berstorff) at slightly above the melting temperature of PES which is 330 ^{0}C and composite pellets were formed. These pellets were preheated to remove moisture and then melted again in a single screw injection molding machine (Milacron 80T) and filled to the molds having dimensions length 120 mm x width 10 mm x thickness 3.85±0.05 mm at a high pressure of 140 bar for a duration of 3 second, after that specimens were allowed to cool at atmospheric pressure for 15 seconds. Injection molded test specimen is shown in figure 1.

Figure 1. Injection moulded composite specimen.

2.3. Free vibration test

Damping properties of the PES/SCF composites were determined by single cantilever free vibration decaying test [11]–[13]. The developed samples were mounted as cantilever beam and 10 mm length was used for clamping purpose. Effective length of the beam was taken as 110 mm for analytical calculations and numerical simulations. Piezoelectric accelerometer, model PCB 356A15 was used to capture the response of the vibrating beam as indicated in figure 2. Mass of the accelerometer used is 10.5 grams and it is capable to measure the frequency in the range 2 to 5000 Hz with ± 5 % accuracy.

Figure. 2. PCB Tri-axial accelerometer sensor.

Accelerometer was mounted at the free end of cantilever beam arrangement with an temporary adhesive (petro wax) as in figure 3. Analog response captured by accelerometer was then digitalized with the help of NI 9234 IEPE input module. Digitalized response was then transferred to the computer system through NI compact data acquisition module cDAQ 9178. Data received by the computer was then analyzed using NI Labview software package.

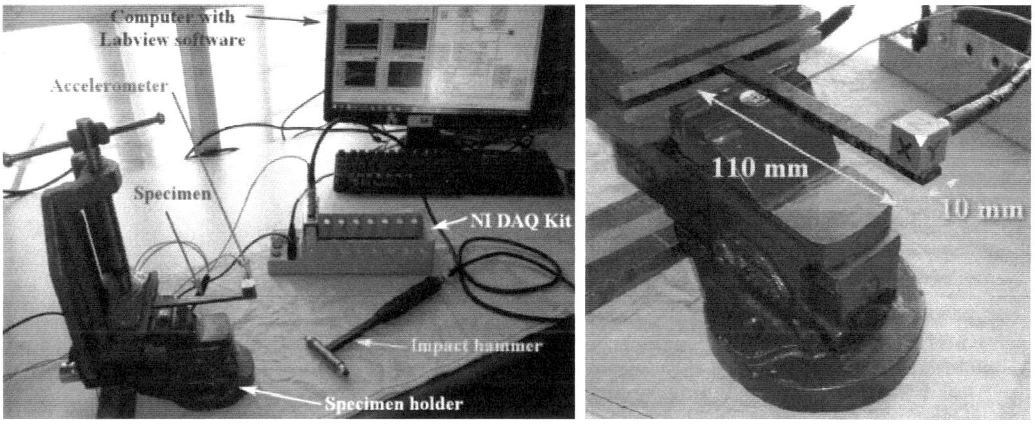

Figure 3. (a) Free vibration test setup (b) Cantilever arrangement

Composite beam was made to vibrate by disturbing it with a small force by an impact hammer. As there was no viscous damper connected, neglecting the air damping, it was assumed that vibrating beam decreases its amplitude of oscillation in every cycle and finally after some finite time attains the equilibrium position. Decrease in the amplitude in every successive cycle is measured using logarithmic decrement parameter.

Logarithmic decrement (δ) is the ratio of natural logarithm of two successive amplitudes and is mathematically expressed as in equation (1).

$$\delta = \left(\frac{x_1}{x_2}\right) = \frac{1}{n}\ln\left(\frac{x_1}{x_{n+1}}\right) \tag{1}$$

Damping ratio (ξ) is related to the logarithmic decrement according to equation (2).

$$\zeta = \frac{\delta}{\sqrt{4\pi^2 + \delta^2}} \tag{2}$$

Since damping offered by the material is very small, the term δ^2 can be neglected.

$$\zeta = \frac{\delta}{2\pi} \tag{3}$$

Damped natural frequency (f_d) is determined by taking the ratio of number of cycles (n) to time taken to complete 'n' number of cycles (t).

$$f_d = \frac{1}{2\pi}\left(\frac{n}{t}\right) \tag{4}$$

Undamped natural frequency (f_n) of the a single degree-of-freedom (DOF) system is related to damped natural frequency as expressed in equation (5).

$$f_n = \frac{2\pi f_d}{\sqrt{1-\zeta^2}} \tag{5}$$

For a single DOF cantilever system with beam mass 'm' and point load 'M' at its tip, natural frequency is given by equation (6).

$$f_n = \frac{1}{2\pi}\sqrt{\frac{3EI/l^3}{M + \frac{33}{140}m}} \tag{6}$$

In the present modal analysis study, point load 'M' indicates the mass of the piezoelectric accelerometer sensor and beam mass 'm' depends on the density and volume of the composite sample [14].

Generally, short fibers are aligned in the direction of melt flow during the injection process based on the theoretical approach. However, short fibers reinforced to composites take the random orientations along and across the thickness due to the fabrication limits and tolerances. Different fiber orientation leads to variation in tensile modulus of the composites. Instead of conducting destructive tensile test, free vibration test can be conducted on specimens without breaking them [9], [13]. From the vibration-time plot, damped natural frequency is calculated using the equation (4). Further, damped frequency is converted into undamped frequency using the relation (5). At last, tensile modulus (E) can be determined with the help of natural frequency using the expression (6), where, I is bending moment of inertia of beam having rectangular cross section and l is the length of beam.

2.4. Numerical analysis

Numerical modal simulation was performed to understand the nature of anisotropic composite material behaviour during the vibration [15], [16]. Finite element based numerical modal analysis was carried out using Abaqus CAE 2017 software package. Mass of the accelerometer sensor (10.5 g) was considered for the analysis by modeling it as a solid volume of dimensions 13 mm x 13 mm x 13 mm with density of 4560 kg/m^3. Geometry of the composite beam was modelled separately as per the dimensions of actual specimens and then interaction was applied between beam and sensor with tie constraint as represented in figure 4(a).

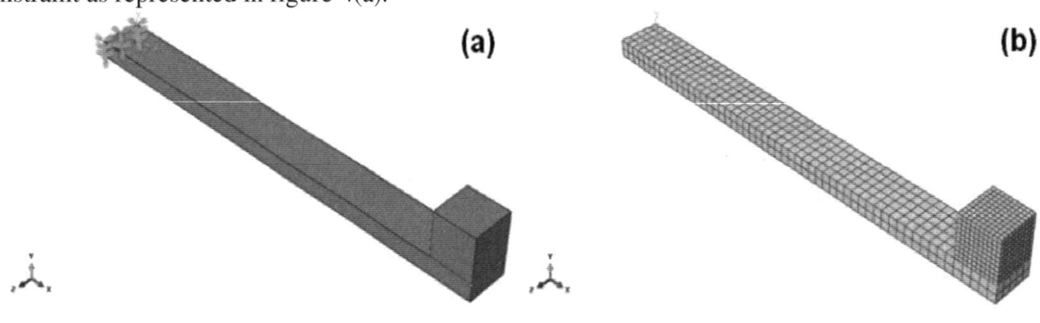

Figure 4. (a) Geometry (b) Mesh created in Abaqus.

Material of the beam was assumed as homogeneous and elastic for numerical analysis. Modulus of elasticity determined by free vibration test was given as material input parameter. Geometry was fixed at one end by arresting all the degrees of freedom of the nodes present at that end. Both the geometries were meshed with C3D8R elements as in figure 4(b) and then linear frequency perturbation analysis was carried out.

3. Results and discussion

3.1. Free vibration test

Acceleration signals recorded for PES/SCF composites were plotted against time using the Matlab tool as shown in figure 5. Amplitudes of acceleration and their corresponding time were labeled for five successive cycles.

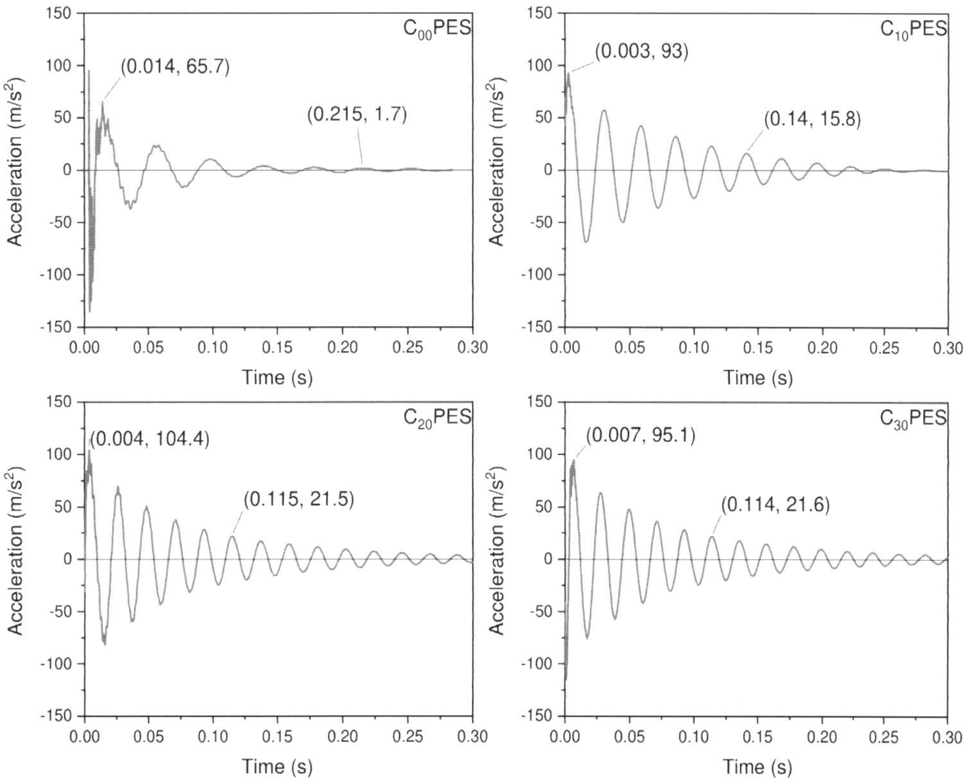

Figure 5. Amplitude of vibration (acceleration) of PES/SCF composites in time domain.

By means of these amplitude and time data, logarithmic decrement for the composites were determined using the equation (1), damping ratio were determined with the help of equation (2) and/or (3). All the computed parameters are presented in table 2.

Table 2. Free vibration test results of PES/SCF composites.

Parameters	Composites			
	C_{00}PES	C_{10}PES	C_{20}PES	C_{30}PES
Time t_1 (s)	0.014	0.003	0.004	0.007
Time t_2 (s)	0.215	0.140	0.115	0.114
Amplitude at time t_1 (m/s^2)	65.7	93.0	104.4	95.1
Amplitude at time t_2 (m/s^2)	1.7	15.8	21.5	21.6
Logarithmic decrement, δ	0.731	0.355	0.316	0.296
Damping ratio, ξ	0.116	0.056	0.050	0.047
Time period (s)	0.040	0.027	0.022	0.021
Damped natural frequency, f_d (Hz)	24.88	36.50	45.05	46.73
Undamped natural frequency, f_n (Hz)	25.04	36.55	45.10	46.78
Tensile Modulus, E (GPa)	2.737	5.847	8.916	9.599

Material damping primarily occurs due to friction interaction between various constituents in the composites and/or within the polymer chains. Materials absorb the vibration energy to create friction interaction between constituents which in turn releases heat energy. This heat needs to be dissipated out from the material [4]. Addition of SCFs into neat PES resulted in improvement in stiffness of the composites. Due to increase in stiffness, vibration energy absorption by the composites decreased.

Neat PES showed the logarithmic decrement of 0.731 resulting in its damping factor as 0.116. Logarithmic decrement values of the composites reduced by 51.4 %, 56. 8% and 59.5% with the reinforcement of 10, 20 and 30 wt.% of SCF into PES. This in turn resulted in lowering of the damping nature of PES/SCF composites. With addition of 10 wt. % of SCF, damping factor decreased drastically by 51.7%. However, with addition of 20 and 30 wt.% of SCF, further reduction in damping factor was very small with respect to $C_{10}P$ composite.

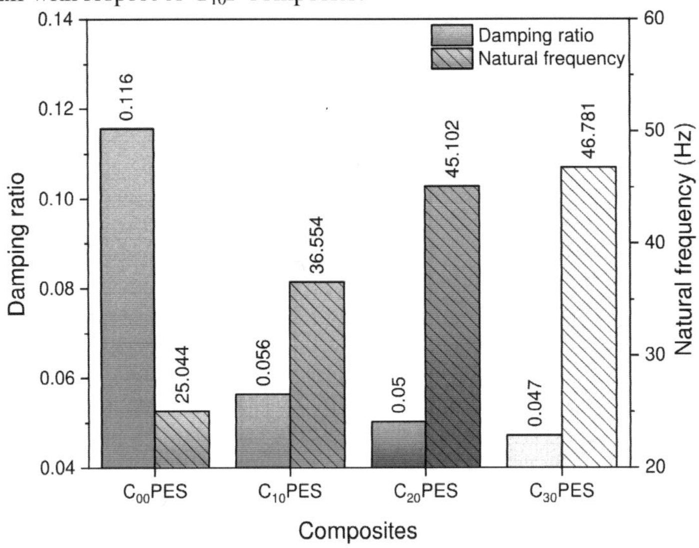

Figure 6. Damping of PES/SCF composites.

Incorporation of SCF into PES resulted in improved stiffness (tensile modulus) of composites. Improvement in tensile modulus was due to good adherence of polymer with fiber surface. Fiber breakage with very few fiber pull-outs were seen on the fractured surface of tensile specimen of C_{30}PES composite shown in figure 7 taken by scanning electron microscopic (SEM), indicating good bonding between fiber and polymer.

Due to good stickiness, the relative motion between the polymer molecules as well as polymer and fiber during the vibration was restricted. This reduced the fiber-matrix friction interaction during vibration and hence, composites lost the ability to absorb the vibration energy and damping of SCF reinforced composites were reduced.

Figure 7. SEM image of fractured tensile specimen of C_{30}PES composite

Damped and undamped natural frequencies were computed according to equations (4) and (5). In addition, Fast Fourier Transform (FFT) analysis was carried out with the help of signal processing tool available in National Instruments Labview software. FFT analysis transformed the acceleration in time domain to frequency domain. Accelerations in frequency domain are shown in figure 8.

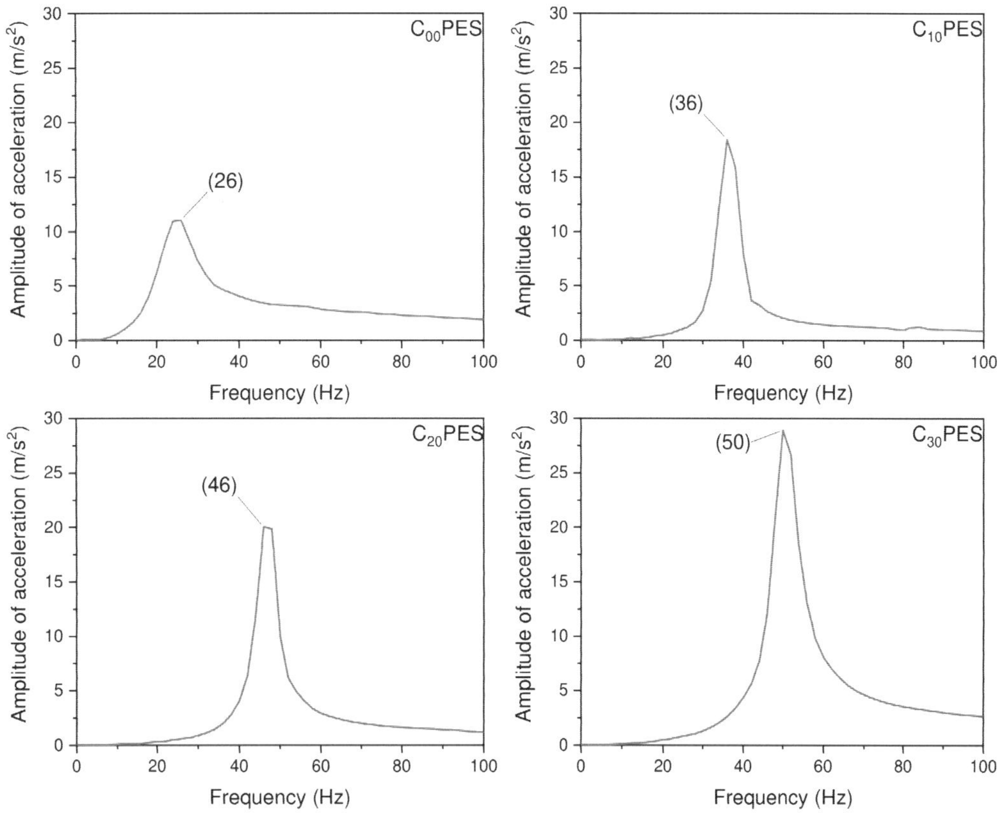

Figure 8. Amplitude of vibration (acceleration) of PES/SCF composites in frequency domain

Frequency corresponding to the first largest peak represents the damped natural frequency of the composite materials. Damped natural frequencies of the composites were in agreement with analytical values computed by equation (4).

Finally, modulus of elasticity of PES/SCF composites were determined using the natural frequency of free vibration test as per equation (6) and the correlated data was mentioned in table 3. It was worthwhile to note that modulus of elasticity of composites determined through free vibration test were found in close match with the tensile test results [10] with less than 5 % error.

Table 3. Tensile modulus of PES/SCF composites.

Composites	Density [5] (kg/m^3)	Voids (%)	Elastic modulus (GPa)		
			Tensile test [10]	Free vibration test	Error (%)
C_{00}PES	1364	0.44	2.760	2.737	0.83
C_{10}PES	1395	0.86	5.886	5.847	0.66
C_{20}PES	1415	2.05	8.798	8.916	1.34
C_{30}PES	1423	4.08	9.191	9.599	4.44

3.2. Numerical simulation

Numerically simulated first mode shape of PES/SCF composites along with the accelerometer mounted on the beam were shown in figure 9. Numerical frequencies were found close to the experimental results as reported in table 4. Since experiments were conducted in air, damping due to air may be one of the factors accounting for the error between numerical and experimental values. In addition, modeling the composite beam with visco-elastic material model may further reduce the error between numerical and experimental frequencies.

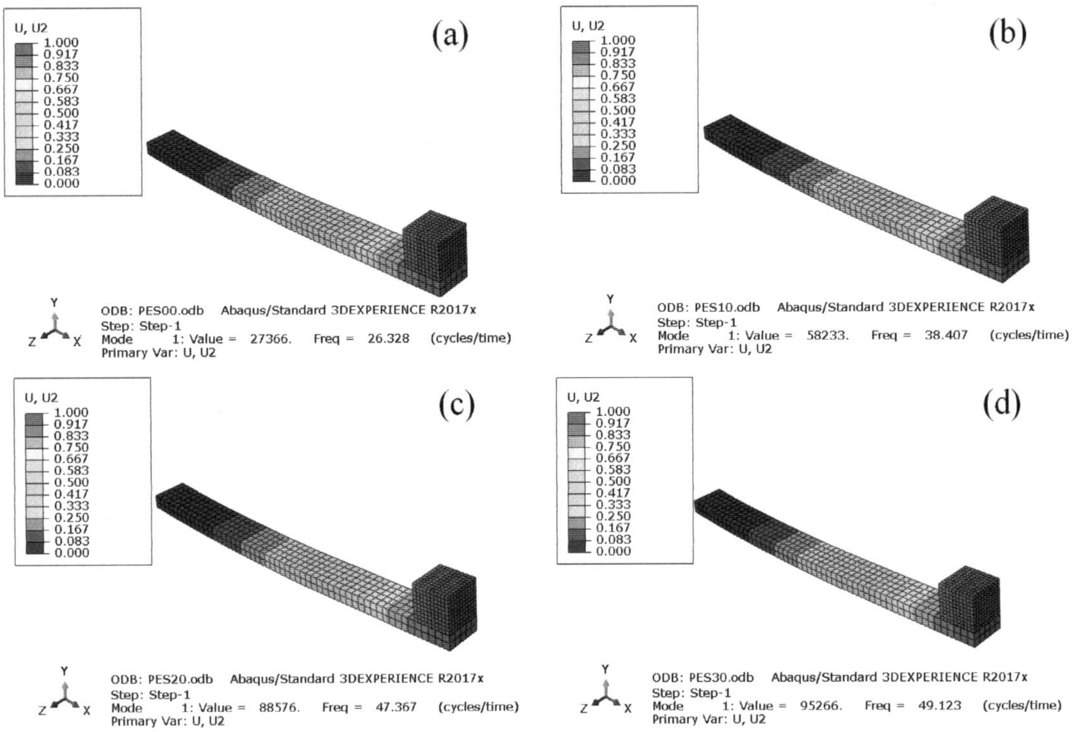

Figure 9. Abaqus CAE simulated frequencies of composites (a) C_{00}PES (b) C_{10}PES (c) C_{20}PES (d) C_{30}PES.

Table 4. Comparison of experimental and numerical natural frequencies.

Composites	Natural undamped frequencies (Hz)		Error
	Experimental	Numerical	(%)
C_{00}PES	25.04	26.328	5.1
C_{10}PES	36.55	38.407	5.1
C_{20}PES	45.10	47.367	5.0
C_{30}PES	46.78	49.123	5.0

4. Conclusion

In conventional tensile test, material is loaded till it breaks and with the help of deformation variation with applied load, tensile modulus is evaluated. Free vibration test is non destructive in which material is subjected to a small impact force and specimen is made to vibrate freely. Experimental results revealed that with increase in SCF loading into PES, damping values of the composites were decreased. Maximum reduction in damping was found in C_{30}PES composite whose tensile modulus enhanced by 267% with respect to neat polymer. Young's moduli of the PES/SCF composites based on free vibration test were in well agreement with tensile test results. In addition, natural frequencies obtained by numerical simulation matched with that of experimental values.

References

[1] S. Erden and K. Ho, 'Fiber reinforced composites', in *Fiber Technology for Fiber-Reinforced Composites*, Elsevier, 2017, pp. 51–79.

[2] N. Forintos and T. Czigany, 'Multifunctional application of carbon fiber reinforced polymer composites: Electrical properties of the reinforcing carbon fibers – A short review', *Compos. Part B Eng.*, vol. 162, pp. 331–343, Apr. 2019, doi: 10.1016/j.compositesb.2018.10.098.
[3] K. Acatay, 'Carbon fibers', in *Fiber Technology for Fiber-Reinforced Composites*, Elsevier, 2017, pp. 123–151.
[4] Y. Li, S. Cai, and X. Huang, 'Multi-scaled enhancement of damping property for carbon fiber reinforced composites', *Compos. Sci. Technol.*, vol. 143, pp. 89–97, May 2017, doi: 10.1016/j.compscitech.2017.03.008.
[5] M. Assarar, W. Zouari, H. Sabhi, R. Ayad, and J.-M. Berthelot, 'Evaluation of the damping of hybrid carbon–flax reinforced composites', *Compos. Struct.*, vol. 132, pp. 148–154, Nov. 2015, doi: 10.1016/j.compstruct.2015.05.016.
[6] S. Han and D. D. L. Chung, 'Mechanical energy dissipation using carbon fiber polymer–matrix structural composites with filler incorporation', *J. Mater. Sci.*, vol. 47, no. 5, pp. 2434–2453, Mar. 2012, doi: 10.1007/s10853-011-6066-7.
[7] Zhi-Fang Fu, Jimin He, 'Modal analysis methods – frequency domain', in *Modal Analysis*, 2001, pp. 159–179.
[8] N. Rajini, J. W. Jappes, S. Rajakarunakaran, and P. Jeyaraj, 'Dynamic mechanical analysis and free vibration behavior in chemical modifications of coconut sheath/nano-clay reinforced hybrid polyester composite', *J. Compos. Mater.*, vol. 47, no. 24, pp. 3105–3121, Nov. 2013, doi: 10.1177/0021998312462618.
[9] P. Cawley and R. D. Adams, 'A Vibration Technique for Non-Destructive Testing of Fibre Composite Structures', *J. Compos. Mater.*, vol. 13, no. 2, pp. 161–175, Apr. 1979, doi: 10.1177/002199837901300207.
[10] B. Harshavardhan, R. Ravishankar, B. Suresha, and S. Srinivas, 'Static and Dynamic Mechanical Performance of Carbon Fiber Reinforced Polyethersulfone Composites', *Appl. Mech. Mater.*, vol. 895, pp. 265–271, Nov. 2019, doi: 10.4028/www.scientific.net/AMM.895.265.
[11] F. Magalhães, Á. Cunha, E. Caetano, and R. Brincker, 'Damping estimation using free decays and ambient vibration tests', *Mech. Syst. Signal Process.*, vol. 24, no. 5, pp. 1274–1290, Jul. 2010, doi: 10.1016/j.ymssp.2009.02.011.
[12] P. P. Hujare and A. D. Sahasrabudhe, 'Experimental Investigation of Damping Performance of Viscoelastic Material Using Constrained Layer Damping Treatment', *Procedia Mater. Sci.*, vol. 5, pp. 726–733, 2014, doi: 10.1016/j.mspro.2014.07.321.
[13] M. M. Wallace and C. W. Bert, 'Experimental Determination of Dynamic Young's Modulus and Damping of an Aramid-Fabric/Polyester Composite Material', p. 4.
[14] S. K. Bhudolia, P. Perrotey, and S. C. Joshi, 'Enhanced vibration damping and dynamic mechanical characteristics of composites with novel pseudo-thermoset matrix system', *Compos. Struct.*, vol. 179, pp. 502–513, Nov. 2017, doi: 10.1016/j.compstruct.2017.07.093.
[15] S. K. Sahu and P. Das, 'Experimental and numerical studies on vibration of laminated composite beam with transverse multiple cracks', *Mech. Syst. Signal Process.*, vol. 135, p. 106398, Jan. 2020, doi: 10.1016/j.ymssp.2019.106398.
[16] P. Das and S. Kr. Sahu, 'Experimental and numerical study on free vibration of cracked woven fiber glass/epoxy composite beam', *Mater. Today Proc.*, p. S2214785320321003, Apr. 2020, doi: 10.1016/j.matpr.2020.03.320.

Acknowledgment
Injection moulding of the specimens were fully supported by the Brakes India Limited, Sundaram Polymers Division, Nanjangud, Mysore District, Karnataka, India. We wholeheartedly thank Mr. P. Krishna Kumar, Vice President–Operations, Brakes India Limited for his support in the fabrication of composites.

Tribological Suitability of aluminium hybrid composite above atmospheric temperature

B.N Sharath*, K.S Madhu, D.G Pradeep, C.V Venkatesh

Department of Mechanical Engineering, Malnad College of Engineering, Hassan, affiliated to VTU, Belagavi, Karnataka, India

*Corresponding author e-mail address: sharathbn04@gmail.com

Abstract. Industries are facing problems with light weight alloys due to tribological impediments in which utmost complicated problem is wear. As a consequence, endeavour has been made to fabricate increased wear resistance materials. In 2014, 2.87 million tons of aluminium was used by the global automotive industry (excluding China). By 2020, an annual demand of 4.49 million tons of aluminium is projected. Key factors of this growth include both increased production of automobiles and broader use of aluminium alloys in modern automobiles.The main advantage of spacecraft aluminum alloys is their ability to withstand hi gh and low temperatures, vibration loads, and radiation. Additionally, aluminium alloys exhibit outstanding strength and durability in cryogenic environments as the temperature drops. In the present investigation Al2618 is used as a Matrix and hard ceramic boron carbide, with amorphous graphite is used as reinforcement. Stir casting is used to synthesis the hybrid composite. Statistical model taguchi is used to assess the dry sliding wear properties of the hybrid Metal matrix composite at different temperature (50°C, 100°C, 150°C). Evaluation of particle distribution of reinforcement in the matrix and worn surface morphology is studied by SEM. Wear resistance of the hybrid composite (reinforced with B_4C & Gr) is enhanced at 150°C than conventional alloy. Glazed layer formation in Al2168 matrix is less than glazing layer formation formed on hybrid composites Adhesive wear is observed during test.

Keywords: Hard ceramic; Boron carbide; graphite; glaze; oxide; tribo layer; worn surface.

1. Introduction

The industrial use of lightweight materials is gaining growing attention from different sectors of manufacturing. The use of aluminium alloys for metal matrix composites is growing rapidly to meet various end uses in automotive components like valve, piston, liners etc. Aluminium alloy, on the other hand, causes high tensile strength and stiffness with lower impact resistance. These composites could be used for components under tensile stress, such as connecting rod [1]. A combination of aluminium alloy and ceramic reinforcement may incorporate the positive tensile characteristics as well as the good impact properties of aluminium alloys. The automobile industry is successfully applying composites reinforced with a variety of Aluminium alloy to replace components such as outer body of car, cylinder head, turbo charger, brake calliper, connecting rod etc. [3]. Metal matrix composites show strong properties, but there is a large volatility in their properties. A vulnerability can and will be solved by introducing more sophisticated manufacturing of aluminium alloys and a composite. Its specific properties will establish a stable foundation for the development of new applications and prospects for hybrid metal matrix composite in the 21st century [2]. Across different applications, the use of hybrid metal matrix composites has opened new opportunities for both academia and industry to build a viable framework for the potential use of hybrid metal matrix composites [4]. Al-Si alloy is a proven alloy with a high wear resistance and low coefficient of thermal expansion, strong tolerance to corrosion and enhanced mechanical properties at high temperature. However, the silicone density is about 3,23g / cm, which is very high compared to Boron Carbide, in fact, Boron Carbide is the third hardest substance used to improve aluminium matrix alloy wear strength. Graphite is typically useful for use at high temperatures and has superior mechanical

IOP Conf. Series: Materials Science and Engineering 1189 (2021) 012018 doi:10.1088/1757-899X/1189/1/012018

properties. Graphite also has a self-lubricating ability to mitigate the wear of tribological component [8]. Therefore, an experiment to examine the results of this study is clearly explained wear behaviour of unreinforced Al2618 and reinforced with B_4C & Gr.

2. Experimentation

2.1 Solidification of composites

There are various routes to synthesize Al2618– B_4C – Gr hybrid metal matrix composites, But in stir casting approaches to its simplicity of fabric and large volume. In order to achieve a high level of mechanical properties in the production of AMCs, a good interfacial bonding between the matrix material and the reinforcement is an important aspect. Care should be taken to preserve optimal casting parameters, such as wettability of the molten metal, homogeneous mixing of the particles, density, and solidification rate. The crucible is made of graphite and the ingot is poured into the crucible by 1.25 kg of matrix material. The melt was held for 1.5 hrs at a temperature between 650 and 700°C. Crucible is made up of graphite and 1.25kg matrix material ingot is poured into the crucible. The melt was maintained at a temperature between 650 to 700°c for 1.5hr, because the slag accumulated on top of the molten metal had been extracted. When the ingot content entered a liquid state, magnesium (4gm) was applied to the molten metal to enable particle dispersion within the molten metal. Weighted Reinforcement 10% of boron carbide was preheated at a required temperature of 550°C to eliminate moisture or any other gas contained in the reinforcement. A fine reinforcement particle (30-50μ) of boron carbide has been added to a molten metal which can afford more nuclei for heterogeneous grain nucleation. Stirring operation at 150 rpm for 10-15 minutes to achieve homogeneous B_4C particle distribution in Al2618. There After the secondary reinforcement, i.e. graphite (3%) has been preheated and poured into the molten metal, constant stirring takes place up to 10-15 min. Hexa-chloro-ethane degassing agent (C2Cl6) is used to reduce porosity, voids and casting defects in molten metal. Once the hot molten metal has been poured into the preheated die, the pouring temperature 650 to 700°C, molten metal must be cooled under the freezing point before solidification begins. Due to the presence of mould surface, the degree of super cooling is reduced as a consequence number of tiny nuclei are formed in molten metal. The rate of cooling in conventional Al2618 is high compared to hybrid composite. Al2618 reinforcement particles in molten metal, the nearest neighbouring gap between nuclei is small, thus suppressing the grain size of the crystal. Compared to the Al2618 alloy grain size, the average grain size is reached in the final solidified composite. The melt is then allowed in the mould to solidify the cast specimens were taken out when it hit room temperature. G99 Cast samples that are synthesized from the stir casting process have been machined to prepare the test specimens in accordance with the ASTM standards. As for this standard, the size of the specimen should be 10 mm in diameter and 30 mm in length. To achieve the uniform flatness on the test specimen buffing was made on the specimen surface, the specimen is then washed with acetone [8].

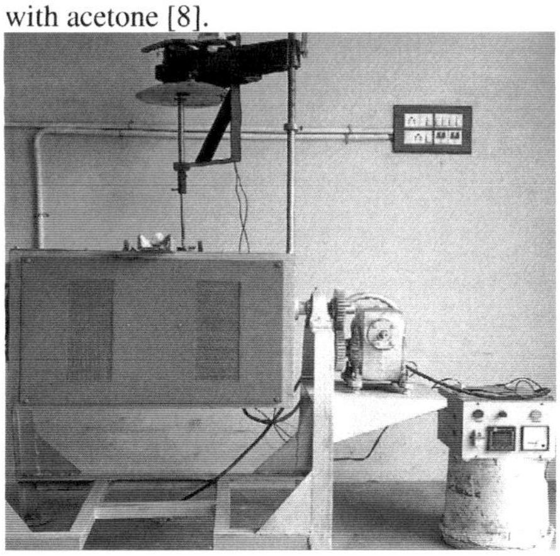

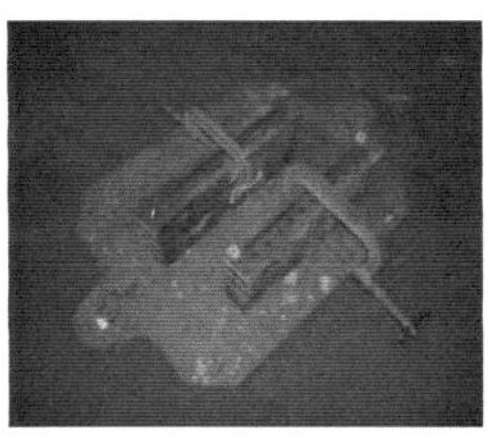

Figure 1a. Preheated Die

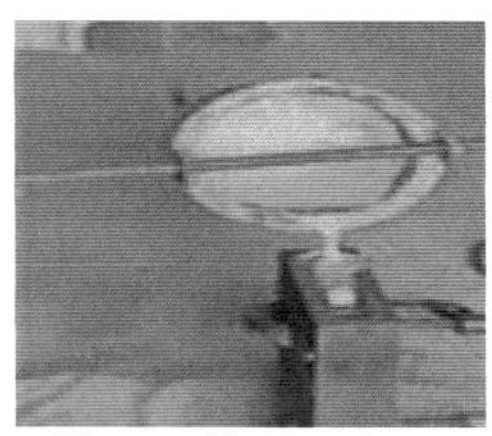

Figure 1b. Pouring

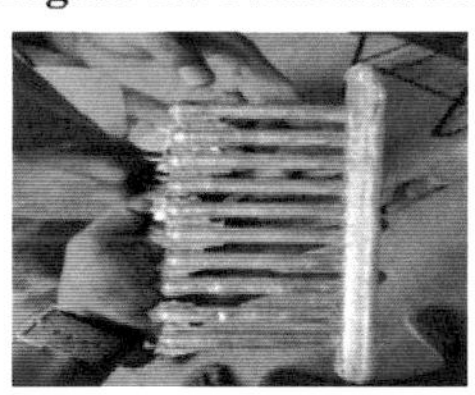

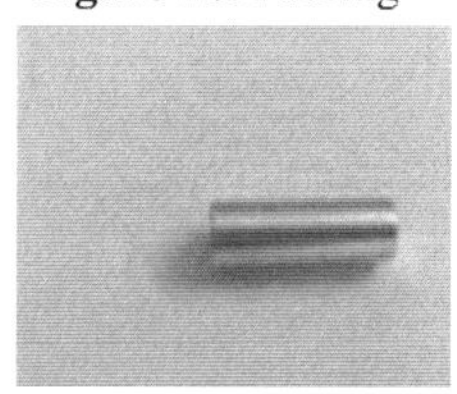

Figure 1. Stir casting setup **Figure 1c.** Cast sample **Figure 1d.** test pin

2.2 Microstructure of composites

The prepared composite is subjected to SEM analysis, allows to see the composite microstructure and to check the distribution of strengthened particles in the matrix as shown in the Figure 3. It is clearly illustrated that B_4C and Gr distribution in Al2618 occurred uniformly. Composite grain structure is fitted with an adequate wear resistance and relies on the presence of reinforced particles which confirmed through XRD spectrum (Figure 5).

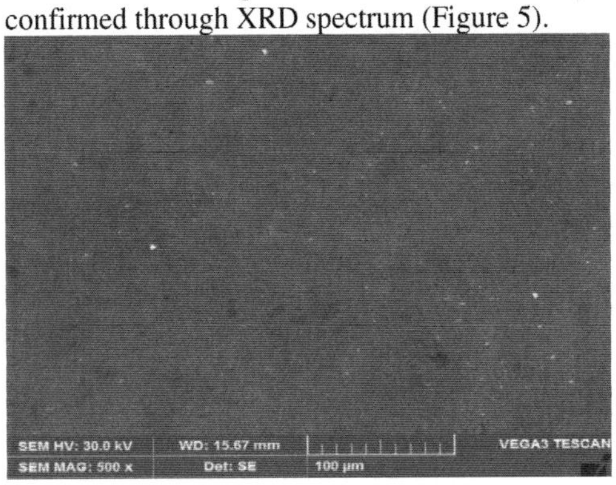

Figure 2. Microstructure of Al2618

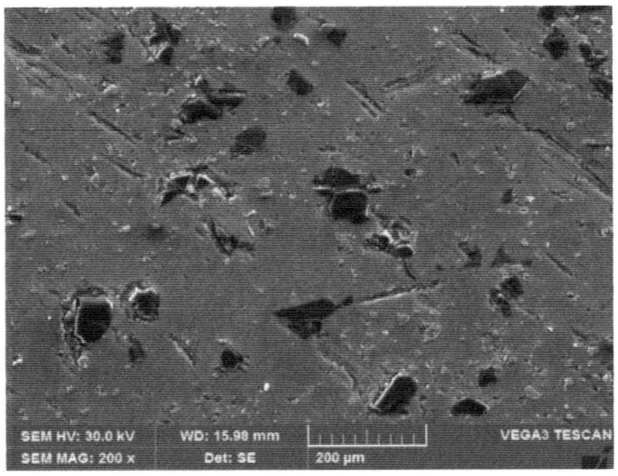

Figure 3. Microstructure of Al2618+B_4C+Gr

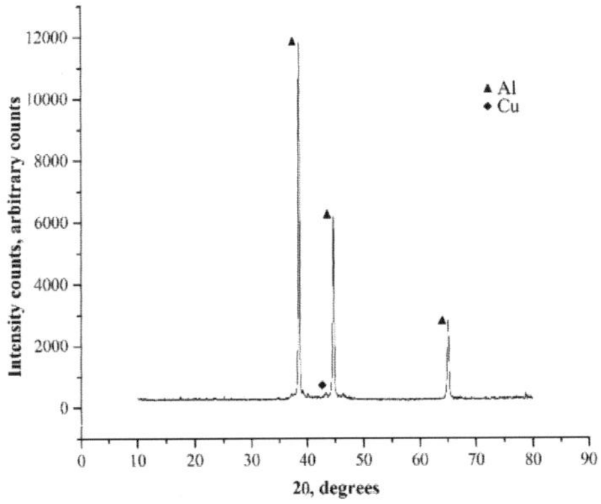

Figure 4. XRD spectrum of Al2618

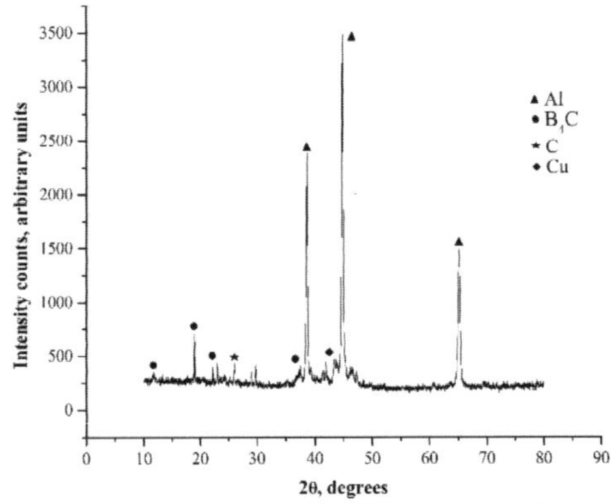

Figure 5. XRD spectrum of Al2618+B_4C+Gr

2.3 Dry sliding wear test at high temperature

The pin-on-test (DUCOM-and friction control TR-20-PHM 400) was used to investigate the attributes of dry sliding wear of hybrid metal matrix composites above Atmospheric temperature. En-32 steel (65HRC) is used when performing the test in compliance with ASTM G99 requirements. Disc track diameter is approximately 140 mm; thickness is 8mm. Disc wear properties have not been studied. Arrangements are rendered as illustrated in Figure 6. Test holder has been placed against the revolving disk that is perpendicular to the disc. The load in step 10N varies from 20N to 40N.and Velocities range from 1.25 m/s to 3.75 m/s, the sliding distance was used in the range from 400 to 800 m and Temperature ranges between 50 to 150°C. During the period of the wear investigation, the standard test procedure was applied to the samples.

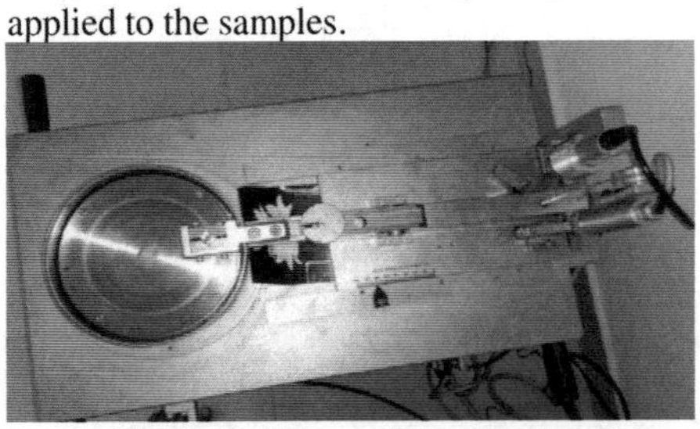

Figure 6. Pin-on-disc apparatus

The surface of both the test samples and the disc were cleaned by soft cotton drenched with acetone prior to the test. Cleaning is required to get rid of any oil and other contaminants found on the disk and sample surface. Specimens are weighed by an electronic measuring system with a precision of 0.0001 mg before and after the test, to measure the difference in weight and is used calculate the rate of wear, to measure the weight difference and use it to compute the wear rate. Using pin heating to describe the wear properties of hybrid metal matrix composite at various temperatures and the experiments were carried out in accordance with the standard orthogonal series L27(taguchi model).

3. Results and Discussions

3.1 High temperature wear analysis

The purpose of this study is to investigate the wear resistance of HMMc (Al2618 + 10% B_4C + 3% Gr as referred with hypo-eutectic alloy [2]. Test temperatures were chosen over ambient temperature i.e. 50,100 & 150 .intention for the selection of that particular range was, generally aluminium alloy allows for oxidation at 50°C. Recrystallization is carried out at 100°C so that these temperatures plays an important role on properties of HMMc particularly when the composite is subject to the temperature described above.

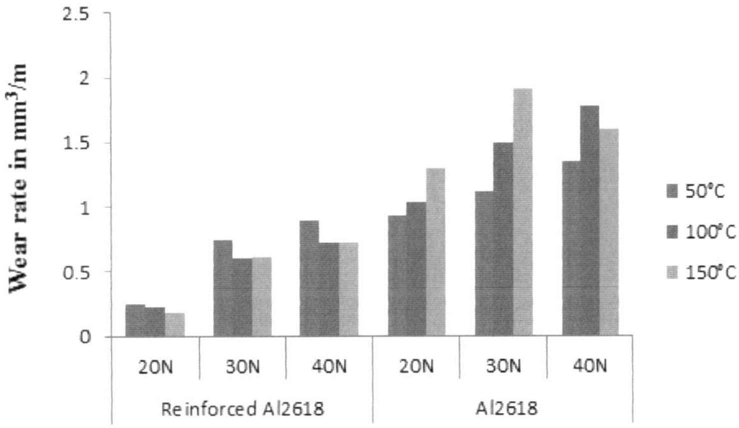

Figure 7. Wear rate v/s Applied load

Prevalence of hard B_4C & Gr particles in the Al matrix leads to expansion in transition load due its strengthening mechanism. Also, they minimize the distance between particles and serve as a strong barrier to dislocation.so that Dislocation movement under the contact load causes the surface to strengthen. The reinforcement particles endorse the lower contact load of 20 N. Hence the load-bearing power of the hybrid composite is increased by the incorporation of B_4C and Gr particles. In addition, the increase in temperature between the pin and disk is the result of by increasing applied load (40 N), it influences to increases in wear rate. Solid ceramic B_4Cp can bear a higher load predominantly because of its potential hardness. In comparison with base alloy (Al 2168), and reinforced alloy is shows maximum wear resistance (5times) at higher load 40 N as shown in Figure 7. In reinforced alloy B_4C particles holds, the applied load and solid lubricant Gr forms a lubricating film, thereby minimizing plastic deformation. The wear rate of A12168+10%B_4C+3%Gr is not significantly exaggerated by load increases, due to the presence of Gr and it is primarily responsible for maintaining wear resistance at all the temperature. The Gr lubricant can quickly flow and squeeze out even at a high load on the surface under contact pressure. Furthermore, this phenomenon is intensified by the distribution of Gr and forms thick tribo film is called glazed layer. This protects the surface against the counter area and reduces the wear rate of A12168+10%B_4C+3%Gr as compared to Al2168. Aluminium and alloy work under heat, their properties must be changed so that such alloys are needed to investigate above the atmospheric temperature. On other hand oxidational & recrystallization temperature is plays a vital role in changing the property of the material that's why (, 100°C, 150°C) these temperatures have been consider for test.

At 50°C wear resistance of the A12168 is less compared to reinforced A12168 because of material softening of unreinforced alloy at the same time oxide layer is begun to form and was not thick in un-reinforced alloy so that wear rate is high at 50°C. At the other hand, hybrid composites have a decreased wear rate by Temperature rise(100°C) mainly due to oxide films formation, it prevents their direct metal-to - metal contact. Oxide films are removed from the surface during continuous sliding. This increases direct metal-to - metal interaction and exposes the fresh area to the atmosphere. When the temperature is. from 50°C to 100°C, during the sliding phase, Shaping and separating oxide layers contribute to reduces in wear rate in hybrid composite as compared to Al2168 alloy. Glazed layer formation in Al2168 matrix is less than glazing layer formation formed on hybrid composites. The formation is less, can be clearly observed from Figure.8 (a, b &c). With the temperature rise, the alloy Al2168 is made smoother. The material transition from metal to steel is improved, leading to an improvement in wear rate. Furthermore, the non-existence of B_4C and Gr particles, thickness of tribo-film is less so that wear rate higher in Al2168 matrix as compared hybridised Al2168.It is also evident from Fig.8 (a-c) and Fig.9 (a-c).

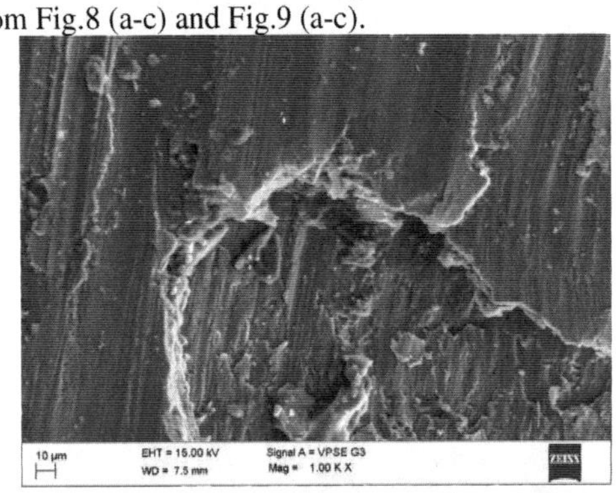

Figure 8a. Hybrdised composite at 50°C

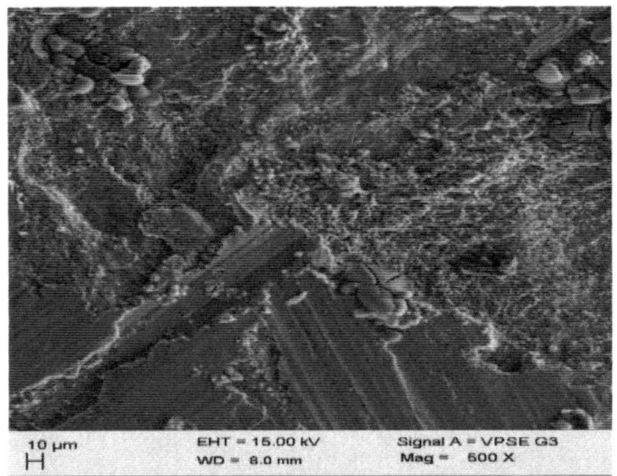

Figure 9a. Al2168 at 50°C

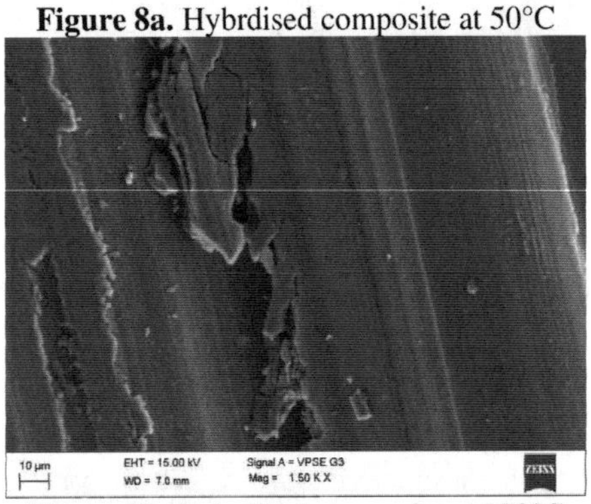

Figure 8b. Hybrdised composite at 100°C

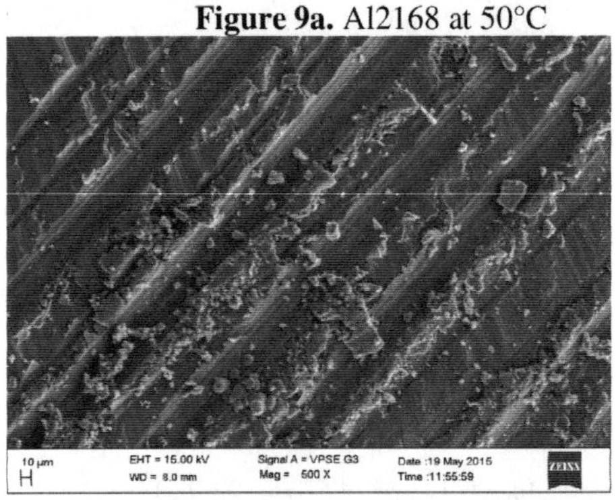

Figure 9b. Al2168 at 100°C.

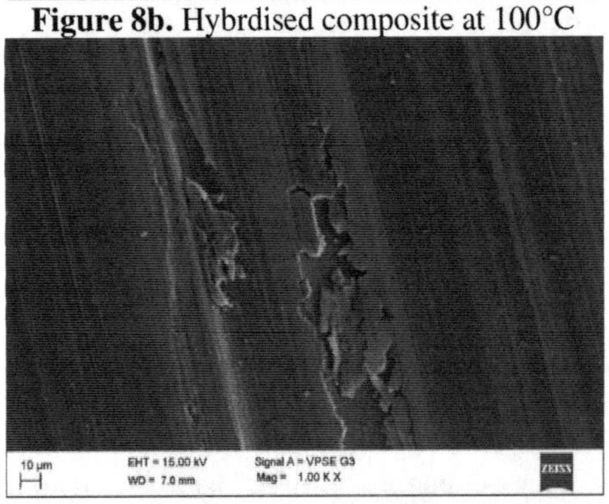

Figure 8c. Hybrdised composite at 150°C

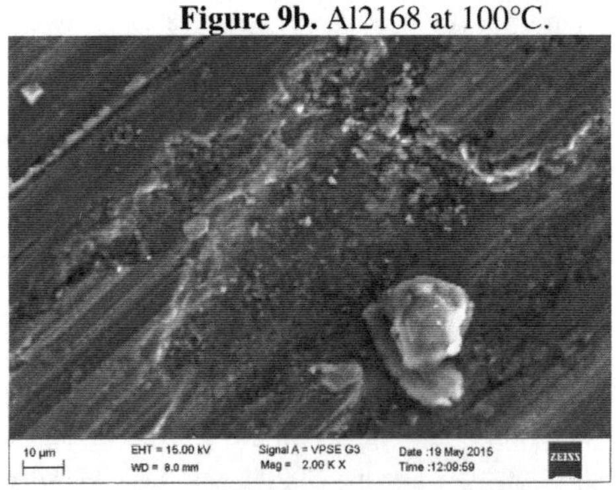

Figure 9c. Al2168 at 150°C

IOP Conf. Series: Materials Science and Engineering 1189 (2021) 012018 doi:10.1088/1757-899X/1189/1/012018

Sintering of superior wear residues occurs at the higher temperature (100° C) with applied load. High compression pressure occurs during the wear test. solid and hard surface known as "glaze." layer is formed because of sintering. The Glaze layer offers long-term wear protection, as breakdown contributes to the formation of additional wear residue as shown in Figure 8b. The damaged Al2168 surface at (50°C) reveals plastic deformation, lesser oxide layer and Al2168 matrix delamination. The large amount of friction heat with 40N loading temperature results in Al2168 being thermally softened. It is Clear from the SEM (Figure 9a.) deeper wear and ploughing grooves are visible. The design in Al Al2168 matrix at the temperature 100°C & 150°C, the deteriorated surfaces are formed on the surface with evidence of oxide and glazing layers. The lack of a glazing layer was clearly detected at a lower temperature (50°C) in Figure 9a.

4. Conclusions

1.Wear resistance is increased by the incorporation of the Boron carbide and Graphite particles in the hybrid metal matrix composites.
2. Glazed layer formation in Al2168 matrix is less than glazing layer formation formed on hybrid composites.
3. Applied load is directly proportional to wear rate
4. Sliding velocity and applied load is more predominated on the wear rate than the sliding distance.
5. SEM micrographs reveal that material removal is micro ploughing and micro pit formation at 50°C

References

[1] Hernandez-Sandoval, J., Garza-Elizondo, G.H., Samuel, A.M., Valtiierra, S. and Samuel, F.H., 2014. *The ambient and high temperature deformation behavior of Al–Si–Cu–Mg alloy with minor Ti, Zr, Ni additions*. J. *Materials & Design*, ***58***, pp. 89-101.
[2] Lee, Jonathan A. "*Cast aluminum alloy for high temperature applications*." (2003).
[3] Surappa MK. *Aluminium matrix composites: Challenges and opportunities*. Sadhana. 2003 Feb 1; **28**(1-2):319-34.
[4] Lee, J.A. and Chen, P.S., National Aeronautics and Space Administration (NASA), 2002. *Aluminum-silicon alloy having improved properties at elevated temperatures and articles cast therefrom*. U.S. Patent 6,399,020.
[5] Ammar, Hany. *Influence of metallurgical parameters on the mechanical properties and quality indices of Al-Si-Cu-Mg and Al-Si-Mg casting alloys*. Université du Québec à Chicoutimi, 2010
[6] Lee, J.A. and Chen, P.S., National Aeronautics and Space Administration (NASA), 2005. *High strength aluminum alloy for high temperature applications*. U.S. Patent 6,918,970
[7] Lee, J. A. *High-Strength Aluminum Casting Alloy for High-Temperature Applications* MSFC Center Director's Discretionary Fund Final Project No. 97-10. 1998.
[8] Sharath, B.N., Madhu, K.S. and Venkatesh, C.V., 2019. *Experimental Study on Dry Sliding Wear Behaviour of Al-B4C-Gr Metal Matrix Composite at Different Temperatures*. J. *Applied Mechanics and Materials*, **895**, pp. 96-101.

Evaluation of Mechanical Properties of Ceramic Reinforced Aluminium-7029 Hybrid Composite

K S Madhu*, B N Sharath, C V Venkatesh, D G Pradeep

Department of Mechanical Engineering, Malnad College of Engineering, Hassan, affiliated to VTU, Belagavi, Karnataka, India

*Corresponding author e-mail address: mks@mcehassan.ac.in

Abstract. Composites are often chosen for structural applications because they have tailored material properties. The production of hybrid metal matrix composites has become a major research area in materials science. With this interest, present paper deals with fabrication of aluminium alloy (Al-7029), boron carbide (B_4C) and graphite (Gr) hybrid composite using conventional stir casting is investigated. Evaluation of mechanical properties such as tensile hardness and impact strength of base alloy with reinforcement [B_4C (2%, 4%, 6%, 8%) with constant 2% Gr] was performed on the samples. After solidification, the samples are prepared according to ASTM standards and tested to find the various mechanical properties like tensile, hardness and impact. It was observed that the hybrid composite tensile strength was increased, with decrease in elongation rate of unreinforced aluminium alloy. The impact strength and hardness property for a hybrid composite was increased compared to base alloy. The composite's microstructure was studied using a Scanning Electron Microscope (SEM).

Keywords: Metal matrix composites; Boron carbide; graphite; Hardness; Impact test; Tensile strength.

1. Introduction

Over the past few years, industries have focused on materials with multiple applications, which are lightweight, durable, and cost effective. Hybrid composites are more popular recently, and they cover multiple material properties. We have a broad range of aluminium alloys to choose from when developing material for automotive applications. Mechanical properties of the aluminium based composite can be enhanced with the addition of reinforcement, when compared to the base alloy. Addition of ceramic reinforcements to aluminium metal matrix composites (AMMCs) exhibit superior strength and rigidity, resistance to creep and wear. Addition of boron carbide (B_4C) improve tensile strength, impact strength, toughness and show stronger interface bonding with the aluminium matrix. Addition of graphite reduces the wear rate of the composite due to its self-lubricating property [1]. The properties of AMCs can be tailored to the requirements of various industrial applications by appropriate combinations of matrix, reinforcement, and processing routes [2-4]. The most used industrial method of producing aluminium-based composites is traditional stir casting. Because of its applicability to mass production and cost-efficiency [5]. Ceramic particle reinforced aluminium composites have shown a significant improvement in their mechanical properties [6]. Al6061 and Al7015 alloy with TiB_2 particulates reinforced composite produced by hot extrusion were studied to evaluate the mechanical properties. Tests were done for room and high temperature condition (500 °C) for both the composites. Al7015+ TiB_2 composites show the better result with hardness up to 148HV and UTS till 400 MPa at room temperature [7]. Mechanical behaviour of B_4C reinforced with Al-7075 matrix composite were studied. It was found that the ultimate tensile strength, compressive strength, and the hardness of the composite increased linearly with increase in volume percentage of B_4C. Also, the wear effect found minimum in the composite with 10 % B_4C [8]. Keeping 3% weight fraction of Gr and by varying the weight percentage (3-12 %) of B_4C Aluminium alloy (LM25) composites were characterised for wear and hardness test. Vicker's hardness value increased with increase in B_4C reinforcement. Aluminium alloy with 12% B_4C+3%Gr found with better characteristics [9]. Dry sliding friction behaviour Al-SiC-Gr particles was studied. Concluded that the load is the most important factor affecting the friction coefficient of the hybrid composite followed by sliding speed. The coefficient of friction increased with increase in load and sliding distance. The author also revealed that the average friction coefficient of the hybrid composite is quite low compared to pure alloy [10]. It was found that wear behaviour of Al 7075/Bagasse ash/Gr hybrid composite the percentage of ceramic phase weight increased and finally suggested wear behaviour of hybrid composites containing graphite showed superior wear resistance [11].

In the present paper, fabrication of aluminium alloy (Al-7029), boron carbide (B_4C) and graphite (Gr) hybrid composite using conventional stir casting is carried. The prepared composites were subjected to mechanical properties such as tensile hardness and impact strength of base alloy with reinforcement [B_4C (2%, 4%, 6%, 8%) with constant 2% Gr]. The composite's micro structure was studied using a Scanning Electron Microscope (SEM).

2. Experimentation

2.1 Material selection

Aluminum-7029 alloy, containing zinc and magnesium as its main alloy elements, selected as the base material due to its good mechanical properties, such as high corrosion resistance, re-crystallization capability and the most economical thermal aluminium alloy. Table 1 and 2 shows the chemical composition of the Al7029 alloy and material composition used for this study.

Table 1. Chemical composition of Al7029 alloy (mass fraction%) [3]

Si	Fe	Cu	Mn	Mg	Zn	Ti	V	Al
0.10	0.12	0.5-0.9	0.03	1.3-2.0	4.2-5.2	0.05	0.05	Remaining

Table 2. Material composition in percentage

Sample	Al7029	Boron carbide	Graphite
A	100	0	0
B	96	2	2
C	94	4	2
D	92	6	2
E	90	8	2

Commercially available boron carbide (partial size average 50 μm [1]) and graphite reinforcements were used for preparing the composites with Al7029 matrix material. The chemical composition of master alloys used in the present study is reported in Table 1. Al7029 alloys was procured from M/s Fen fee Metallurgical, Bangalore, India. Boron carbide was procured from Parshwamani metals, Mumbai, India, and graphite was procured from Graphite India, Bangalore, India.

2.2 Material preparation

Hybrid Al7029-B_4C-Gr composites were fabricated by stir casting technique using electrical furnace of 5 kW capacity. Figure.1a and figure.1c shows photograph of casting furnace and cast specimen used in the present study. Graphite crucible is used because Al7029 matrix material melt was held for 90 minutes at a temperature between 600 and 700 ºC, at this temperature graphite crucible does not react with Al7029. Magnesium (3%) was added to the molten metal as the Al7029 ingot material reached a liquent condition to enable the dispersion of particles inside the molten metal. 2% boron carbide has been preheated at temperature of 500 °C to remove moisture or any other gases. Boron carbide fine-reinforcing particles (50μ) has been applied to a molten metal, which can have heterogeneous nuclei. Stirring is carried for 15-20 minutes at 150 rpm to acquire consistent distribution of B_4C particulate in Al7029. Graphite (2%) was pre-heated and poured into the premixed molten metal (Al7029 with B_4C). Stirring is further carried up to 10-15 minutes to ensure proper mixing of Gr with molten metal. Degasification agent hexa-chloroethane (C_2Cl_6) is used to eliminate porosity, voids, and casting defects. Progressive solidification will begin when hot molten metal has been poured into the pre-heated die. The molten pool is then allowed to solidify and are extracted at room temperature [1].

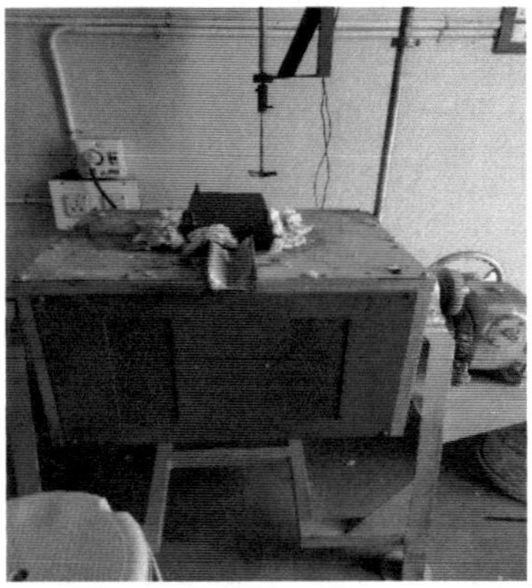

Figure 1. Stir casting setup

Figure 1a. Preheated Die

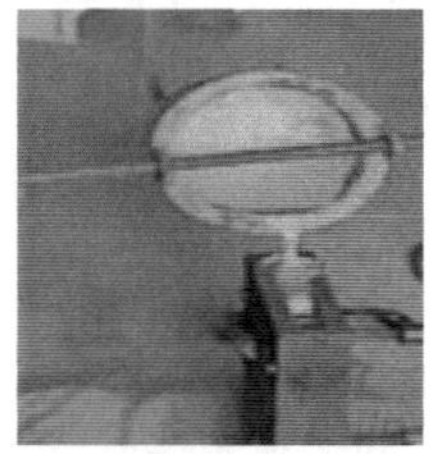

Figure 1b. Pouring

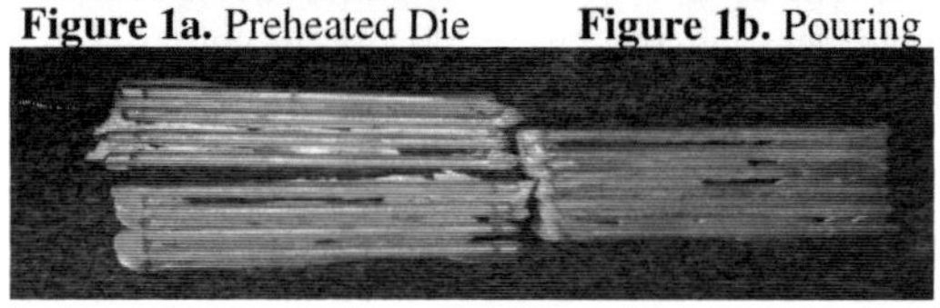

Figure 1c. Cast sample

2.3 Test procedure

Cast samples are machined as per ASTM standards accordingly. Tensile behaviour of the sample is conducted as per ASTM E8 standard. As per ASTM E10 standard, Brinell hardness test was carried out by applying load of 50 g for a period of 10 seconds. Readings were taken at five various places on the sample and an average value was used for study. Charpy impact test was carried to measure the samples' ability to resist high rate loading and samples were prepared according to the ASTM E23 standard. Metallographic investigation was carried out through scanning electron microscope (SEM). Etching was done using 10% NaOH solution on the surface of polished sample.

3. Results and Discussions

3.1 Tensile strength

Purpose of this study is to investigate tensile strength of Al7029 and HMMc (Al7029-B_4C-Gr). Tensile test is carried out in Universal testing machine. Tensile strength increases progressively with a rise of up to 6% in the composition of B_4C. Results of the tensile tests also revealed significant improvement in (2 % wt of B_4C & Gr) reinforced hybrid composite. Increase in tensile strength (reinforcement up to 6% B_4C & 2%Gr) is also attributed to increase in grain boundary area due to grain refinement and effective propagation of applied tensile load to well-bound reinforcement with a standardized distribution. Strength of ceramic materials lies much higher than metallic materials and higher than that of the matrix alloy. Under the applied stress, grain boundaries serve as an obstruction to the dislocation movement and ultimately results in a dislocation stack on the grain boundaries. Again, multi-directional stress induced during elongation easily starts multi-gliding system under applied stress so that dislocations were found developing and moving in several directions. These multi glide planes agglomerate with reinforcement particles under the applied stress forms grain boundary ledges. As applied load increases, these ledges act as obstacle to dislocation movement resulting in pile-ups. The cumulative impact of these two barriers enhances the hybrid composite 's strength.

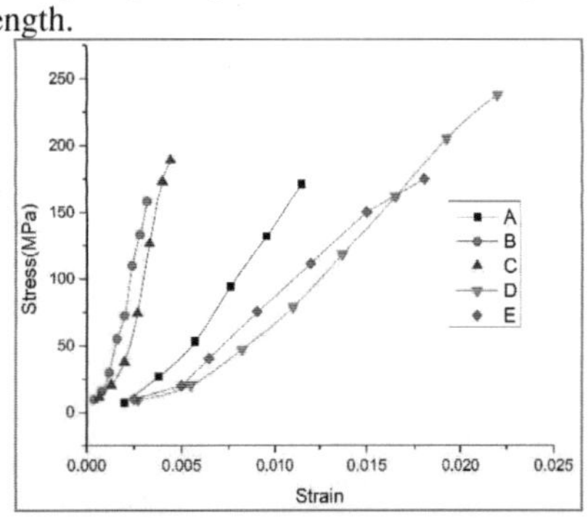

Figure 2. Stress v/s strain

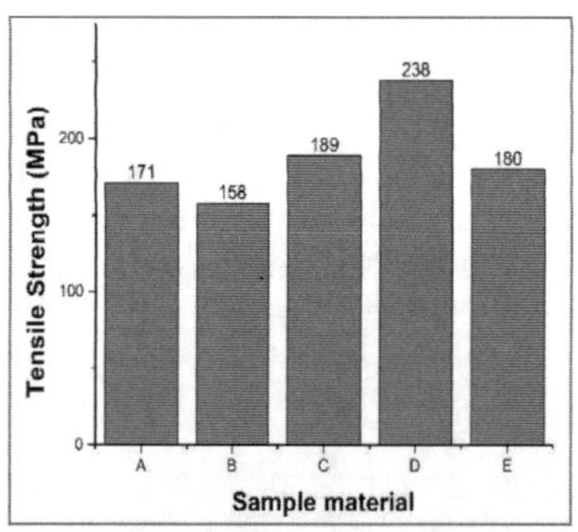

Figure 2a. Tensile strength

3.2 Hardness test

Experiments are done with the constant Gr (2%), changing weight fraction B_4C (2%, 4%, 6%, 8%). The graph reveals that Al7029 alloy surface hardness rises steadily, with rise in the weight percentage of strengthening particulates (B_4C & constant 2 % Gr). The results of hardness test conducted on unreinforced Al7029 and hybridised Al7029 samples have shown that matrix hardness levels are growing with reinforcement material up to 6% (Figure 3.). Effects of hardness measurements shows that the increase in reinforcement content results in a considerable increase in hardness which can be mainly due to; the existence of hard ceramic B_4C & Gr particles scattered in the matrix, the deformation restricted due to their inclusion during indentation and the reduced size of grain due to the chilling phase.

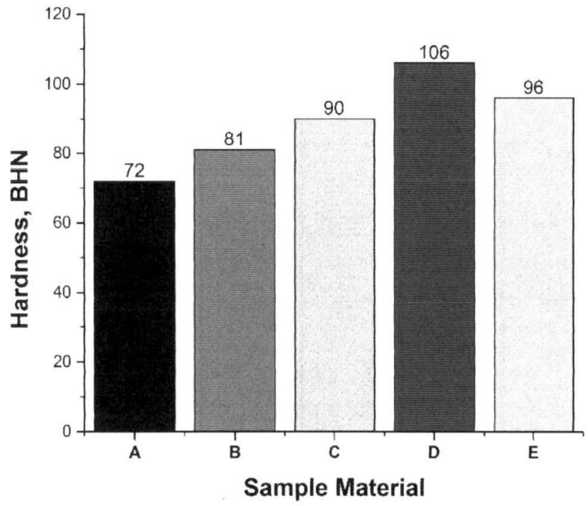

Figure 3. Hardness

3.3 Impact test

Figure 4 portrays the variation in the impact strength of the specimen for varying wt. % of reinforcements. Energy observed by the 6 wt. % primary reinforced composite is superior to the other composites. The increase in impact strength is due to the reduction in the porosity. Experiment is carried out on Al7029 Alloy with different weight divisions B_4C (0%, 2%, 4%, 6%, 8%) and Gr (2%). At the beginning, the impact strength increases with increase in percentage composition of B_4C and Gr. It has been concluded that (Al7029+6%B_4C-2%Gr) hybrid composite has the highest impact capacity. It is concluded that highest impact strength is for Al6082 hybrid composite with 8% B_4C and 2% Al_2O_3 reinforcements, this indicates that maximum amount of energy is absorbed during fracture.

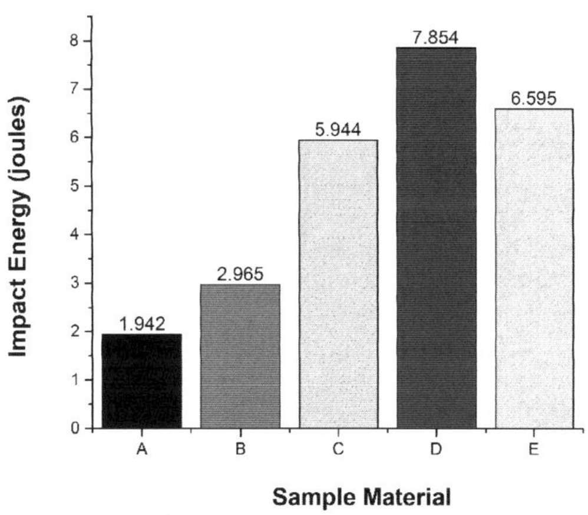

Figure 4. Impact energy

3.4 SEM Analysis

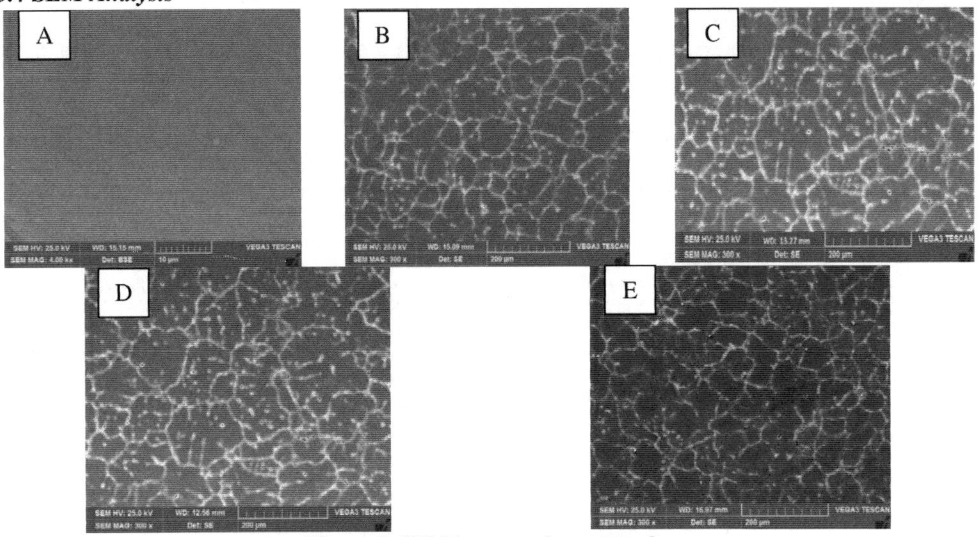

Figure 5. SEM images of cast samples

SEM micrographs of base alloy and hybrid composites is shown in figure 5. The distribution of the B_4C and Gr particles in the Al7029 is noticeably uniform. Micrographs discloses the uniform particles distribution of the cast composites. The uniform particles distribution enhances the mechanical properties of the reinforced Al7029. B_4C particles are observed to be accumulated within the grain boundary regions and geometrical trapping by dendrites is observed [1]. B_4C and Gr particles have located at grain boundary region, so that it increases the bonding between matrix and the reinforcement. Hence there is increase in the grain strength, which results in increase in mechanical properties as revealed from mechanical tests. Also, both the reinforcements act as obstacles to the dislocations, at the boundary region which improves the mechanical properties.

4. Conclusions

1. Mechanical properties of hybrid composites increase with an increase in wt. % of B_4C & Gr. Tensile strength maximization is obtained by 6% B_4C and 2% Gr.
2. From hardness test, the better effects will be seen if the B_4C is used at 6 % with 2 % Gr.
3. It is inferred that for (Al7029+6 % B_4C+2 % Gr) hybrid composite has highest impact strength which indicates that maximum energy is absorbed during fracture.
4. B_4C and Gr were found to be distributed reliably. It was confirmed, the interfacial analysis indicated that the reinforcement particulate is related closely to the matrix strengthening.

References

[1] Sharath, B.N., Madhu, K.S. and Venkatesh, C.V., 2019. *Experimental Study on Dry sliding Wear Behaviour of Al-B_4C-Gr Metal Matrix Composite at Different Temperatures*. J. *Applied Mechanics and Materials,* **895**, pp. 96-101.

[2] Jones, R. M., *Mechanics of Composite Materials, Tailor Francis,* 2nd Edition, 1999.

[3] Davis, Joseph R. *Aluminum and aluminum alloys.* ASM international, 2001: p351-416.

[4] Surappa, M.K., *Aluminium matrix composites: Challenges and opportunities. Sadhana,* **28** (1-2), 2003: pp.319-334.

[5] Surappa, M.K. and Rohatgi, P.K., *Preparation and properties of cast aluminium-ceramic particle composites. Journal of materials science,* **16**(4), 1981: pp.983-993.

[6] Kala, H., Mer, K.K.S. and Kumar, S., *A review on mechanical and tribological behaviors of stir cast aluminum matrix composites.* Procedia Materials Science, **6**, 2014: pp.1951-1960.

[7] Onoro, J., Salvador, M.D. and Cambronero, L.E.G., *High-temperature mechanical properties*

of aluminium alloys reinforced with boron carbide particles. Materials Science and Engineering: A, **499**(1-2), 2009: pp.421-426.
[8] Baradeswaran, A.E.P.A. and Perumal, A.E., *Influence of B_4C on the tribological and mechanical properties of Al 7075–B_4C composites. Composites Part B:* Engineering, **54**, 2013: pp.146-152.
[9] Thirumalai, T., Subramanian, R., Kumaran, S., Dharmalingam, S. and Ramakrishnan, S.S., *Production and characterization of hybrid aluminum matrix composites reinforced with boron carbide (B_4C) and graphite*, 2014.
[10] Suresha, S. and Sridhara, B.K., *Friction characteristics of aluminium silicon carbide graphite hybrid composites.* J. Materials & Design, **34**, 2012: pp.576-583.
[11] Imran, M., Khan, A.A., Megeri, S. and Sadik, S., 2016. *Study of hardness and tensile strength of Aluminium-7075 percentage varying reinforced with graphite and bagasse-ash composites.* Resource-Efficient Technologies, **2**(2), pp.81-88.

IOP Conf. Series: Materials Science and Engineering 1189 (2021) 012020 doi:10.1088/1757-899X/1189/1/012020

Elastic modulus measurement of thermoplastic composites through modal analysis

Harshavardhan B[1,2], R Ravishankar[1], B Suresha[2], and Arun C Dixit U[3]

[1]Department of Mechanical Engineering, Sri Jayachamarajendra College of Engineering, Mysuru-570 006, India

[2]Department of Mechanical Engineering, The National Institute of Engineering, Mysuru-570 008, India

[3]Department of Mechanical Engineering, Vidyavardhaka College of Engineering, Mysuru-570 002, India

Abstract. Conventional tensile testing of materials employ pulling of a specimen till it breaks to determine the elastic modulus and other tensile properties. Modal analysis is one of the ways to measure the elastic modulus without destructing the sample. The present study was carried out to evaluate the elastic modulus of short carbon fiber (SCF) reinforced polyethersulfone (PES) composites filled with potassium titanate whisker (PTW) and ultra high molecular weight polyethylene (UHMWPE) through modal analysis. Specimens were prepared by extrusion and injection moulding processes. Modal analysis was done with cantilever type free-vibration test rig on the developed composites using an accelerometer sensor. Based on the acceleration response of the composites, their natural frequencies and damping values were determined. Further, elastic moduli of the composites were computed and compared with that of actual tensile test. Elastic modulus of PES/SCF composites determined by free vibration test was well in agreement with that of tensile test results. In addition, numerical modal analysis simulation was performed for by considering the tensile modulus and density of each of the composite material. Natural undamped frequencies of the composites found by numerical modal analysis were closer to that of the experimental values.

1. Introduction

Carbon fiber reinforced composites are widely used in automobile, marine and aerospace industries due their high load carrying capacity[1]–[3]. However, high stiffness and high strength materials exhibit poor damping properties. Damping properties of fiber reinforced composites can be enhanced by hybridization with filler materials. Fillers will introduce more number of internal interfaces and forms a way to dissipate the vibration energy [4]. Vibration techniques can be used as non-destructive tests to detect and locate the flaws in fiber reinforced composites [5]. Modal analysis is an important tool to determine the vibration parameter such as natural frequency and corresponding mode shapes of a system [6]. Neglecting air damping, a vibrating system is expected to attain the zero velocity if the material constituting the system absorbs all the vibration energy and dissipates it to surrounding. Based on the amount of reduction in amplitude of the vibrating body in the successive cycles, logarithmic decrement can be evaluated. Logarithmic decrement will be useful in computing the damping ratio[7]–[9]. Natural damped frequency is determined by taking the reciprocal of the time period from the acceleration-time plot. Natural undamped frequency is computed through the relation between damping ratio and damped frequency.

Modal analysis was done with cantilever type free-vibration test rig on the developed composites using an accelerometer sensor. Acceleration response during the vibration of the composites due to the initial applied impact was captured till amplitude of acceleration decay to zero. Damped time period is measured from the time plot. For cantilever beam system, its natural frequency [10] is determined using the relation (1).

$$f_n = \frac{1}{2\pi}\sqrt{\frac{3EI/l^3}{M + \frac{33}{140}m}} \quad (1)$$

where E is the tensile modulus of the material, I is the moment of inertia of beam cross section in bending, l is the length of the beam, m is the mass of beam and M is the mass of the accelerometer. Equation (1) can be further used to evaluate the Young's modulus of the composite material [11]. Numerical modal analysis can assist in understanding the experimental results better[12], [13].

2. Materials and methods

2.1. Materials

The polyethersulfone (PES) was procured from the company Solvay, USA and was used as matrix material. The short carbon fiber (SCF) based on polyacrylonitrile (PAN) was purchased from the company Teijin, Japan and was used as reinforcement material. Potassium titanate whisker (PTW) was supplied by Nobel Alchem, India and Ultra high molecular weight polyethylene (UHMWPE) was procured from Spectra Plast, India. PTW and UHMWPE were used as fillers to provide the self lubrication. The properties of PES, SCF, PTW and UHMWPE are stated in the previous work [14]. Materials formulation and designations adopted in the present work are presented in table 1. All the composites were fabricated for a dimension of 120 mm x 10 mm x 3.85 mm by extrusion process followed by injection moulding as stated in elsewhere[14].

Table 1. Material formulations and designations

Composites	Matrix wt. % PES	Fiber wt. % SCF	Filler wt. % PTW	Filler wt. % UHMWPE
$C_{00}P$	100	00	-	-
$C_{25}P$	75.0	25	-	-
$C_{25}W_{2.5}P$	72.5	25	2.5	-
$C_{25}W_{5.0}P$	70.0	25	5.0	-
$C_{25}U_{2.5}P$	72.5	25	-	2.5
$C_{25}U_{5.0}P$	70.0	25	-	5.0

2.2. Methods

2.2.1. Free vibration test. Damping properties of the PES/SCF composites were determined by single cantilever free vibration decaying test [10]. The developed samples were mounted as cantilever beam and 10 mm length was used for clamping purpose. Effective length of the beam was taken as 110 mm for analytical calculations and numerical simulations.

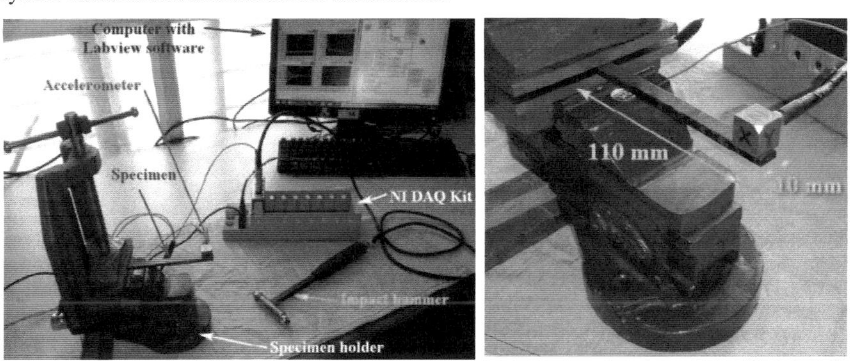

Figure 1. (a) Free vibration test rig with computer interface (b) Cantilever composite beam

Piezoelectric accelerometer, model PCB 356A15 was used to capture the response of the vibrating beam as indicated in figure 1. Mass of the accelerometer used is 10.5 grams and it is capable to measure the frequency in the range 2 to 5000 Hz with ± 5 % accuracy. Accelerometer was mounted at the free end of cantilever beam arrangement with a temporary adhesive (petro wax) as in figure 1. Analog response captured by accelerometer was then digitalized with the help of NI 9234 IEPE input module. Digitalized response was then transferred to the computer system through NI compact data acquisition module cDAQ 9178. Data received by the computer was then analyzed using NI Labview software package.

2.2.2. Numerical simulation of modal analysis. Numerical modal analysis was performed with Abaqus CAE 2017 finite element software. Steps involved in the simulation process are illustrated in figure 2. Geometry of the beam was modeled as rectangular cross section bar (10 mm x 3.85 mm) of 125 mm length and properties obtained from the experiments were assigned to created beam geometry. Accelerometer sensor was modeled as cube of 13 mm and cube was assigned with customized material of density 4560 kg/m^3 so that mass of the sensor equals to its actual value of 10.5 gram.

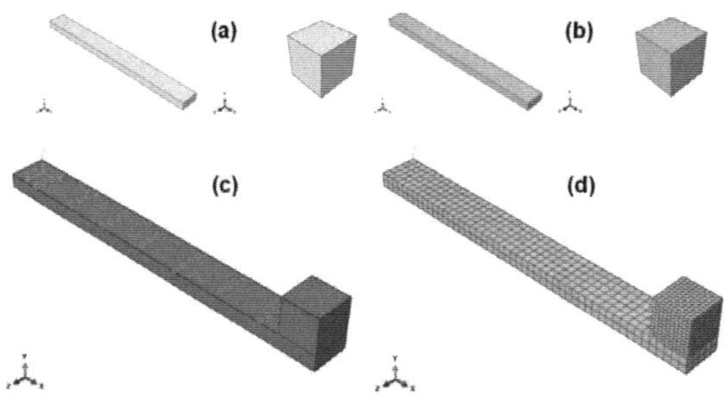

Figure 2. Numerical simulation steps (a) Geometry (b) Materials assigned (c) Assembly (d) Meshing

Beam and cube were assembled together followed by applying the tie constraint between them. Both the geometries were meshed with C3D8R elements. Mesh convergence was achieved with minimum of 500 elements. Linear perturbation frequency analysis was carried to extract the first natural frequency and mode shapes.

3. Results and discussion

3.1. Test results

3.1.1. Free vibration test. Responses of the composites during the free vibration are presented with respect to time as shown in figure 3. From the acceleration of vibration versus time plots, time taken for five successive cycles of vibration and corresponding reduction in amplitudes were noted as presented in table 2.

Table 2. Free vibration test results of PES/SCF composites

Parameters	Composites					
	$C_{00}P$	$C_{25}P$	$C_{25}W_{2.5}P$	$C_{25}W_{5.0}P$	$C_{25}U_{2.5}P$	$C_{25}U_{5.0}P$
Time t_1 (s)	0.014	0.003	0.003	0.015	0.004	0.019
Time t_2 (s)	0.215	0.113	0.112	0.121	0.123	0.148
Amplitude at time t_1 (m/s^2)	65.7	101.6	94.0	124.2	87.8	128.4
Amplitude at time t_2 (m/s^2)	1.7	21.8	19.5	29.4	16.7	20.1
Logarithmic decrement, δ	0.731	0.308	0.315	0.288	0.332	0.371
Damping ratio, ξ	0.116	0.049	0.050	0.046	0.053	0.059
Time period (s)	0.040	0.022	0.022	0.021	0.024	0.026
Damped natural frequency, f_d (Hz)	24.88	45.45	45.87	47.17	42.02	38.76
Undamped natural frequency, f_n (Hz)	25.04	45.51	45.93	47.22	42.08	38.83

Damping ratio of each composite was determined based on vibration amplitude decayed in mode-1 vibration. Addition of SCF into neat PES resulted in reduction of damping by 57.8% due to stiffening of the $C_{25}P$ composite. Addition of 2.5 wt.% of PTW did not exhibit much change in the damping with respect to $C_{25}P$ composite. However, damping value of $C_{25}P$ composite reduced further by 6.1 % with addition of 5 wt. % of PTW. This attributed to the slight improvement in the tensile modulus due to PTW addition.

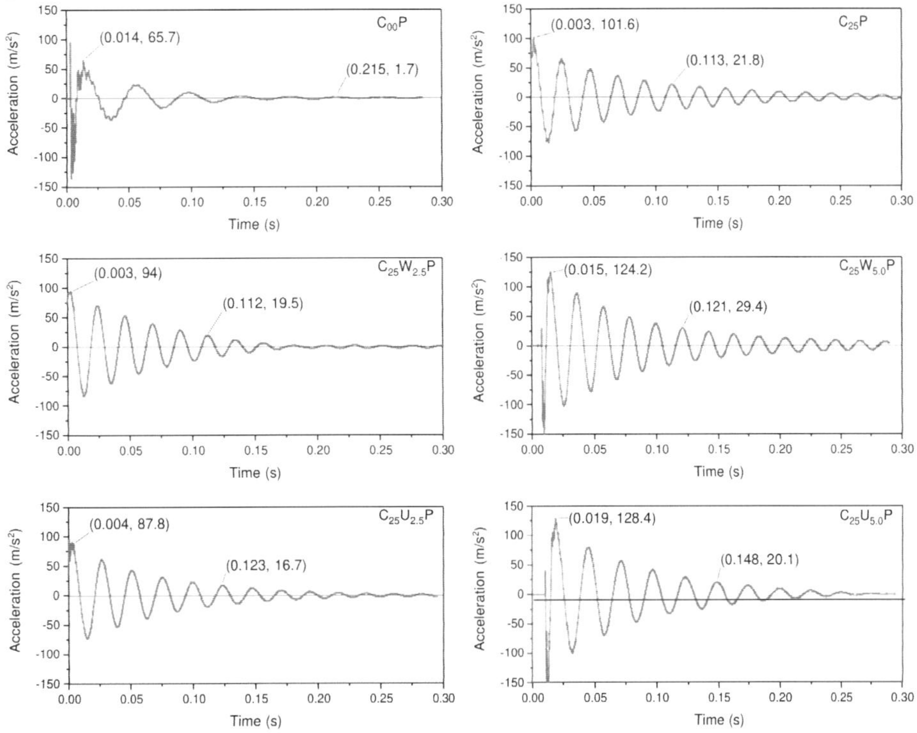

Figure 3. Vibration response of composites in time domain (a) $C_{00}P$ (b) $C_{25}P$ (c) $C_{25}W_{2.5}P$ (d) $C_{25}W_{5.0}P$ (e) $C_{25}U_{2.5}P$ (f) $C_{25}U_{5.0}P$

UHMWPE filled composites showed improvement in damping ratio by 8.2 % and 20.4 % with addition of 2.5 and 5 wt. % of UHMWPE into $C_{25}P$ composite. This indicated slight softening and enhanced ductility in the UHMWPE filled composites.

Neat PES showed maximum time period of vibration of 0.04 seconds and minimum damped frequency of 24.88 Hz. With addition of SCF and PTW fillers, lowering of time period and enhancement of natural frequency was observed. However, time periods of UHMWPE filled composites found higher than PTW filled composites and represented decrease in their natural frequencies.

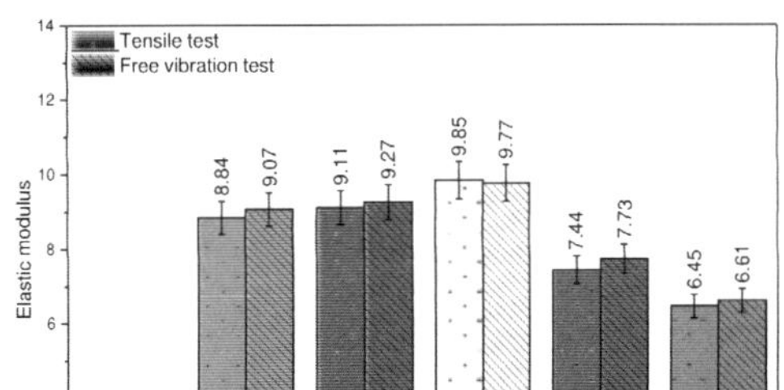

Figure 4. Elastic modulus of PES/SCF composites

As natural frequency is directly proportional to the stiffness which in turn depends on elastic modulus of the material, tensile modulus of all the developed composites were determined as higher than unfilled PES due to their vibrations at higher natural frequencies. Computed elastic modulus of PES/SCF composites filled with PTW and UHMWPE measured by free vibration test are printed in figure 4 and also comparison with that of tensile test results are shown. Instead of computing the natural frequencies of the composites based on time period consideration, Fast Fourier Transform (FFT) analysis was also performed in which acceleration signal in time domain was transformed to acceleration in frequency domain as shown in figure 5. Frequency corresponding to the maximum peak in the frequency domain curves represents the natural frequencies of the composites which were exactly same as reciprocal of the time period.

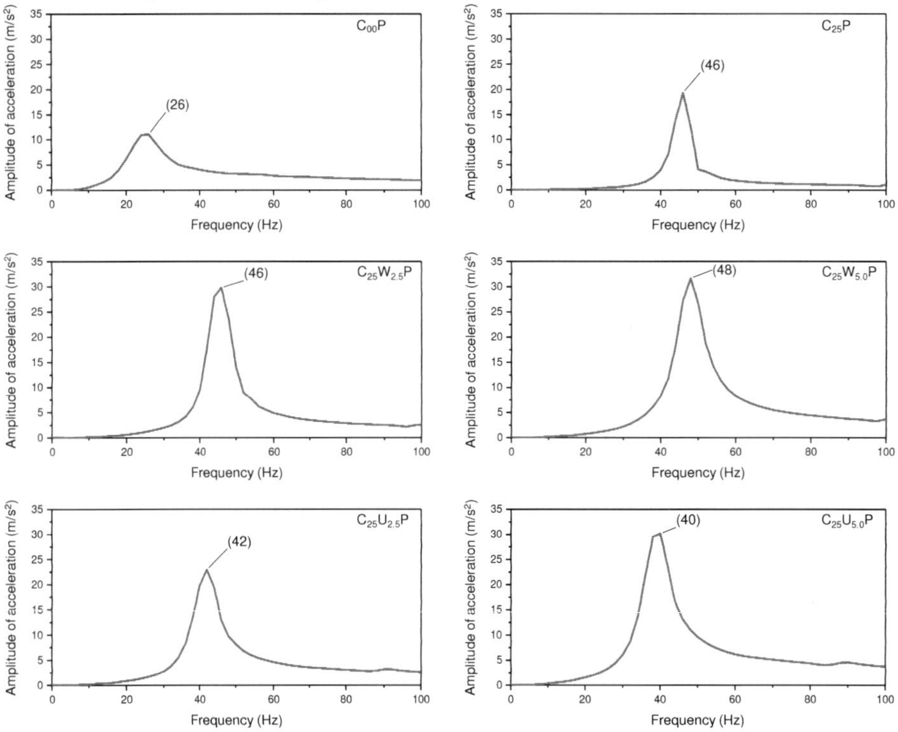

Figure 5. Acceleration-time plots of PES/SCF composites

3.1.2. Numerical modal analysis. Simulated mode-1 shapes of each composite along with their natural frequencies were presented in figure 6. Error between the numerical and experimental natural frequencies of the PES/SCF composites was around 5% as reported in table 3. This error was due to the assumption of homogeneous and elastic material model. More accurate simulation results can be achieved by assigning the geometry with visco-elastic material model parameters obtained from stress relaxation and/or creep test.

Table 3. Comparison between numerical and experimental results

Composites	Experimental	Numerical	Error (%)
$C_{00}P$	25.04	26.33	5.1
$C_{25}P$	45.51	47.80	5.0
$C_{25}W_{2.5}P$	45.93	48.31	5.2
$C_{25}W_{5.0}P$	47.22	49.59	5.0
$C_{25}U_{2.5}P$	42.08	44.23	5.1
$C_{25}U_{5.0}P$	38.83	40.78	5.0

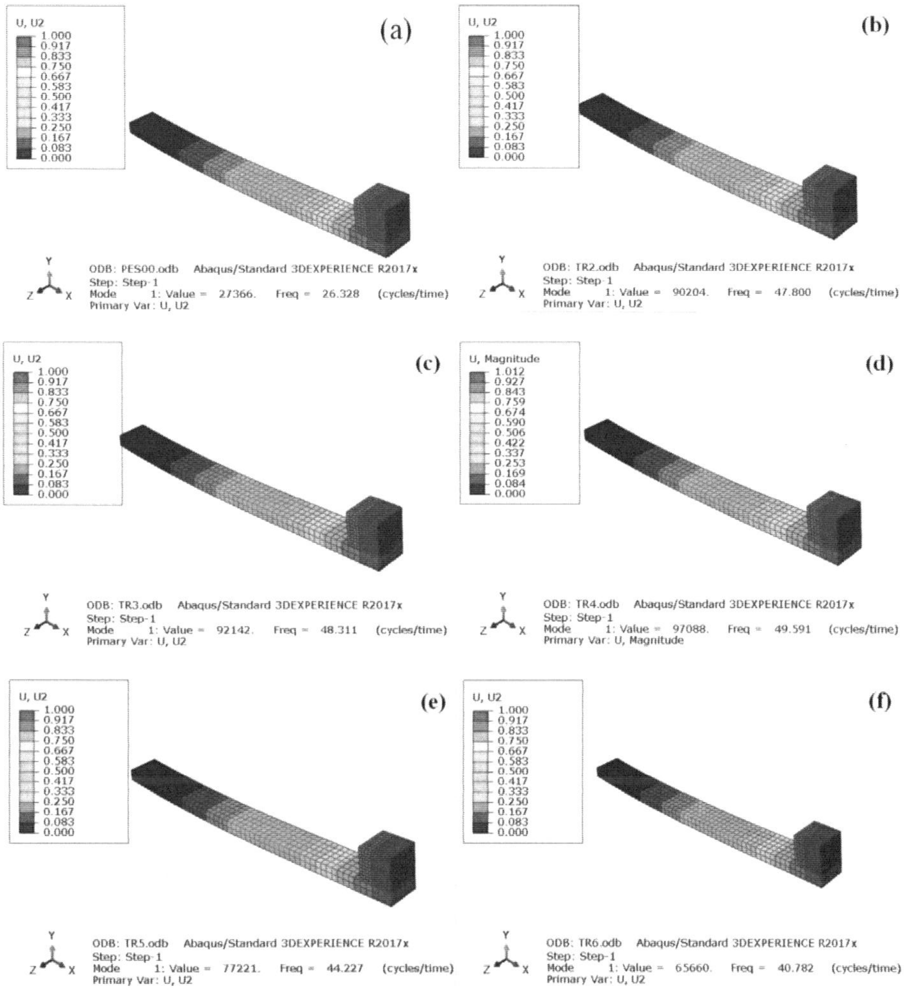

Figure 6. Numerical modal analysis results (a) $C_{00}P$ (b) $C_{25}P$ (c) $C_{25}W_{2.5}P$ (d) $C_{25}W_{5.0}P$ (e) $C_{25}U_{2.5}P$ (f) $C_{25}U_{5.0}P$

4. Conclusion

Elastic moduli of PES/SCF composites determined by free vibration test were well in agreement with that of tensile test results. Error between modulus obtained by free vibration test and tensile test was attributed to the anisotropic nature of composites due to random distribution and orientation of fibers in the matrix material. In conventional tensile test, even if five replicate tests were done on each composite as per ASTM standard, usually ±10 % variation in tensile properties from their mean is observed. Hence, when tensile modulus is the property of interest for designing, free vibration test can be used as an alternate method to estimate the tensile modulus through natural frequencies of composite materials without any material destruction. Numerical simulation results were validated with the experimental values with minimum error.

5. References

[1] N. Forintos and T. Czigany, 'Multifunctional application of carbon fiber reinforced polymer composites: Electrical properties of the reinforcing carbon fibers – A short review', *Compos. Part B Eng.*, vol. 162, pp. 331–343, Apr. 2019, doi: 10.1016/j.compositesb.2018.10.098.

[2] S. Erden and K. Ho, 'Fiber reinforced composites', in *Fiber Technology for Fiber-Reinforced Composites*, Elsevier, 2017, pp. 51–79.
[3] K. Acatay, 'Carbon fibers', in *Fiber Technology for Fiber-Reinforced Composites*, Elsevier, 2017, pp. 123–151.
[4] Y. Li, S. Cai, and X. Huang, 'Multi-scaled enhancement of damping property for carbon fiber reinforced composites', *Compos. Sci. Technol.*, vol. 143, pp. 89–97, May 2017, doi: 10.1016/j.compscitech.2017.03.008.
[5] P. Cawley and R. D. Adams, 'A Vibration Technique for Non-Destructive Testing of Fibre Composite Structures', *J. Compos. Mater.*, vol. 13, no. 2, pp. 161–175, Apr. 1979, doi: 10.1177/002199837901300207.
[6] Zhi-Fang Fu, Jimin He, 'Modal analysis methods – frequency domain', in *Modal Analysis*, 2001, pp. 159–179.
[7] F. Magalhães, Á. Cunha, E. Caetano, and R. Brincker, 'Damping estimation using free decays and ambient vibration tests', *Mech. Syst. Signal Process.*, vol. 24, no. 5, pp. 1274–1290, Jul. 2010, doi: 10.1016/j.ymssp.2009.02.011.
[8] K. Senetakis, A. Anastasiadis, and K. Pitilakis, 'A comparison of material damping measurements in resonant column using the steady-state and free-vibration decay methods', *Soil Dyn. Earthq. Eng.*, vol. 74, pp. 10–13, Jul. 2015, doi: 10.1016/j.soildyn.2015.03.009.
[9] P. P. Hujare and A. D. Sahasrabudhe, 'Experimental Investigation of Damping Performance of Viscoelastic Material Using Constrained Layer Damping Treatment', *Procedia Mater. Sci.*, vol. 5, pp. 726–733, 2014, doi: 10.1016/j.mspro.2014.07.321.
[10] S. K. Bhudolia, P. Perrotey, and S. C. Joshi, 'Enhanced vibration damping and dynamic mechanical characteristics of composites with novel pseudo-thermoset matrix system', *Compos. Struct.*, vol. 179, pp. 502–513, Nov. 2017, doi: 10.1016/j.compstruct.2017.07.093.
[11] M. M. Wallace and C. W. Bert, 'Experimental Determination of Dynamic Young's Modulus and Damping of an Aramid-Fabric/Polyester Composite Material', p. 4.
[12] S. K. Sahu and P. Das, 'Experimental and numerical studies on vibration of laminated composite beam with transverse multiple cracks', *Mech. Syst. Signal Process.*, vol. 135, p. 106398, Jan. 2020, doi: 10.1016/j.ymssp.2019.106398.
[13] P. Das and S. Kr. Sahu, 'Experimental and numerical study on free vibration of cracked woven fiber glass/epoxy composite beam', *Mater. Today Proc.*, p. S2214785320321003, Apr. 2020, doi: 10.1016/j.matpr.2020.03.320.
[14] B. Harshavardhan, R. Ravishankar, B. Suresha, and S. Srinivas, 'Static and Dynamic Mechanical Performance of Carbon Fiber Reinforced Polyethersulfone Composites', *Appl. Mech. Mater.*, vol. 895, pp. 265–271, Nov. 2019, doi: 10.4028/www.scientific.net/AMM.895.265.

Acknowledgment
Injection moulding of the specimens were fully supported by the Brakes India Limited, Sundaram Polymers Division, Nanjangud, Mysore District, Karnataka, India. We wholeheartedly thank Mr. P. Krishna Kumar, Vice President–Operations, Brakes India Limited for his support in the fabrication of composites.

IOP Conf. Series: Materials Science and Engineering 1189 (2021) 012021 doi:10.1088/1757-899X/1189/1/012021

A Review on the Effect of Surface Roughness and Liquid Slip on fluid flow in PDMS Microchannels

Rakesh T G [1], Girisha C [2], and Satish V T [3]

[1]Assistant professor, Department of Mechanical, AMCEC, Bangalore
[2] HOD, Department of Mechanical, AMCEC, Bangalore
[3]Assistant professor, Department of Mechanical, MCE, Hassan.

Abstract: Microfluidics is about studying flows with characteristic length scales of the orders of microns (typically sub millimeters). The fluid dynamics at the microscale can vary from "macro fluidic" behavior in those factors such as surface tension and energy dissipation. The fluidic resistance will keep on developing to dominate the system. Polydimethylsiloxane (PDMS) is a polymer commonly used for microfluidic chip manufacture and prototyping. PDMS becomes a hydrophobic elastomer after cross linking. Small dimensions make it impossible to produce a microchannel with smooth surface. Hence entrance effect, surface roughness, turbulence and viscous effects have a reasonable effect on fluid flow. This paper aims to look into characterization techniques of surface roughness in a channel and liquid slip phenomenon considering changes in elementary geometry, flow conditions and different roughness elements.

Keywords: Surface Roughness, LBM, Microchannel.

1. Introduction

Microfluidics has been an essential part in "lab on chip" device. The complications of micro-hydrodynamics were considered in different conditions. As length scale becomes smaller, surface effects tend to dominate. PDMS polymer can be used for the fabrication of microchannels using soft lithography technique developed by Whitesides.. Soft lithography has the constraints of fabricating different geometries of the channel. This method could only create a cross section in rectangles. Theoretically, circular channel is the best shape for fluid flow mechanism. The reasons are, First of all, channels of circular geometry offer least surface area exhibited to the fluid that could limit the friction between the fluid and the wall surface. So, the energy required to pump the water for a given flow rate is less. Secondly, internal pressure of the fluid is efficient to handle in circular shape. Pressure distribution is uniform around the entire circumference of the channel. PDMS formula is (C2H6OSi)n and its fragmented formula is CH3[Si(CH3)2O]nSi(CH3)3, n is the number of monomers repetitions. PDMS is optically transparent within the range of optical frequencies (240 – 1100), which helps to visualize the flow using a microscope. Using a simple plasma treatment PDMS can be adhered thoroughly to glass or another PDMS layer. Hence the production of multilayers PDMS devices is more advantageous than that of a glass substrate. A wide range of patterns and channel dimensions can be easily prepared, including straight, T-shapes, X-, Y-, and S-curve which may replicate the vascular network.

IOP Conf. Series: Materials Science and Engineering 1189 (2021) 012021 doi:10.1088/1757-899X/1189/1/012021

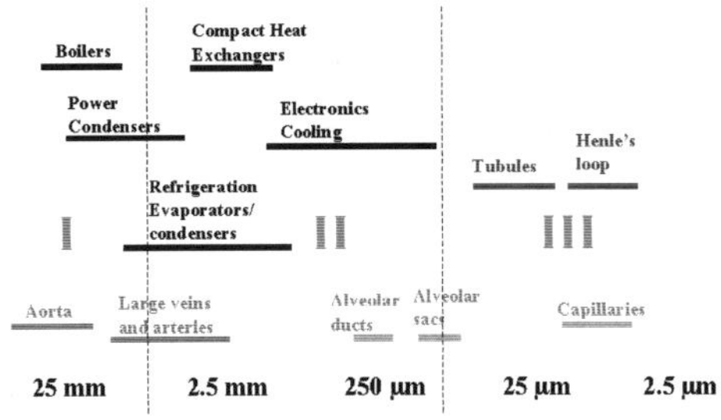

Figure 1 [1]: Classification of channels.

The rudimentary flow behavior in microchannels has been broadly studied and phenomenal advancements are achieved in experimental and theoretical investigations. Therefore, for liquid flows in microchannels Navier-Stokes equations and use of the continuum assumption are generally assumed to be valid and, in many studies, are experimentally in good agreement. Rigid microchannels have reported exceptionally higher values of friction constant (f_{Re}).). Reduction of aspect ratio (α) and the ratio of channel height to width in rectangular geometry microchannels increased the friction factor considerably. Huge deviation is observed in the forecasted values with respect to the friction factor (f) due to experimental variability, channel geometry, and surface roughness. Among these the roughness of the surface should be carefully examined. Lattice Boltzmann method (LBM) [1] has emerged as productive alternative simulations for fluid flow.

2. Evaluation of Flow and Theory

These are with respect to Rectangular microchannels
Mean hydraulic diameter was calculated as follows:

$$D_h = 4A/P$$

$$D_h = \frac{4 \times W_{ch} \times H_{ch}}{2\,(W_{ch} + H_{ch})}$$

The numerator is volume of microchannels. Denominator is the surface area of rectangular microchannels.
Reynolds number for the laminar flow in microchannel should be less than 2000, which is achieved here [2]. Expression for Reynolds number is,

$$Re = (\rho \times V \times D_h)/\mu$$

Reynolds number is directly proportional to the velocity of water through the microchannel, ass the Density, hydraulic diameter and dynamic viscosity of water remains constant,

The cumulative drop in pressure in the microfluidic system, Psys, is calculated using the equation.
$P_{sys} = \Delta P_{tube} + \Delta P_{ent} + \Delta P_{needle} + \Delta P_{interference} + \Delta P_{capillary} + \Delta P_{exit}$

Across the microchannel length L in steady-state laminar flow the pressure drop is defined as

$$\Delta p = Q\,(12\mu L)\,/\,(h^3 w) = v\,(12\mu L^2)\,/\,h^3$$

To understand the flow phenomenon Normalized friction constant (C*) is used widely to observe a deviation from the laminar theory. This value of C* is a non dimensional number.

$$C* = fRe_{experiment} / fRe_{theory}$$

A measure of the departure from the continuum is introduced through the Knudsen number, Kn, defined as,

$$k_n = \alpha/D_h$$

In liquid flow Knudsen number is much less than unity in the channels with hydraulic diameter ranging from 10^{-6} m to 10^{-3} m. Occurrence of deviation between the results are due to specific flow characteristics which are unaccounted in conventional theory.

3. Effect of Surface Roughness

The presence of wall roughness leads to increase the friction between the fluid and channel due to which there is a decrease in the value of the critical Reynolds number, during which the flow transition takes place from laminar to turbulent. In laminar flow the drag coefficient on the Reynolds number remains the same for both smooth and rough micro-channels. More surface roughness leads to increase in surface area resulting in adsorption effect and possibility of trapped air bubbles during the flow. Effect of Surface roughness is crucial with respect to fluid flow because of four reasons: (i) Controlling surface roughness at micro scale in a realistic manner is not easy, (ii) Roughness of the surface causes interfacial properties such as interfacial forces, wettability and trapped air. (iii) Rough surface generates swollen boundaries that make accurate interpretations of specific boundary positions more complicated, and (iv) No theoretical characterization for the inherent randomness and nonperiodic roughness of the surface. A good approach for rough contours or surfaces and their random behavior is very important in some engineering problems. Fractal modeling [2] can provide an alternative to generate rough geometries based on statistically self-similar geometries. Table 1 shows the important parameters to be studied and investigated to understand the flow of fluid in microscale.

The following are the best analytical characterization techniques to understand the effect of surface roughness in a PDMS microchannel.

3.1 Fast Fourier Transform (FFT) method:

Fast Fourier Transform (FFT) [3] can be used to simulate the surface roughness. In this method the roughness can be measured as a function which represents the height of surface at each point in xy plane.

$$f(x) = \sum_{k=0}^{M-1} \sum_{l=0}^{N-1} \sqrt{S}\,\bar{m},n \exp\left(i2\pi\left[\emptyset\, m,n + \frac{mx}{M} + \frac{ny}{N}\right]\right)$$

Where x = 0, 1, 2 . . . (M-1), and y = 0, 1, 2 . . . (N-1) are the coordinate of the base plane are taken as integer, $\emptyset m,\ n$ is the phase angles ranging over 0 and 2π. Sm,n is spectral density used to generate surface and m and n are integers with m = 0, 1, 2 . . . M/2 and n = 0, 1, 2, . . . N/2. Autocorrelation function is used for plotting the surface.

spectral density is calculated by following equation:

$$S(m,n) = 1/MN \sum_{k=0}^{M-1} \sum_{l=0}^{N-1} \mathrm{Rrs}\ \exp\left(-i2\pi\left[\frac{mr}{M} + \frac{ns}{N}\right]\right)$$

IOP Conf. Series: Materials Science and Engineering 1189 (2021) 012021 doi:10.1088/1757-899X/1189/1/012021

3.2 Mandelbrot–Weierstrass function:

fractal dimension (D) can quantitatively describe surface microscopic roughness[4]. Mandelbrot–Weierstrass function is used to model the rough surface represented by following equation.

$$f(x) = \sum_{l=0}^{N-1} p^{-n(2-f)} \left(1 - k\cos[p^{n}x + \emptyset]\right)$$

f(x) is the rough profile, f is a non integer value known as fractal dimension, varies between 1 and 2, P is the multiplier frequency value, which usually ranges between 1.1 which 3.0. and $\emptyset$ is a randomly generated phase. k can be used to alter the profile, For regular surfaces usually these values will be fixed k = 1 and p = 2. Routine measurements used in fabrication can detect highly irregular surfaces. In fractal models, surface roughness can thus be described by fractal dimension f. The major advantage of fractal modelling is it is scale independent. Once the topography is generated we can find all the parameters such as RMS (Root Mean Square Error) values. Since some of the experimental results are reported in terms of RMS or any other parameters, we can change the model parameters. Any profile can be developed and fine-tuned by setting the parameter D and k. Fig.2[3] shows the Mandelbrot – Weierstrass equation generated surfaces and plotted using wire-mesh models.

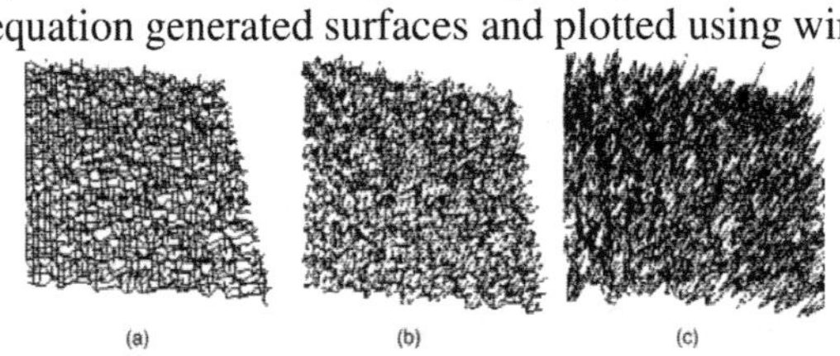

Figure 2 [3]: Surfaces drawn using M – W for various D values: (a) D = 1.3, (b) D = 1.5 and (c) D = 1.8.

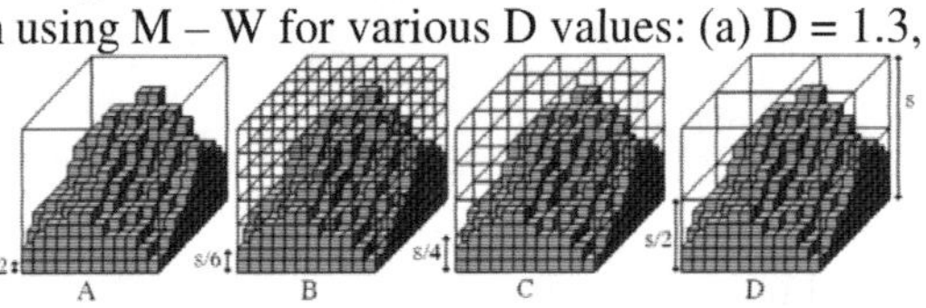

Figure 3 [3]: Fractal dimensions of surface roughness using LBM.

D is varied between 1 to 2, in order to know the impact of fractal dimension on speed. Velocity is computed at a steady state across the microchannel height from the lowest lattice point to the topmost lattice point, and plotted as shown in Fig. 4. As fractal dimension increases there will be increase in drag coefficient. Corresponding drop in fluid velocity is observed due to this reason near surface. For further more roughness surface (D=2.0), It can also partially block the flow in the channel [4]. Partial channel blockage for the roughest surface is observed up to 28.5 per cent of the total channel height. Remember that only the bottom surface with determined roughness is modelled in simulation. Velocity at the wall tend to oscillate does not remain constant and due to the roughness of the surface.

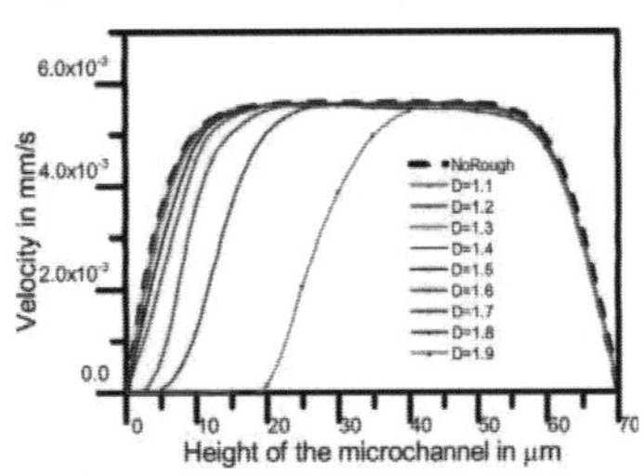

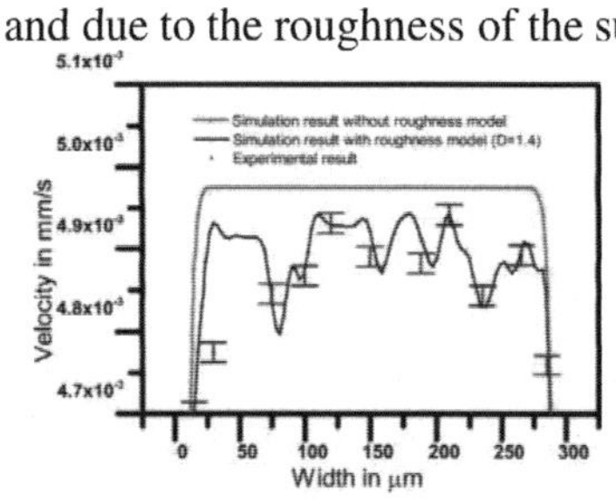

Figure 4. [6]: Effect of surface roughness **Figure 5.** [6]: experimental and simulation results

Simulation can be performed with and without surface roughness to compare the velocity profile between these two conditional parameters. In Fig. 6 it is clearly visible that the plot is flat and ideal when no roughness is considered, the velocity indicator is consistent throughout the channel. Experimental results however reveal that speed is oscillating and not constant. The simulation without ruggedness did not notice the velocity oscillatory behaviour. These oscillations occur because of the presence of ruggedness in the channel surface. Implementing a surface roughness model of MW significantly improved the accuracy by capturing the oscillatory behavior in the middle of the tube. This method is thus proven successful in capturing the real phenomenon in the numerical simulations. This model also offered an analysis of the near-surface vorticity behavior. Local oscillations can be seen to be produced because of the Presence of surface ruggedness. It is possible to use this oscillatory activity for microfluidic tasks such as mixing. In addition, by optimizing the manufacturing process one can also improve the mixing by imparting the necessary roughness.

4. Liquid Slip

Knowing basic mechanisms which control liquid slip is still evasive. Without slip velocity, the wall slip length can be perceived as an increase in channel width. The wall slip velocity for liquid flow was first proposed by Navier in 1823, generally expressed as $U_s=\beta\delta u/\delta n$. The parameter β is called Slip length ranging from several nanometers to several microns, and has been thought to be a material property parameter and the working fluid. The wall slip velocity is illustrated in following figure.

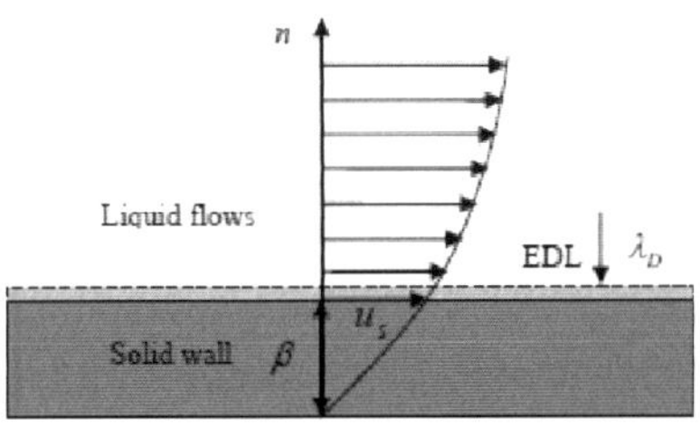

While few studies suggest the reasons behind the liquid slip due to either (i) a depleted water layer or (ii) an efficient air gap created by the wall nanobubbles. Another approach to hypothesizing liquid slip is based on intermolecular forces that influence the transportation of momentum from the liquid to wall.

Figure 6.: Wall slip velocity between wall material and working fluid

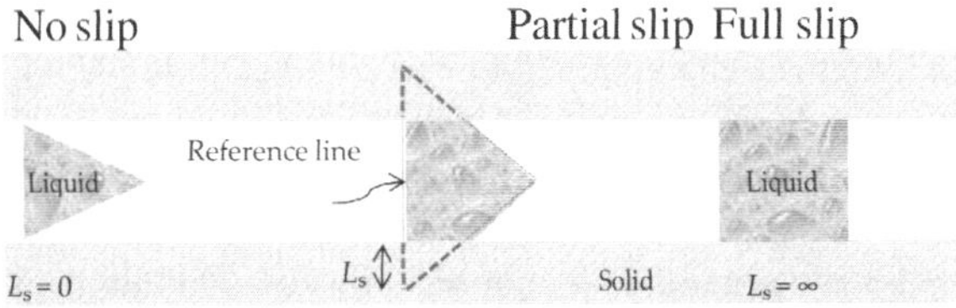

Figure 7. [9]: Slip length for various slip conditions.

Lund et al. derived a definition of effective slip length on a surface with periodic roughness in terms of the intrinsic slip length and surface contact area, using the Stokes flow and homogenization method of steady state. Its derived effective slip length [7] reflects the harmonic mean weighted by the surface-to-fluid contact area and is given as follows:

$$L_{eff} = \left[\frac{\sqrt{1+s^2}}{Ls}\right]^{-1}$$

Where Ls is intrinsic slip length, s is slope.

The surface wettability of the material is significantly impacted at nanoscale due to liquid slip. Macroscopically, depending on the contact angle produced by a liquid droplet with the surface, the wettability of a surface is defined as wetting, nonwetting, or neutral. Contact angles greater than 90 ° with a liquid is called non-wetting while contact angles smaller than 90 ° with a liquid are considered wetting. A contact angle of 90 ° with a liquid that forms a surface is termed neutral. Nonwetting surfaces are known to be strongly inclined towards liquid slip introduction.

Conclusion

The roughness of the surface depends on the packaging and thus relies directly on the mold packaging. PDMS precisely copies even the tiniest holes and gets the same mold geometry. Soft lithography combined with a thick photoresist reflow phenomenon is a new approach to PDMS microchannel manufacturing to minimize surface roughness. [5]. We could easily change the circular channel diameter and the channel routes. It is much needed to investigate the effect of surface roughness on the microchannel fluid flow for a PDMS-based microfluidic application. The results show that the surface velocity is not constant, but shows oscillating behaviour. This action can be clarified by accounting for ruggedness information on the channel surfaces. Local velocity oscillations in microchannel were recorded because of the ruggedness of the wall. The function of Weierstrass-Mandelbrot was used to develop surface roughness based on a fractal. Fractal dimension of the device constructed can be determined experimentally using AFM (Atomic Force Microscopy) and substituted in model's WM function. Using 3D LBM method the fluid flow can be simulated. The findings demonstrate fluid flow oscillatory behaviour. The result also shows that an increase in the fractal dimension increases the ruggedness and thus the fluid velocity is reduced. Thus, the turbulence in the microchannel can be tuned by synthesizing rough surfaces, and the mixing efficiency can be improved. With respect to liquid slips it affects in terms of nominal scale, but when flow is droplet induced this parameter is very significant.

REFERENCES:

[1] Liu Yanfeng, Li Hongwei, Zhang Jingwei and Xue Jin. “Liquid Flow Characteristics in Microchannels “*Applied Mechanics and Materials Vols 130-134 (2012) pp 1484-1490.*

[2] Johana Kuncová-Kallio, and Pasi J. Kallio “PDMS and its Suitability for Analytical Microfluidic Devices” *28th IEEE EMBS Annual International Conference New York City, USA, Aug 30-Sept 3, 2006.*

[3] Young-Min Kim', Woo-Seung Kim, Sang-Hoon Lee & Ju-Yeoul Baek “Effects of Surface Roughness on the Flow Characteristics in PDMS Microchannels”. *3rd Annual International IEEE EMBS Special Topic Conference on Microtechnologies in Medicine and Biology 2005.*

[4] Harry Garg, Vipender Singh Negi, “Numerical analysis of different shapes of microchannel for miniature cooling system”. *Proceedings of 2014 RAECS UIET.*

[5] Jong Seob Choi, Yunxian Piao, Tae Seok Seo, “Fabrication of a circular PDMS microchannel for constructing a three-dimensional endothelial cell layer.” *Springer April 2013.*

[6] Wang Lei, Wu Jiankang and Chen Bo “Analytic Solution of Liquid Flow in Rectangular PDMS-GLASS Microchannel with Wall Slip and Electro-Viscous Effects” *Applied Mathematical Sciences, Vol. 3, 2009, no. 44, 2195 – 2214*

[7] H Lin, Michel V “Electrokinetic instability in microchannel flows: A review”. *mechrescom.2008.07.012*

[8] Whiteside., Scott Ganz, Thomas Dickey et al “Foldable printed circuit boards on paper substrates “ *Advanced Functional Materials 2009.*

[9] Ageev, Grasegger, Morita et al. “Design, Simulation and Fabrication of Polydimethylsiloxane (PDMS) Microchannel for Lab-on-Chip (LoC) Applications.” *Applied Mechanics and Materials (Volume 819) January 2016*

IOP Conf. Series: Materials Science and Engineering 1189 (2021) 012022 doi:10.1088/1757-899X/1189/1/012022

Experimental investigation on the machinability characteristics in drilling of syntactic foams

Ashrith H S[1,4], Jeevan T P[2] and Mrityunjay Doddamani[3]
[1]Mechanical Engineering, Malnad College of Engineering, Hassan, India
[2]Mechanical Engineering, National Institute of Technology, Karnataka, Surathkal, India
[4]ashuhiriyal@gmail.com

Abstract. Syntactic foams (SFs) finds its application in automobiles, aircrafts and underwater vehicles. Drilling is most frequently used machining operation for assembly of these composite structures. In the present investigation, an effort is made to analyse the effect of cutting speed (C_s), feed rate (f_r), drill diameter (D) and filler content (F) on machinability characteristics in drilling glass microballoon (GMB) reinforced epoxy SFs. Machinability characteristics analyzed includes thrust force (T_f) and surface roughness (S_r). SFs are fabricated by dispersing 25, 35 and 45 vol.% of GMBs in epoxy resin. Experiments are carried out based on full factorial design (FFD) using TiAlN coated carbide drills. Experimental results show that T_f and S_r increases with increasing C_s. As f_r increases, T_f increases but decreases the S_r. Increasing GMB content significantly reduces the T_f and S_r of developed SFs.

1. Introduction

Lightweight materials synthesized by dispersing hollow microballoons in matrix material are called syntactic foams [1]. SFs are widely used for structural and buoyancy modules of submerged vehicle structures, where drilling holes is essential for installing various components. The quality of the drilled hole strongly depends on the input factors and their levels. Nearly 60% of composite parts are rejected due to poor hole quality which in turn substantially increasing the overall production cost [2].

Several researchers evaluated the effects of different input factors on drilling polymer composites. Studies have focused only on understanding the drilling process of fiber reinforced polymer composites (FRPC). Results shows that the increase in C_s leads to increased T_f in drilling FRPC and the twist drills performed better than multifacet drills [3]. Basavarajappa et al. [4] analysed the influence of C_s and f_r on T_f, S_r, and specific cutting coefficient in drilling of FRPC with and without silicon carbide filler. Results indicate that f_r has a significant influence on T_f and S_r. A study on the impact of C_s, drill geometry and f_r on T_f, hole diameter and circularity error found that the optimum conditions were different in drilling of unreinforced and reinforced polyamides [5].

Even though a comprehensive literature on drilling of FRPC is available, drilling of syntactic foams is les explored, which is the focus of current study. In the current investigation an effort has been made to examine the influence of Cs, f_r, F and drill diameter (D) on T_f and S_r in drilling of GMB/epoxy SFs using CNC vertical machining center and using coated carbide twist drills.

2. Materials and methods

2.1. Constituent materials
Epoxy resin (LAPOX L-12) and hardener (K-6) obtained from Atul Ltd., India is used as resin. Hollow GMBs (SID-350 grade) procured from Trelleborg Offshore, USA are used as fillers.

2.2. Sample preparation
SFs are prepared by dispersing 25, 35 and 45 by vol.% of GMBs. GMBs are added into the resin matrix and stirred slowly until homogeneous slurry is formed. Hardener by 10 wt.% is then added to the slurry, which is then stirred for additional 5 minutes followed by a degassing period of 5 minutes prior to pouring to the molds (ϕ 35×16 mm). Silicone gel is smeared to the mould surfaces. Specimens are cured for 24 h at room temperature. Experimentation

2.3. *Experimentation*
In the present investigation, T_f and S_r are chosen as responses. Input process parameters (*Cs*, f_r, *F* and *D*) and their levels are chosen based on previous investigations (Table 1). Overall 27 experiments with three trials are performed for each drill diameter based on FFD (Table 2).

Table 1. Input factors and levels

Parameters	Level		
	1	2	3
Cutting speed (C_s), m/min	25	75	125
Feed (f_r), mm/rev	0.04	0.08	0.12
Filler content (F), %	25	35	45

Table 2. Experimental layout plan and measured value of responses

Exp. No.	F	C_s	f_r	D_8		D_{16}	
				T_f	S_r	T_f	S_r
1			0.04	42.294	2.367	112.46	2.264
2		25	0.08	57.373	2.559	138.99	2.033
3			0.12	62.641	2.670	155.70	1.961
4			0.04	39.683	3.248	113.12	2.748
5	25	75	0.08	58.576	3.152	143.46	2.452
6			0.12	67.659	3.102	163.99	2.374
7			0.04	40.341	3.261	117.05	2.911
8		125	0.08	63.050	3.445	151.20	2.670
9			0.12	75.948	3.543	175.55	2.587
10			0.04	33.983	2.711	99.56	2.271
11		25	0.08	50.742	2.538	124.45	1.982
12			0.12	54.375	2.324	139.53	1.851
13			0.04	34.006	3.211	99.54	2.695
14	35	75	0.08	51.264	3.081	128.25	2.401
15			0.12	58.712	2.962	147.14	2.265
16			0.04	37.298	3.585	102.79	2.919
17		125	0.08	55.056	3.423	135.31	2.620
18			0.12	66.320	3.320	158.02	2.479
19			0.04	34.210	2.603	88.57	2.200
20		25	0.08	44.859	2.439	111.83	1.852
21			0.12	48.017	2.673	125.27	1.663
22	45		0.04	30.236	3.107	87.87	2.625
23		75	0.08	46.018	2.981	114.94	2.272
24			0.12	51.673	2.914	132.20	2.078

25		0.04	29.532	3.500	90.44	2.850
26	125	0.08	48.971	3.322	121.32	2.492
27		0.12	58.599	3.549	142.39	2.292

Drilling experiments are performed according to FFD with coated carbide twist drills (D_8 and D_{16}) fitted on a vertical CNC machining centre (MAX MILL PLUS+, MTAB, Chennai, India). Strain gauge type of dynamometer (Syscon Instruments Pvt. Ltd., India) is used to measure the T_f and S_r of hole is examined using Mitutoyo surftest (SJ 301, Japan). Input factors (I) and their levels (L) are coded together as I_L. For example, C_{s25} represents 25 m/min cutting speed.

3. Result and discussion

3.1. Analysis of thrust force

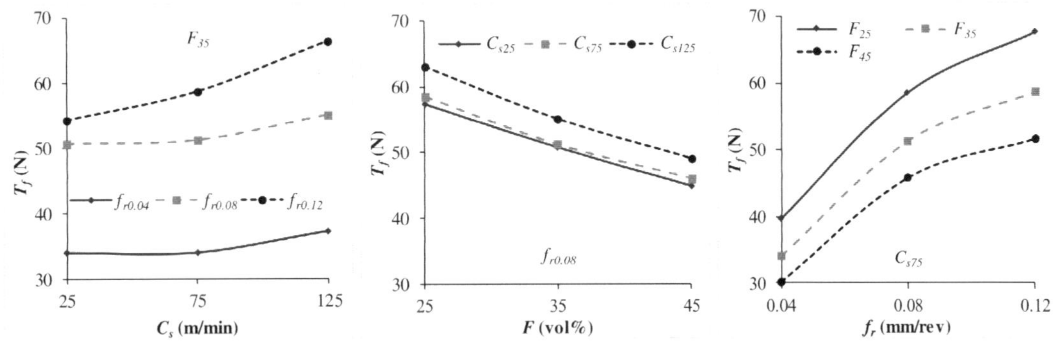

(a) (b) (c)

Figure 1. Effect of (a) C_s and f_r, (b) F and C_s and (c) f_r and F on T_f for D_8.

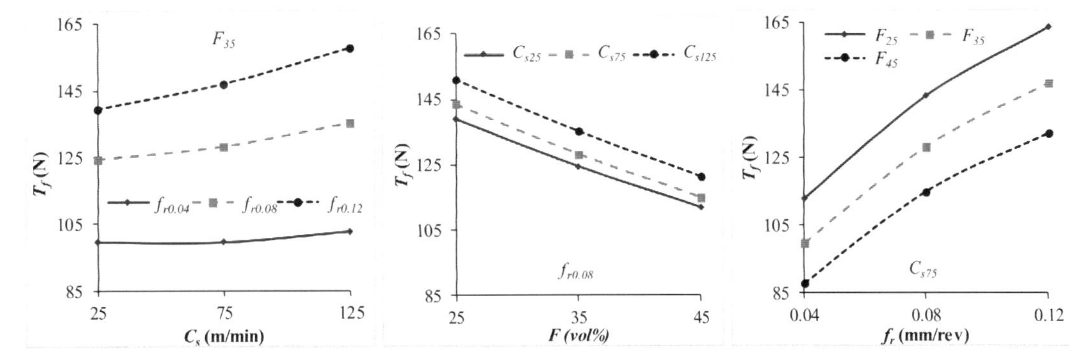

(a) (b) (c)

Figure 2. Effect of (a) C_s and f_r, (b) F and C_s and (c) f_r and F on T_f for D_{16}.

Experimentally measured values of T_f are presented in Table 2. The variation of T_f with Cs at different f_r for D_8 and D_{16} are shown in Figure 1a and Figure 2a respectively. T_f increases with increasing cutting speed and feed. Increased tool wear at higher Cs might result in such an observation [3]. T_f as a function of F and Cs for D_8 and D_{16} are shown in Figure 1b and Figure 2b respectively. T_f decreases with increasing F, while increases with increasing Cs. Decrease in T_f is due to the increased brittle behaviour of the SFs and also the presence of void inside GMBs. Figure 1c and Figure 2c shows the increasing T_f with increasing f_r, while decreases with increasing F for D_8 and D_{16} respectively. As f_r increases resistance offered by the SFs increases leading to increased friction between tool and SFs resulting higher T_f [4, 6].

3.2. Analysis of surface roughness

Surface roughness increases with increasing cutting speed for both D_8 and D_{16} as shown in Figure 3a and Figure 4a respectively. Heat increases at the drill and SFs interface with increasing *Cs* leading to rough surface [7, 8].

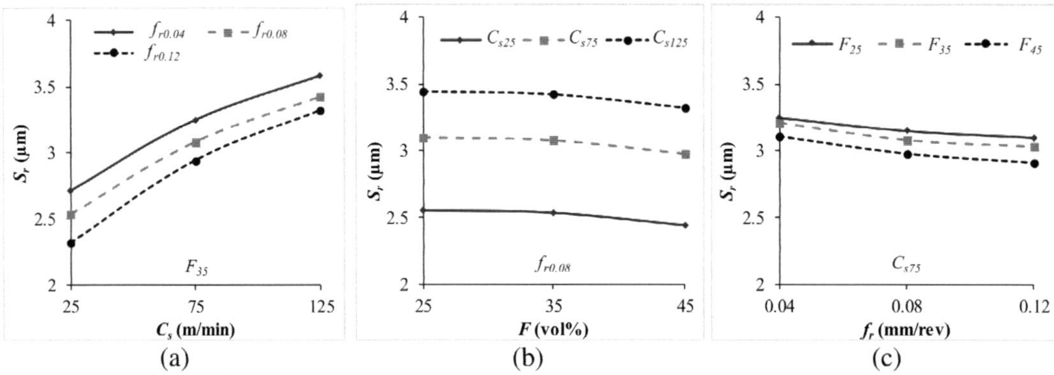

Figure 3. Effect of (a) C_s and f_r, (b) F and C_s and (c) f_r and F on S_r for D_8.

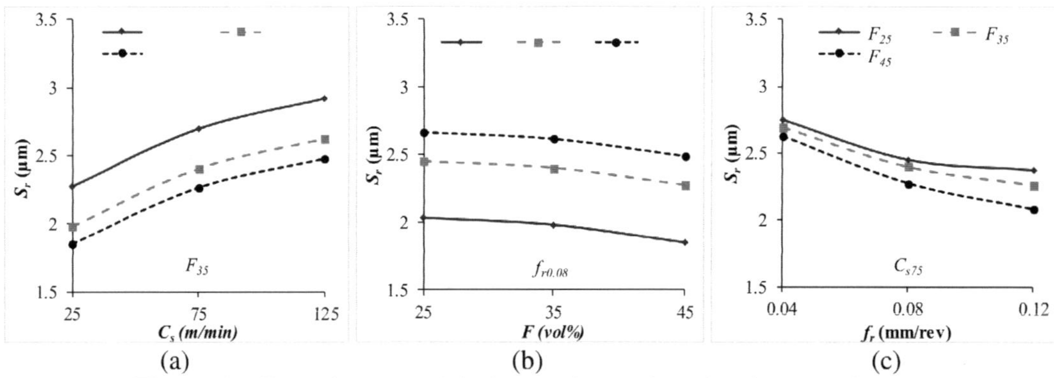

Figure 4. Effect of (a) C_s and f_r, (b) F and C_s and (c) f_r and F on S_r for D_{16}.

Figure 3b and Figure 4b shows decrease in S_r with increasing F, while increasing trend is observed with increasing *Cs*. Decrease in S_r with increasing F is owing to the burnishing and honing effect produced by abrasive fillers [4]. Additionally, lower T_f with increased F results in reduced S_r [7, 9]. The variation of S_r with f_r and F for D_8 and D_{16} are shown in Figure 3c and Figure 4c respectively. S_r decrease with increasing F and f_r. At higher f_r, temperature decreases due to reduced contact time between tool and SFs leading to lower f_r [8].

3.3. Influence of drill diameter on responses

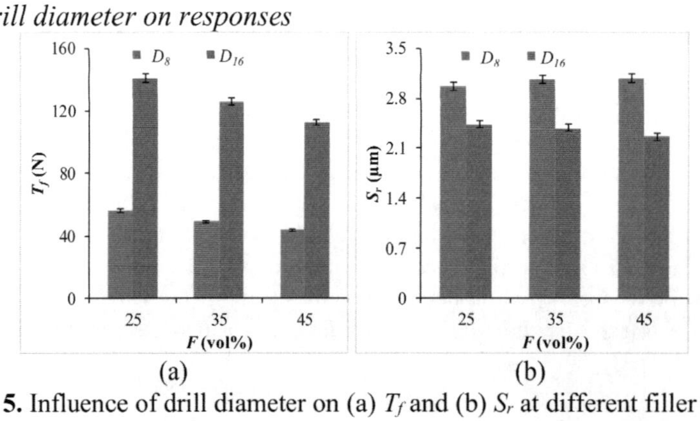

Figure 5. Influence of drill diameter on (a) T_f and (b) S_r at different filler content.

Figure 5a shows the effect of increasing D at different F on the T_f developed in drilling GMB/epoxy SFs. It is observed that the T_f increases by 2.5 times with increasing D from D_8 to D_{16}. As D increases, the contact area of the drilled hole increases leading to higher T_f [10, 11]. Increasing D from D_8 to D_{16} decreases the S_r of the drilled hole as shown in Figure 5b. S_r decreases by 17.84 and 26.88% with increasing D for F_{25} and F_{45} respectively. At any given Cs, D_{16} has a lower spindle speed than D_8 which results in lower S_r [7, 12].

From experimental investigation it is confirmed that GMB content strongly impacts machinability characteristics of SFs. Higher filler percentage is preferred in the SFs from drilling operations perspective, which is also beneficial for weight sensitive applications.

Conclusions

GMB/epoxy SFs are studied for drilling characteristics, such as T_f and S_r. Cs, f_r and F are varied at different levels to measure the responses. The effect of each variable on machinability characteristics are analyzed and presented. The following conclusions are drawn based on the experimental investigation:

- Increase in Cs, f_r and D increases T_f while decreasing trend is obtained with increasing GMB content.
- Minimum T_f is observed at a combination of $C_{s125}f_{r0.04}F_{45}D_8$.
- S_r increases with increasing Cs but decreases with the increase in f_r, F and D.
- S_r is found to be minimum at a combination of $C_{s25}f_{r0.12}F_{45}D_{16}$.
- Increasing GMB content from F_{25} to F_{45} reduces T_f and S_r by 20.92 and 3.92% respectively.
- SFs with high GMB content exhibits better machinability in terms of T_f and S_r.

References

[1] Gupta, N., Kishore, E. Woldesenbet, and S. Sankaran, *Studies on compressive failure features in syntactic foam material.* Journal of Materials Science, 2001. **36**(18): p. 4485-4491.

[2] Capello, E., A. Langella, L. Nele, A. Paoletti, L. Santo, and V. Tagliaferri, *Drilling polymeric matrix composites.* Machining. Fundamentals and recent advances, 2008. **10**: p. 167-194.

[3] Lin, S. and I. Chen, *Drilling carbon fiber-reinforced composite material at high speed.* Wear, 1996. **194**(1-2): p. 156-162.

[4] Basavarajappa, S., A. Venkatesh, V. Gaitonde, and S. Karnik, *Experimental investigations on some aspects of machinability in drilling of glass epoxy polymer composites.* Journal of Thermoplastic Composite Materials, 2012. **25**(3): p. 363-387.

[5] Rubio, J.C.C., L.J. da Silva, W. de Oliveira Leite, T.H. Panzera, S.L.M. Ribeiro Filho, and J.P. Davim, *Investigations on the drilling process of unreinforced and reinforced polyamides using Taguchi method.* Composites Part B: Engineering, 2013. **55**: p. 338-344.

[6] Gaitonde, V., S. Karnik, F. Mata, and J.P. Davim, *Study on some aspects of machinability in unreinforced and reinforced polyamides.* Journal of Composite Materials, 2009. **43**(7): p. 725-739.

[7] Gaitonde, V., S. Karnik, J.C. Rubio, A. Abrão, A.E. Correia, and J.P. Davim, *Surface roughness analysis in high-speed drilling of unreinforced and reinforced polyamides.* Journal of Composite Materials, 2012. **46**(21): p. 2659-2673.

[8] Campos Rubio, J.C., A.M. Abrão, P. Eustaquio Faria, A.E. Correia, and J.P. Davim, *Delamination in high speed drilling of carbon fiber reinforced plastic (CFRP).* Journal of composite materials, 2008. **42**(15): p. 1523-1532.

[9] Palanikumar, K., *Experimental investigation and optimisation in drilling of GFRP composites.* Measurement, 2011. **44**(10): p. 2138-2148.

[10] El-Sonbaty, I., U. Khashaba, and T. Machaly, *Factors affecting the machinability of GFR/epoxy composites.* Composite structures, 2004. **63**(3): p. 329-338.

[11] Rajamurugan, T., K. Shanmugam, and K. Palanikumar, *Analysis of delamination in drilling glass fiber reinforced polyester composites.* Materials & Design, 2013. **45**: p. 80-87.

[12] Khashaba, U., I. El-Sonbaty, A. Selmy, and A. Megahed, *Machinability analysis in drilling woven GFR/epoxy composites: Part I–Effect of machining parameters.* Composites Part A: Applied Science and Manufacturing, 2010. **41**(3): p. 391-400.

A Taguchi Approach in Investigating the Effects of Process Parameters in Drilling of Glass Fibre Reinforced Plastic Composite

RAVI SHANKAR S. N.[1] SURJYA K. PAL[2] AND ARUN K. SAMANTARAY[3]

[1]DEPARTMENT OF MECHANICAL ENGINEERING, MALNAD COLLEGE OF ENGINEERING, HASSAN, KARNATAKA, INDIA
[23]DEPARTMENT OF MECHANICAL ENGINEERING, INDIAN INSTITUTE OF TECHNOLOGY KKHARAGPUR, WEST BENGAL, INDIA
Corresponding author:[1]ravishivara@yahoo.com

Abstract. Glass fibre reinforced plastic (GFRP) composite has found an important place in the industries replacing many metals and alloys. The joints formed are found to be the weakest portions of the assembled GFRP parts. Conventional drilling is the most widely adapted manufacturing process for an efficient assembly. Moreover, drilling of glass fibre reinforced plastics is difficult to carry out due to the anisotropic, non-homogeneous structure and high abrasiveness of the glass fibre and the complex geometry of the drill. For these reasons, the damage gets induced to the hole/workpiece and the cutting tool. In this study, Taguchi approach is employed to analyse the drilling process. An L25 orthogonal array, signal to noise S/N ratio and analysis of variance (ANOVA) are used to identify the effects of process input parameters on the output factors. The process input parameters are drill diameter, cutting speed and feed rate. The output factors/responses are the thrust force, torque, power, temperature, tool life, surface roughness, deviation from dimensional parameters of the holes, peel-up and push-out delaminations. In every set of experiment, an exhaustive work has been carried out till each drill bit reached its tool life. Optimal process parameters are available from the S/N ratio plots whereas ANOVA has identified the significant input factors and their percentage contribution on the responses.
Keywords: Drilling, Glass fibre reinforced plastic (GFRP), Delamination, Design of experiments (DOE), Taguchi approach, Analysis of variance (ANOVA)

Nomenclature

D	Drill diameter, mm	T	Average chip temperature, ^oC
f	Feed rate, mm/rev	t_l	Tool life, s
F_{cal}	Test of significance, calculated	V_b	Tool flank wear, µm
F_d	Push-out delamination	v_c	Cutting speed, m/min
F_p	Peel-up delamination	V_f	Fibre volume fraction
F_{tab}	Test of significance, tabulated	Δ_c	Deviation from cylindricity, mm
F_z	Thrust force during drilling, N	Δ_d	Deviation from nominal diameter, mm
M_z	Drilling torque, Nm	Δ_o	Deviation from circularity, mm
P	Power, W	θ	Fibre orientation angle
R_a	Surface roughness, µm		

IOP Conf. Series: Materials Science and Engineering 1189 (2021) 012023 doi:10.1088/1757-899X/1189/1/012023

1. Introduction

Glass fibre reinforced plastic (GFRP) composite material is used extensively in the industries nowadays due to its high strength to weight ratio, stiffness to weight ratio, rigidity, low thermal expansion, high specific modulus of elasticity, flexibility in design, parts consolidation, dimensional stability and low tooling costs. Their application is in the area of aerospace, marine, automotive, electronic industries, transportation industries, military vehicles and sports equipments. Because of electrical insulating properties, GFRP is widely used in appliances, tools and machinery. Corrosion resistant glass fibre reinforced tanks and pipes offer extended service life as compared to metals. A GFRP composite provides protection against biological agents, nuclear radiation and resists the degradation caused by chemicals. Hole making is the most important secondary manufacturing process to give final shape to the assembled GFRP component. Efforts have been made by various researchers in the recent past to optimize the process parameters to minimize the drilling induced damages. Among the damages, peel-up and push-out delaminations are found to be the most critical, since they reduce the structural integrity and long term performance of the GFRP material.

Godwin et al. [1] have reviewed the strength of the joints in fibre reinforced plastics (FRPs). Koniget al. [2-3] have produced keynote papers on machining of FRPs and reviewed quality definition and assessment in drilling of fibre reinforced thermosets. Teti [4] has reviewed machining of composite materials. Singh et al. [5] attempted to review the status of the work done in drilling of composite materials with emphasis on damage mechanisms and reported initial experimental findings on drilling of GFRP composite laminates. Hochenget al. [6] have reviewed the path towards delamination-free drilling of composite materials. Abraoet al. [7], in their review, have found that the phenomena associated with shearing of polymeric composite materials require additional studies, in order to allow a better understanding of the behavior of polymeric composites when subjected to cutting.

Tagliaferriet al. [8] has worked on the effect of drilling parameters on the finish and mechanical properties of GFRP composites. They have found that the damage zone width is correlated to the ratio between drilling speed and feed rate, Vr/ Vt; in particular, higher the Vr/ Vt value, better is the cut quality. For a definite value of Vr/ Vt, the damage zone decreases to a minimum value, beyond which the damage stays constant. Malhotra [9] found that the flank wear is higher than chisel edge wear. Drilling was interrupted at preset depths to study the quality of the cut surface and the damage development during drilling GFRP by Caprino and Tagliaferri [10]. According to that study, damage induced in the form of delamination is strongly dependent on the feed rate. Khashaba [11] experimentally investigated the influence of drilling and material variable on thrust force, torque and delamination of GFRP composites. Variable feed technique was used to achieve delamination free drilling. Sonbatyet al. [12] made an experimental study on drilling glass fibre reinforced epoxy composites.They found that cutting speed has insignificant effect on the thrust force and surface roughness of epoxy resins whereas increase in feed, drill size, and fibre volume fractions led to increase in the thrust force and torque. Drill diameter combined with feed has a significant effect on surface roughness. Experimental investigation by Arul et al. [13] on the influence of tool material on the dynamics of drilling GFRP composites has revealed that defect-tolerant drilling can be attained by proper selection of cutting parameters and tool material. Monitoring of flank wear, hole shrinkage and acoustic emission during drilling substantiates this finding. Khashabaet al. [14] analysed drilling of chopped composites. Their observations revealed no clear effect of the cutting speed on delamination size, while the delamination size decreased with the decrease in feed. They commented that delamination free drilling of chopped composites, with high fibre volume fraction, remained as a problem for further investigation. A finite element (FE) approach was proposed to study the drilling characteristics of UD-GFRP composite laminates in Singh et al. [15]. Raoet al. [16] have studied the effect of drilling induced damage on notched tensile and pin bearing strengths of woven GFR-epoxy composites. They concluded that the feed rate is found to be the major contributing factor on drilling damage. The effect of cutting tools with four different geometries and composition on thrust force and delamination were studied by Abraoet al. [17]. In contrast to other reports, a direct relationship between thrust force and delamination was not observed, since the drill responsible for the highest thrust force was also responsible for the smallest delamination. Hence it is concluded that the relationship between thrust force and delamination is not straightforward.

Mathew et al. [18] have used DOE technique to investigate the effect of geometry of a trepanning tool on the thrust and torque during drilling of GFRP. It is revealed that the performance of the trepanning tool is superior to that of conventional twist drills in terms of thrust, torque and hole quality. Low production cost and ease of regrinding are its major additional advantages due to its simple geometry. Davimet al. [19] evaluated the cutting parameters (cutting velocity and feed) and the influence of the matrix under specific cutting force, delamination factor, and surface roughness in two types of matrix (Viapal VUP 9731 and ATLAC 382-05) by DOE technique using carbide drills. From their investigation it was concluded that the specific cutting force decreases and the delamination factor increases with both the cutting parameters (cutting velocity and feed); whereas the surface roughness increases with the feed and decreases with the cutting velocity. A correlation between the cutting velocity and the feed rate with the specific cutting pressure, thrust force, damage factor and surface roughness in a GFRP material was established by Davimet al. [20]. The analysis of variance (ANOVA) was performed for the investigation of the cutting characteristics of GFRPs by using cemented carbide (K10) drills with appropriate geometries. The objective of the experiment was to establish a correlation between the cutting velocity and the feed rate with the specific cutting pressure, thrust force, damage factor and surface roughness. A study on the influence of process parameters on the cutting force and torque during drilling of glass fibre polyester reinforced composites by using Taguchi method has been attempted by Mohan et al. [21]. They reveal that the speed and drill (carbide) size have more significant influence on the cutting thrust force than specimen thickness and feed rate. Study of the response table indicated that the specimen thickness and drill size are significant and dominant parameters on torque than any other combination. Al-Sulaimanet al. [22] investigated the use of electrical power for online monitoring of tool condition and found that the differential electrical power is a better indicator of tool wear than conventional mechanical power method. Delamination analysis in the drilling process of GFRP using carbide drills by Mohan et al. [23] reveals that the specimen thickness and cutting speed have significant influence on peel up delamination and the specimen thickness and feed have more significant influence on push out delamination. DOE, Taguchi method and ANOVA have been adapted for the study. The residual strength of drilled UD-GFRP has been investigated by using Taguchi method inKishore et al. [24]

2. Plan of experiments

Design of experiments is a statistical technique used to study the effects of multiple variables (also called factors or parameters) simultaneously. DOE is highly effective wherever and whenever it is suspected that the performance of a part or process is controlled by more than one factor. The Taguchi technique or Taguchi approach is an experimental strategy in which a modified or standardized form of DOE is utilized. The Taguchi approach, which is a form of DOE with special application principles, is chosen for the present study [25].

Table 1. Different Levels Assigned to the Factors

Process parameters	*Factors assigned*	*Level 1*	*Level 2*	*Level 3*	*Level 4*	*Level 5*
Drill diameter (*D*), mm	A	6	7	8	9	10
Cutting speed (v_c), m/min	B	1.6	9.2	16.8	24.4	32
Feed rate (*f*), mm/rev	C	0.02	0.09	0.16	0.23	0.3

The process parameters selected for the present study were chosen from the range of process parameters opted by earlier researchers. A few trial experiments were conducted initially. The results were compared with earlier reports for confirmation. The planning of experiment for the drilling of GFRP composite was based on Taguchi L_{25} orthogonal array (Table 3). The orthogonal array was formed by selecting three factors/parameters and five levels as shown in Table 1. Accordingly, a total number of 25 sets of experiments were carried out.In each set of experiment, the drill bit was examined for the wear measurement in the Leica inverted microscope at regular intervals after drilling.

The regular intervals were after drilling the first hole, third hole, fifth hole, eighth hole, eighteenth hole and so on, further maintaining a constant interval of ten holes till the drill bit reached the end of its tool life; the tool life criteria being 300 μm flank wear (V_b). Some drill bits had short life, i.e., they reached their tool life criteria after drilling only 18 or less holes. On the other hand one of them showed a long life by reaching its tool life criteria after drilling in between 228 to 238 holes. Hence the number of experiments carried out, varied between 18 and 238. A total number of 2175 experiments were conducted in the present work.The input factors (process parameters) are the drill diameter (D), cutting speed (v_c) and feed rate (f). The responses measured are the thrust force (F_z), torque (M_z), power (P), temperature (T), tool flank wear (V_b), tool life (t_l), peel-up delamination factor (F_p), push-out delamination factor (F_d), arithmetic mean value of the surface roughness (R_a), deviation from the nominal diameter (Δ_d), deviation from circularity (Δ_o) and deviation from cylindricity (Δ_c).

3. Experimental setup and procedure

The schematic diagram of the experimental setup along with various sensors and the data acquisition system are shown in Fig. 1. Glass fibre reinforced plastic (GFRP) whose mechanical properties shown in Table 2, was supplied by Sunrise FRP Ltd, Bangalore, India. It was prepared by passing glass fibre mat (woven roving of Vetrotex) through a hot press lamination process using two rollers. Then a hot cure epoxy was applied on the glass mat. The epoxy resin used was Araldite LY-556 with hardenerHT-972. The whole set was cured under high temperature ranging from 120 - 140 °C. The laminate so cured was cut to the required size of 100mm x 50mm and 9 mm thickness. The fibre orientation (θ)was 0/90 with a fibre volume fraction (V_f)being 65-68%.

The workpiece was held in the fixture to avoid any displacement due to shock and vibration during drilling. A CNC vertical machining center BFW Agni (BMV 45 T20) of 16 kVA capacity having a maximum spindle speed of 10000 rpm and a maximum feed rate of 10000 mm/min was used for drilling the GFRP work material. Standard HSS twist drills of regular geometry and point angle of 118° were used to generate holes in the GFRP composite. The work which was held in the fixture was mounted on the KISTLER® dynamometer (type 9272A). The thrust force and the torque during drilling were measured from the drilling dynamometer which was connected to the charge amplifier (5070A). Machining center's main power supply cables were made to pass through the arrangement made in the Hall effectMontronix power sensor PS 100-DGM for measurement of power.

Table 2. Mechanical Properties of GFRP Work Material

Property	*Value*	*Unit*
Specific gravity	1.65-1.75	
Glass content	65-68	% weight
Water absorption	0.07-0.08	% weight
Tensile strength	2.35-2.74	N/mm^2
Flexural strength	2.94-3.09	N/mm^2
Compressive strength	3.33-3.72	N/mm^2
Impact strength	300-310	KJ/m^2
Fiber orientation	0/90	deg

Temperature in the machining zone was measured by an Omega temperature sensor (model OS 523-3). The charge amplifier, power sensor and temperature sensor were all connected to the data acquisition card (NI 6210 card 2). This NI DAQ card, which gave the output signals in volts, was interfaced with an IBM PC. The acquired data was processed and stored with the help of LabVIEW software (Version 7.1). Actual values of thrust force, torque, power and temperature were later converted to their respective units of measurement by the corresponding conversion factors.

Drill wear was monitored by withdrawing the drill bit at regular intervals between the experiments. The drill flank wear was measured in the Leica DMILM type invertedmicroscope. Once the drill bit reached its tool life, the corresponding hole drilled in the GFRP workpiece was examined for its peel-up and push-out delaminations in the Leicastereo zoom microscope. Thus the delamination factors were determined. Surface roughness of the hole was measured three times for each test in the Mitutoyo SJ-201P surface roughness tester. The cut-off length selected was 0.8 mm with 5 samplenumbers. The deviations from the nominal diameter, circularity and cylindricity, all related to the geometrical parameters of the hole were measured in the TESA micro-hite DCC 4-7-4 coordinate measuring machine.

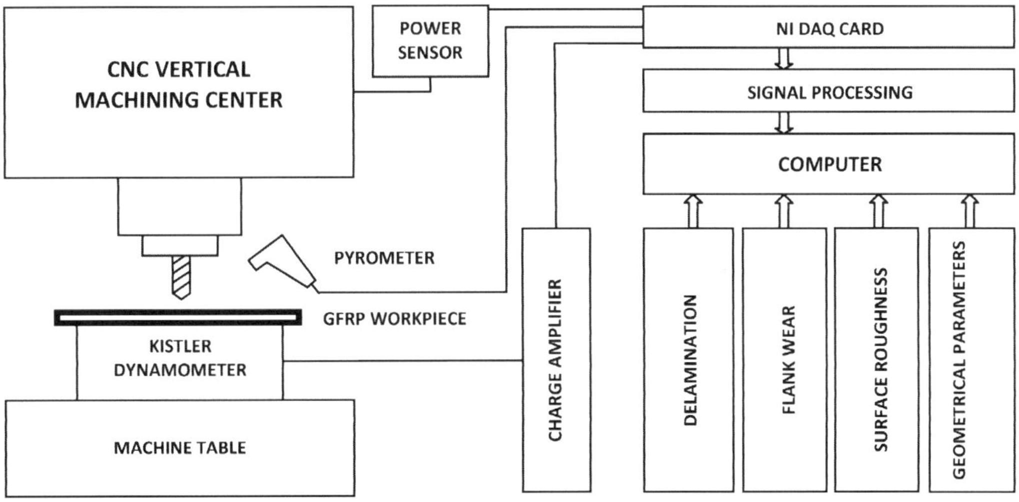

Figure 1. Schematic diagram of the experimental set up.

4. Analysis of experimental results

Taguchi L_{25} orthogonal array and experimental results are given in Table 2. A statistical analysis was performed using the commercial software package MINITAB 15®on each response to study the effect of the drill diameter, cutting speed and the feed rate. Hence the analysis consists of calculation of signal to noise (*S/N*) ratio values and analysis of variance (ANOVA) test for each response factor. The *S/N* ratio is the mean square deviation of the data set plotted in a log (to the base 10) scale with a -10 multiplier which changes the desirability from *smaller is better* to *larger is better* model.Higher the *S/N* ratio better is the result.

In the smaller the better model,

$$S/N ratio = -10 \, log \frac{1}{n}[\sum_{i=1}^{n} Y_i^2], \quad (1)$$

where, for Y observed data, n is the number of observations.

For larger the better model,

$$S/N ratio = -10 \, log \frac{1}{n}\left[\sum_{i=1}^{n} \frac{1}{Y_i^2}\right]. \quad (2)$$

The designed parameter is considered as a significant factor if test$F_{cal}$is greater thanF_{tab}. Test F_{cal} is the ratio of mean of squared deviations (MS) to the MS error and $F_{tab}$and istaken from the statistical table. The percentage contribution of the factors (shown in Table 4) in influencing the response is the percentage ratio of sum of squares to the total sum of squares, where the used variable

Degrees of freedom (DF) = Number of levels – 1 (3)

The delamination factor is the ratio of the maximum diameter (D_{max}) of the damage zone measured, to the nominal hole diameter (D). Peel-up delamination (F_p) and push-out delaminations (F_d)were determined by usingEq. (4) after measuring the outer diameters (D_{max}) of both the delaminations, respectively.

$$\text{Delamination factor} = \frac{D_{max}}{D} \quad (4)$$

5. Results and discussion

Main effect plots of signal to noise ratios of the entire responses verses the process parameters are shown in Fig. 2. For better productivity, it is desirable to have *minimum* thrust force, torque, power, temperature, surface roughness, deviations from the geometrical parameters, peel-up and push-out delaminations but *maximum* tool life. Therefore the calculation of *S/N* ratio follows *smaller the better* model for all the responses except tool life for which the calculation of *S/N* ratio follows *larger the better* model. An increase in *S/N* ratio indicates the decrease in thrust force, torque, power, temperature, surface roughness, deviation from nominal diameter, deviation from circularity, deviation from cylindricity, peel-up and push-out delaminations. But for tool life, an increase in *S/N* ratio indicates longer life. From the results, it is found that increase in tool diameter has increased the thrust force. This may be due to the increase in the length of the chisel edge and the increase in depth of cut. Since test F_{cal}>F_{tab}, as seen in the ANOVA Table 4 the drill diameter is a significant factor influencing the thrust force. But the percentage contribution is only 11.49. There is a slight reduction of thrust force with increase in cutting speed.

Table 3.Orthogonal Array L_{25}

L_{25} *test*	*A*	*B*	*C*
1	1	1	1
2	1	2	2
3	1	3	3
4	1	4	4
5	1	5	5
6	2	1	2
7	2	2	3
8	2	3	4
9	2	4	5
10	2	5	1
11	3	1	3
12	3	2	4
13	3	3	5
14	3	4	1
15	3	5	2
16	4	1	4
17	4	2	5
18	4	3	1
19	4	4	2
20	4	5	3
21	5	1	5
22	5	2	1
23	5	3	2
24	5	4	3
25	5	5	4

IOP Conf. Series: Materials Science and Engineering 1189 (2021) 012023 doi:10.1088/1757-899X/1189/1/012023

The matrix material may have been softened due the rise of cutting temperature with the increase in the cutting speed. But the contribution is a mere 4.17 % and is not a significant factor. Feed rate is highly significant in influencing the thrust force with a 77.97% contribution. This may be due to the increase in shear area as the tool encounters more material per revolution.

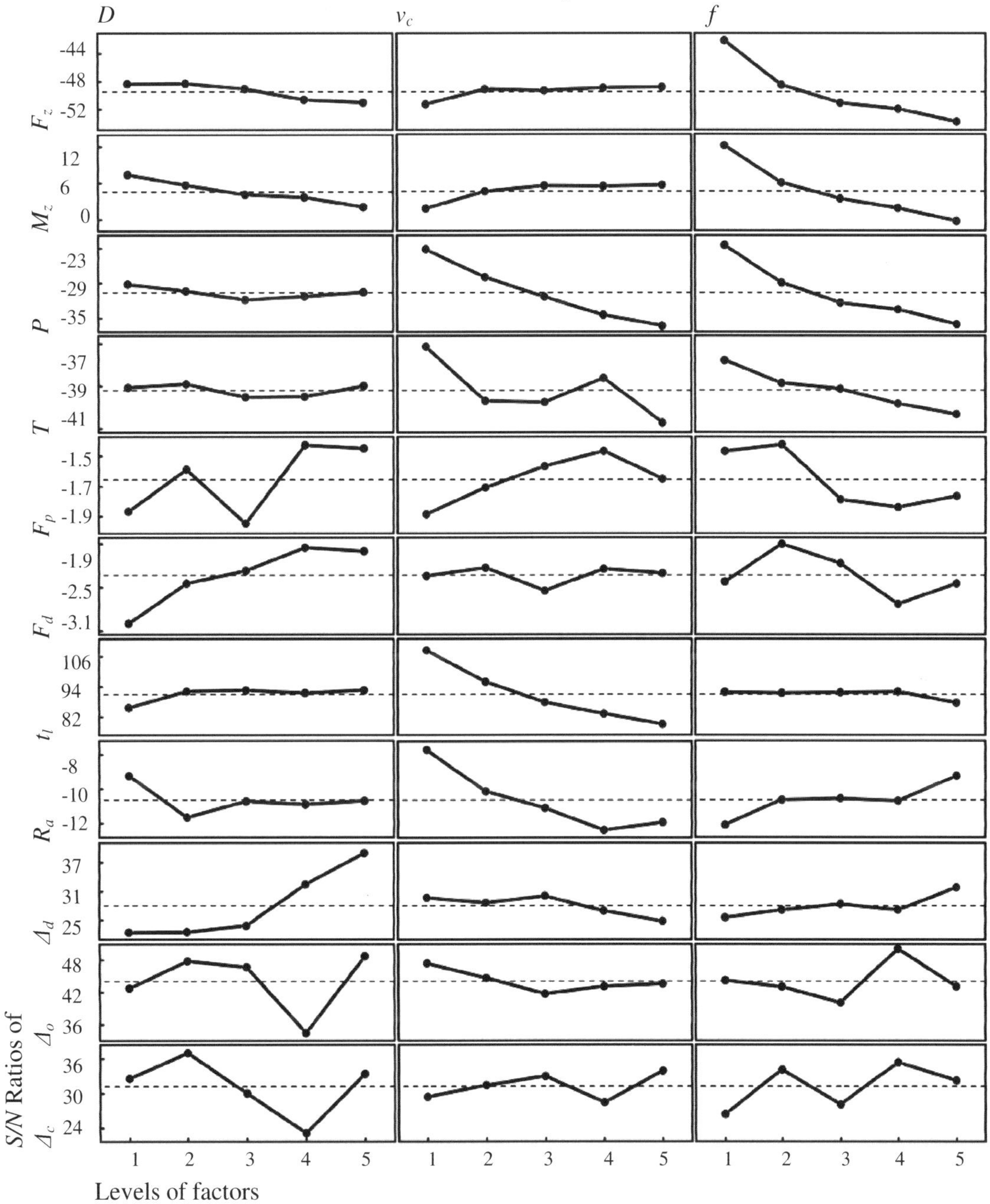

Figure 2 Signal to Noise Ratio Plots

The trend observed in the effects of process parameters on torque is almost similar to the thrustforce. Though all the process parameters are significant, feed rate is the more significant factor with 64.04% contribution. This is due to the increase in cross-sectional area of the undeformed chip [8].There is a steep rise in power consumption with the increase in both the cutting speed and feed rate, as can be

IOP Conf. Series: Materials Science and Engineering 1189 (2021) 012023 doi:10.1088/1757-899X/1189/1/012023

seen in Fig.2. They are very significant in influencing the power consumption with a contribution of 44.2% and 42.81%, respectively. The measurement of electrical power has indicated the increase in power with the increase in spindle motor rpm. The increase in the feed rate increases the power consumption because the tool encounters more work material (with the increase in shear area) during cutting.

Table 4. Analysis of Variance (ANOVA) for the Responses

Response	*Source*	*DF*	*Seq SS*	*Adj MS*	F_{cal}	F_{tab}	*% Contribution*
F_z	*D*	4	55912	13978	5.41	3.26	11.49
	v_c	4	20316	5079	1.97	3.26	4.17
	f	4	379522	94880	36.74	3.26	77.97
	Error	12	30990	2582			
	Total	24	486739				
M_z	*D*	4	0.61462	0.15365	10.73	3.26	19.13
	v_c	4	0.36868	0.09217	6.44	3.26	11.48
	f	4	2.05711	0.51428	35.92	3.26	64.04
	Error	12	0.17180	0.01432			
	Total	24	3.21221				
P	*D*	4	95.5	23.9	0.11	3.26	0.47
	v_c	4	9033.6	2258.4	10.60	3.26	44.20
	f	4	8748.6	2187.1	10.26	3.26	42.81
	Error	12	2557.3	213.1			
	Total	24	20435.0				
T	*D*	4	520.0	130.0	0.47	3.26	4.36
	v_c	4	5440.3	1360.1	4.90	3.26	45.63
	f	4	2631.1	657.8	2.37	3.26	22.07
	Error	12	3332.0	277.7			
	Total	24	11923.4				
F_p	*D*	4	0.011110	0.002778	2.33	3.26	29.83
	v_c	4	0.004697	0.001174	0.98	3.26	12.61
	f	4	0.007134	0.001783	1.50	3.26	19.15
	Error	12	0.014308	0.001192			
	Total	24	0.037249				
F_d	*D*	4	0.095819	0.023955	3.97	3.26	42.27
	v_c	4	0.008694	0.002174	0.36	3.26	3.84
	f	4	0.049753	0.012438	2.06	3.26	21.95
	Error	12	0.072417	0.006035			
	Total	24	0.226683				
t_l	*D*	4	4.9488E+10	1.2372E+10	0.78	3.26	5.72
	v_c	4	5.9776E+11	1.4944E+11	9.48	3.26	69.12
	f	4	2.8412E+10	7102933832	0.45	3.26	3.29
	Error	12	1.8921E+11	1.5767E+10			
	Total	24	8.6486E+11				
R_a	*D*	4	4.6543	1.1636	1.25	3.26	12.33
	v_c	4	15.8939	3.9735	4.26	3.26	42.11
	f	4	6.0064	1.5016	1.61	3.26	15.91
	Error	12	11.1921	0.9327			
	Total	24	37.7466				
Δ_d	*D*	4	0.0222776	0.0055694	6.97	3.26	53.14
	v_c	4	0.0070614	0.0017654	2.21	3.26	16.84
	f	4	0.0030001	0.0007500	0.94	3.26	7.16
	Error	12	0.0095835	0.0007986			
	Total	24	0.0419226				
Δ_o	*D*	4	0.0038669	0.0009667	2.03	3.26	32.51
	v_c	4	0.0009548	0.0002387	0.50	3.26	5.00
	f	4	0.0013532	0.0003383	0.71	3.26	11.38
	Error	12	0.0057209	0.0004767			
	Total	24	0.0118958				
Δ_c	*D*	4	0.0105614	0.0026403	3.82	3.26	41.53
	v_c	4	0.0018466	0.0004616	0.67	3.26	7.26
	f	4	0.0047330	0.0011832	1.71	3.26	18.61
	Error	12	0.0082897	0.0006908			
	Total	24	0.0254306				

Temperature rises with the increase in cutting speed, which is the only significant factor with a contribution of 45.63%. This may be due to high friction at higher cutting speeds and more heat generation while shearing more work material with high feed rate. The chips produced at low feed rates were of ribbon type and were winding themselves to the drill bit for a while before getting separated. This sometimes blocked the new chips from getting exposed to the sensor. Hence the average temperature recorded may not be consistent for few readings.

It is seen that an increase in drill diameter and cutting speed reduces the peel-up and push-out delaminations. But an increase in feed rate has increased the delaminations. This may be due to the increase in thrust force with the increase in feed rate. The trend in both the cases appears to be the same. In the case of peel-up delaminations, as the test $F_{cal}$is less thanF_{tab} for all the process parameters, they are less significant in influencing the delaminations. But the test $F_{cal}$is greater thanF_{tab} for drill diameter in the case of push-out delaminations and is a very significant factor in influencing the delaminations.

Larger the better model is chosen for the tool life *S/N* ratios. Figure 2 shows an increase in tool life with increase in drill diameter, but decrease in tool life with increase in cutting speed and feed rate. Though the drill diameter and feed rate are not the significant factors, the cutting speed appears to be very significant in influencing the tool life. This may be due to the increase in temperature at the elevated speeds which may have affected the tool material as well as the matrix in the composite. Increase in drill diameter and cutting speed increased the surface roughness, whereas increase in feed rate decreased the surface roughness. From the ANOVA results, cutting speed seems to be a very significant factor in influencing the surface roughness, making the other two factors less significant. Increase in surface roughness may be attributed to the thermal softening of the matrix which becomes sticky because of rise in temperature at higher speeds.

Deviation from the nominal diameter has decreased with increase in drill diameter and feed rate. This may be due to the same amount of fibre pullouts irrespective of the drill diameter since it is a very significant factor. Though there is an increase in deviation with increase in cutting speed, both the cutting speed and feed rate are not significant in influencing the deviation from the nominal diameter. With the increase in the cutting speed, the deviation from circularity increases and the deviation from cylindricity decreases. But this is the least significant factor for both the deviations. Though the trends of drill diameter and feed rate are similar in both the cases, they are inconsistent. Though none of the factors are significant in the case of deviation from circularity, drill diameter is a significant factor in influencing the deviation from cylindricity.

Table 5. Optimal Parameter Settings of Input Factors

Physical requirement	***Optimal combinations***		
	Drill diameter	***Cutting speed***	***Feed rate***
Minimum thrust force	7	32	0.02
Minimum torque	6	32	0.02
Minimum power	6	1.6	0.02
Minimum temp	7	1.6	0.02
Maximum tool life	10	1.6	0.23
Minimum peel-up delamination	9	24.4	0.09
Minimum push-out delamination	9	9.2	0.09
Minimum surface roughness	6	1.6	0.3
Minimum deviation from nominal hole diameter	10	16.8	0.3
Minimum deviation from circularity	10	1.6	0.23
Minimum deviation from cylindricity	7	32	0.23

6. Conclusions

In the present work, Taguchi method with ANOVA identified the significant input factors and their percentage contribution on the responses. The results are obtained after all the drill bits reached their tool life. The increase in thrust force is mainly due to the feed rate and drill diameter to a smaller extent. The torque also increases considerably with increase infeed rate but effects of drill diameter and cutting speed were also found to be significant. The power increases significantly with the increase in cutting speed and feed rate whereas the drill diameter has the least significance. Cutting speed and feed rate are the most significant factors in influencing the temperature. Moreover, cutting speed is also a significant factor in decreasing the tool life and interestingly increasing the surface roughness. None of the factors are significant in case of peel-up delamination but drill diameter is a significant factor for the push-out delamination. Drill diameter is a significant factor for both deviation from the nominal diameter and deviation from cylindricity whereas there are no significant factors for deviation from circularity. The optimal process parameters available from the *S/N* ratio plots are shown in Table 5.

References

[1] E. W. Godwin, F. L. Matthews, (1980), "A review of the strength of joints in fibre-reinforced plastics," Composites, pp. 155-160.
[2] W. Konig, Ch. Wulf, P. Grab, H. Willerscheid, (1985), "Machining of fibre reinforced plastics," Annals of the CIRP, Vol. 34, no.2, pp. 537-548.
[3] W. Konig, P. Grab, (1989), "Quality definition and assessment in drilling of fibre reinforced thermosets," Annals of the CIRP, Vol. 38, no. 1, pp. 119-124.
[4] R. Teti, (2000), "Machining of Composite Materials," CIRP Annals – Manufacturing Technology, Vol. 51, pp. 611-634.
[5] I. Singh, D. Nayak, R. Saxena, N. Bhatnagar, (2004), "Drilling induced damage in FRP composite laminates," Institution of Engineers, Vol. 85, pp. 37-40.
[6] H. Hocheng, C. C. Tsao, (2005), "The path towards delamination-free drilling of composite materials" Journal of Materials Processing Technology, Vol. 167, pp. 251-264.
[7] A. M. Abrao, P. E. Faria, J. C. Campos Rubio, P. Reis, J. Paulo Davim, (2007), "Drilling of fiber reinforced plastics: A review," Journal of Materials Processing Technology, Vol. 186, pp.1-7.
[8] V. Tagliaferri, G. Caprino, A. Diterlizzi, (1990), "Effect of drilling parameters on the finish and mechanical properties of GFRP composites," Int J Mach. Tools Manufact, Vol. 30, no.1, pp. 77-84.
[9] S. K. Malhotra, (1990), "Some studies on drilling of fibrous composites," Journal of Materials Processing Technology, Vol. 24, pp. 291-300.
[10] G. Caprino, V. Tagliaferri, (1995), "Damage development in drilling glass fibre reinforced plastics," Int J Mach Tools Manufact, Vol. 35, no. 6, pp. 817-829.
[11] U. A. Khashaba,(2004), "Delamination in drilling GFR-thermoset composites," Composite Structures, Vol. 63, pp. 313–327.
[12] E. Sonbaty, U. A. Khashaba, T. Machaly, (2004), "Factors affectting the machinability of GFR/epoxy composites," Composite structures, Vol. 63, pp. 329-338.
[13] S. Arul, L. Vijayaraghavan, S. K. Malhotra, R. Krishnamurthy, (2006), "Influence of tool material on dynamics of drilling of GFRP composites," Int J AdvManufTechnol, Vol. 29, pp. 655-662.
[14] U. A. Khashaba, M. A. Seif, M. A. Elhamid, (2007), "Drilling analysis of chopped composites, Composites: Part A, Vol. 38, pp. 61-70.
[15] I. Singh, N. Bhatnagar, P. Viswanath, (2008), "Drilling of uni-directional glass fiber reinforced plastics: Experimental and finite element study," Materials and Design, Vol. 29, pp. 546-553.
[16] B. SrinivasaRao, R. Rudramoorthy, S. Srinivas, B. NageswaraRao, (2008), "Effect of drilling induced damage on notched tensile and pin bearing strengths of woven GFR-epoxy composites," Materials Science and Engineering A, Vol. 472, pp. 347–352.

[17] A. M. Abrao, J. C. Campos Rubio, P. E. Faria, J. P. Davim, (2008), "The effect of cutting tool geometry on thrust force and delamination when drilling glass fibre reinforced plastic composite," Materials and Design; Vol. 29, pp. 508–513.
[18] J. Mathew, N. Ramakrishnan, N. K. Naik, (1999), "Investigatio ns into the effect of geometry of a trepanning tool on thrust and torque during drilling of GFRP composites," Journal of Materials Processing Technology, Vol. 91, pp. 1-11.
[19] J. P. Davim, P. Reis, C. C. António, (2004), "Drilling fiber reinforced plastics (FRPs) manufactured by hand lay-up: influence of matrix (Viapal VUP 9731 and ATLAC 382-05)," Journal of Materials Processing Technology,
Vol. 155-156, pp. 1828-1833.
[20] J. Paulo Davim, P. Reis, C. António, (2004), "Experimental study of drilling glass fiber reinforced plastics (GFRP) manufactured by hand lay-up,"Composites Science and Technology, Vol. 64, pp. 289–297.
[21] N. S. Mohan, A. Ramachandra, S. M. Kulkarni, (2005), "Influence of process parameters on cutting force and torque during drilling of glass–fiber polyester reinforced composites," Composite Structures, Vol. 71, pp. 407–413.
[22] F. A. Al-Sulaiman, M. A. Baseer, A. K. Sheikh, (2005), "Use of electrical power for online monitoring of tool condition," Journal of Materials Processing Technology, Vol. 166. pp. 364–371.
[23] N. S. Mohan, S. M. Kulkarni, A. Ramachandra,(2007), "Delamination analysis in drilling process of glass fiber reinforced plastic (GFRP) composite materials," Journal of Materials Processing Technology, Vol. 186, pp. 265–271.
[24] R. A. Kishore, R. Tiwari, A. Dvivedi, I. Singh, (2009), "Taguchi analysis of the residual tensile strength after drilling in glass fibre reinforced epoxy composites," Materials and Design, Vol. 30, Issue 6, pp. 2186-2190.
[25] Ranjit K Roy, (2002), "Design of experiments using the Taguchi approach", John Wiley & Sons. Inc.

IOP Conf. Series: Materials Science and Engineering 1189 (2021) 012024 doi:10.1088/1757-899X/1189/1/012024

Synthesis of eggshell based hydroxyapatite using hydrothermal method

Shahid Hussain[1] **and Kazi Sabiruddin**[1,2]

[1]Discipline of Mechanical Engineering, Indian Institute of Technology Indore, Simrol, Indore 453552, India
[2]Email: skazi@iiti.ac.in

Abstract. Eggshell are rich sources of calcium carbonate and calcium oxide, but treated as bio-waste material. The natural origin and wide availability of eggshell are suitable to synthesize hydroxyapatite having chemical composition and crystalline structure similar to human teeth and bones. In present work, eggshell powders are sintered at different temperature 200°C, 400°C, 600°C, 800°C and 900°C for 1 h time duration in muffle furnace. The sintered powders are cold compacted at a fixed load of 15 tons for 10 min duration in a hydraulic pellet press. A suitable die set having 13 mm diameter is used to prepare compacted powder samples of 6 mm height. The sintered eggshell pellets are characterized using X-ray diffraction (XRD), Field Emission Scanning Electron Microscope (FESEM) and Vickers hardness tester. Chemical composition of eggshell pellet changes from calcium carbonate to calcium oxide at 900°C for 1 h time duration and noticed through XRD. Increased hardness of the sintered eggshell pellet is noticed at more than 600°C temperature. The waste eggshells can be utilized to synthesize hydroxyapatite with the help of ball milling and subsequently heat treatment. A straight forward and cost effective method (i.e. hydrothermal method) is proposed for converting eggshell into hydroxyapatite. In this method, a solution of calcium oxide powder obtained from eggshell, tri calcium phosphate powder and demineralized water in stoichiometric proportions is employed as the precursor. Now, this solution is heat treated at 1050°C for 3 h time duration in muffle furnace to complete the reaction. X-ray diffraction result confirms the formation of hydroxyapatite, apart from little amount of other compounds in final product.

Keywords: Eggshell; Hydroxyapatite; XRD; Micro hardness; FESEM

1. Introduction

Hydroxyapatite (HA) is a bioactive material means ability to produce bone apatite-like material on their surfaces. The chemical composition and structure of hydroxyapatite is similar to mineral constituent of natural teeth and human bones. They can be used in dental, orthopaedic, and maxillofacial applications as its bioactivity helps in bone osseointegration and ingrowth [1]. Various techniques have been developed to synthesize hydroxyapatite powders. The synthesis techniques can be divided into two methods: wet techniques and dry techniques. In wet techniques, the reaction is carried out in solution using various solvent at different temperature. On the other hand, no solvent is used in dry methods. The wet methods can be further categorized into hydrothermal, chemical precipitation, emulsion, sol-gel, sonochemical and hydrolysis. The dry techniques include mechanochemical method and solid state method. The more convenient and effective method is hydrothermal method to produce hydroxyapatite among various synthetic routes [2]. The

IOP Conf. Series: Materials Science and Engineering 1189 (2021) 012024 doi:10.1088/1757-899X/1189/1/012024

hydroxyapatite can be synthesized by different biogenic materials and synthetic calcium. The hydroxyapatite obtained through biogenic materials is different in response to that of synthetic calcium based hydroxyapatite. Different biogenic materials like eggshell, seashell, coral, and cuttlefish shell have successfully been focused to obtain hydroxyapatite. In biogenic based hydroxyapatite, the properties like pore structure and chemical composition are preserved as in precursor material [3]. In this work, a solution of calcium oxide powder obtained from eggshell, tri calcium phosphate powder and demineralized water in stoichiometric proportions is employed as the precursor to carry out chemical reaction at 1050°C for 3 h time duration to synthesize hydroxyapatite [4].

2. Materials and Methods

Waste eggshells were collected from Bhopal Caterer Mess, IIT Indore, Madhya Pradesh, India. Tap water was used to clean the collected eggshell and eggshell membranes were removed manually by peeling off. Further, eggshells were rinsed with demineralized water. Then, washed eggshell were dried and crushed into very small pieces by using hammer. The powder form of eggshells particle was obtained by using planetary ball milling process (Retsch PM 100). The ratio of grinding media to eggshell was 10:1 (mass ratio). Calcium oxide was synthesized from eggshell powder by two stage heat treatment process. In the first stage, eggshell powder was placed in muffle furnace to heat at 450°C for 2 hours. Any organic residues are expected to be destroyed at this temperature. In the second stage, the obtained powder was heat treated at 900°C for another 2 hours. By releasing carbon dioxide, the eggshell transform into calcium oxide at this temperature according to the following equation (1).

$$CaCO_3 \rightarrow CO_2 + CaO \qquad (1)$$

The obtained calcium oxide powder from eggshell and tri-calcium phosphate (manufacturer: Loba Make) was mixed homogeneously with demineralized water in an alumina crucible. The obtained calcium oxide powder of 0.72 g and tri-calcium phosphate of 11.62 g were mixed in 30 g demineralized water to carry out reaction according to equation (2). After mixing, this solution was heat-treated in muffle furnace (*BIO-TECHNICS INDIA*) and maintained at temperature 1050°C for time duration 3 h.

$$3Ca_3(PO_4)_2 + CaO + H_2O \rightarrow Ca_{10}(PO_4)_6(OH)_2 \qquad (2)$$

The powders is cold compacted at 15 tons and 10 min holding time in a hydraulic pellet press (25 tons, Kimaya engineering). A suitable die set having 13 mm diameter is used to prepare compacted powder samples of 6 mm height. The X-ray diffraction (XRD) analyses of the pellets were carried out using an X-ray diffractometer (Rigaku Smart Lab 3kW) with scanning speed 2°/min operating at 40 mA and 30 kV. Microstructural characterization of the pellets was conducted using a field emission scanning electron microscope (Zeiss Supra-55) under secondary electron mode. Surface hardness of the pellets were carried out using Vickers hardness tester at an indentation load of 100 g, dwell time 15 sec and indentation speed 25 micron/sec.

3. Results and discussion

The XRD patterns of the eggshell powder at room temperature, 200°C, 400°C and 600°C are shown in figure 1. It is found that in all four figures most of the characteristics diffraction peaks of eggshell powders at room temperature and different temperature are coinciding with calcite polymorphic phase of calcium carbonate. Additionally, these XRD patterns indicate that eggshell at room temperature and upto temperature 600°C are mainly composed of calcium carbonate ($CaCO_3$) with little amount of impurities like organic matter, magnesium carbonate and calcium phosphate [5].

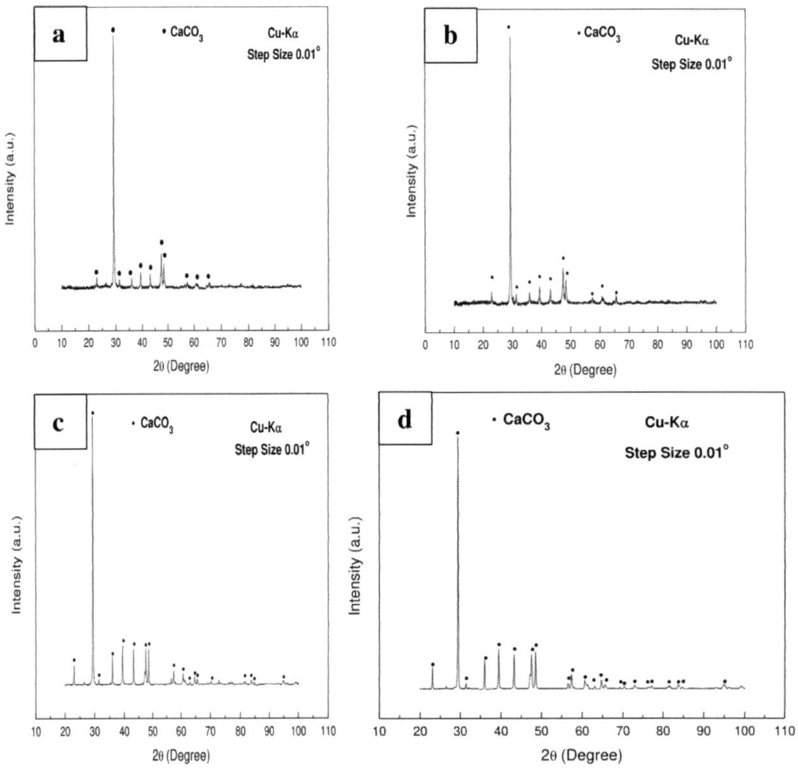

Figure 1. XRD pattern of eggshell at (a) room temperature, (b) 200°C, (c) 400°C, (d) 600°C.

When eggshell powder is heated at 800°C or above, calcium carbonate is converted into calcium oxide (CaO) by releasing of carbon dioxide as depicted in figure 2. Physical appearance of eggshell powder also changes as temperature increases from room temperature to 900°C. The color of eggshell powders at room temperature, 200°C, 400°C, 600°C and 900°C are white, light brown, brown, black and bright white respectively.

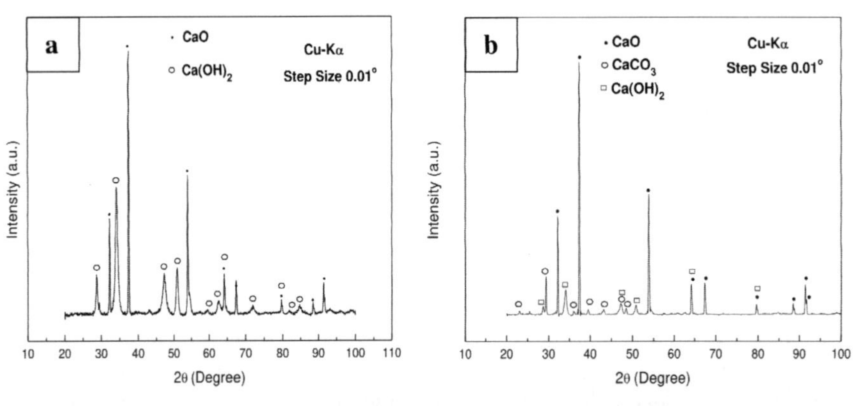

Figure 2. XRD pattern of eggshell at (a) 800°C, (b) 900°C temperature.

Micro images of the eggshell powder at room temperature, 200°C, 400°C, 600°C and 900°C are shown in figure 3. Furthermore, it is noticed that remarkable difference in surface morphology of eggshell powder with an increase in heating temperature from room temperature to 900°C. Microstructure of powder at 900°C temperature is quite different than that of remaining powders.

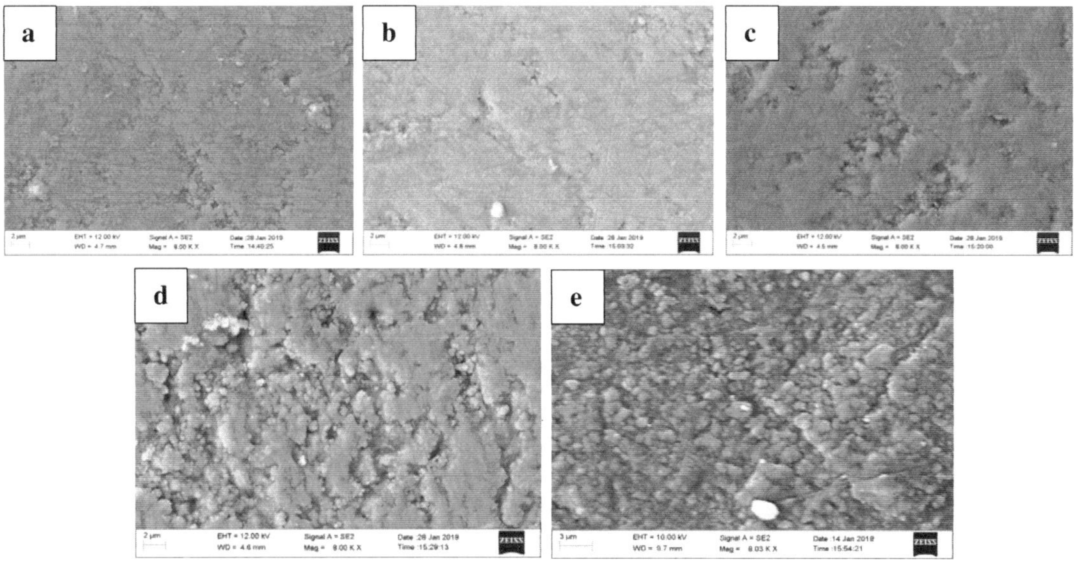

Figure 3. FESEM image of eggshell at (a) room temperature, (b) 200°C, (c) 400°C, (d) 600°C, (e) 900°C.

The micro hardness of room temperature eggshell powder and heat treated eggshell powder at different temperature calculated by Vickers hardness tester is shown in figure 4. The hardness of eggshell powder at room temperature, 200°C, 400°C and 600°C are almost same. But hardness of eggshell powder at 900°C is almost double than that of remaining powders. This increment in hardness value may be due the transformation of calcium carbonate into calcium oxide. Figure 5 shows X-ray diffraction pattern of final product which confirms the peaks of hydroxyapatite mainly and few peaks are also belong to the other compound which is minute in amount.

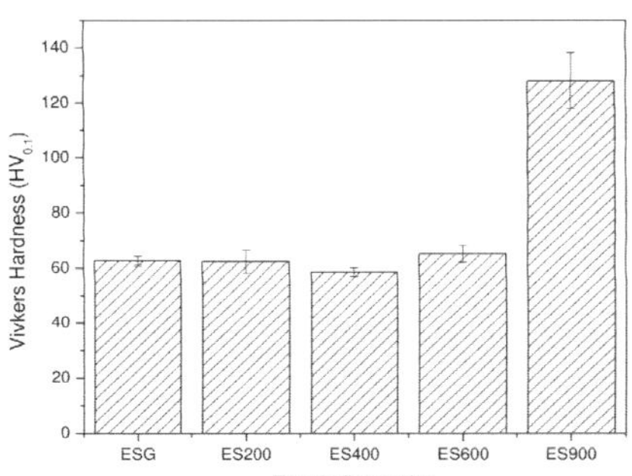

Figure 4. Hardness of eggshell at different temperature.

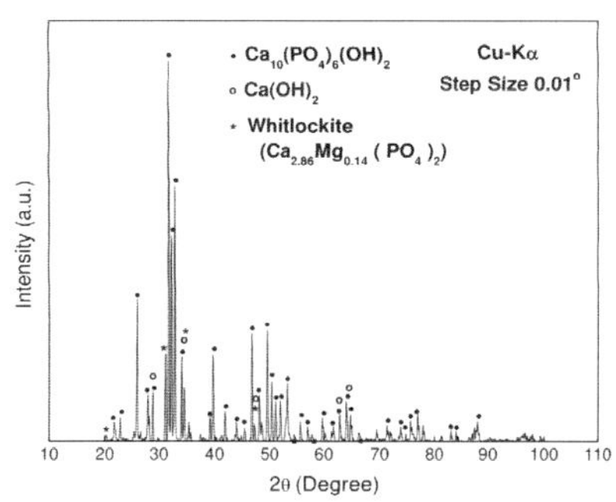

Figure 5. XRD pattern of eggshell based hydroxyapatite.

IOP Conf. Series: Materials Science and Engineering 1189 (2021) 012024 doi:10.1088/1757-899X/1189/1/012024

Conclusions

a) With an increase in temperature from room temperature to 600°C, eggshell powder retaining its chemical composition i.e. calcium carbonate. But, further increase in temperature results in mainly calcium oxide.
b) The hardness value of eggshell powder is almost same, with an increase in temperature from room temperature to 600°C. With further increase in temperature leads to transformation of calcium carbonate into calcium oxide which in turn result in higher hardness value.
c) Hydroxyapatite powder has been synthesized successfully using eggshell powder.

References

[1] Dorozhkin S V 2010 *Biomaterials* **31** 1465– 85.
[2] Rhee S H 2002 *Biomaterials* **23** 1147–52.
[3] Akram M, Ahmed R, Shakir I, Ibrahim W A W and Hussain R 2014 *J. Mater. Sci.* **49** 1461–75
[4] Rivera E M, Araiza M, Brostow W, Castano V M, Dıaz-Estrada J R, Hernandez R and Rodrıguez J R 1999 *Mat. Lett.* **41** 128–34.
[5] Balaz Matej 2018 *J. Colloid Interface Sci.* **256** 256-75.

IOP Conf. Series: Materials Science and Engineering 1189 (2021) 012025 doi:10.1088/1757-899X/1189/1/012025

Experimental Investigation of effect of hybridization of UHMWPE fibers on the mechanical and physical properties of Basalt fiber reinforced polymer matrix composites

R Raghavendra Rao*, S Pradeep, S Karthik, K R Gopi, B S Shivashankara

Department of Mechanical Engineering, Malnad College of Engineering, Hassan – 573202, VTU, Belagavi, Karnataka, India

*Corresponding author e-mail address: rrr@mcehassan.ac.in

Abstract. Polymer matrix composite materials have attracted the researchers over a period due to its superior specific strength compared to conventional materials. Bio based fibers started to emerge rapidly for industrial components, encompassing the technological/scientific aspects and the economic, environmental and social issues. Amongst the variety of bio-based fibers, Basalt fiber occupies a prominent position, due to its overall production and its unique chemistry-related features. It is derived from volcano rocks. UHMWPE having higher specific strength, non-reactive, chemically inert, and hence finds its applications in biomedical field. In the present investigation, UHMWPE fiber (plain woven fabric) along with the basalt fiber (plain woven fabric) are added as reinforcement to LY556 epoxy to bring in hybrid characteristics to the material and comparative studies are conducted with non-hybrid composites. Fabrication is carried out by hand lay-up technique. Mechanical tests such as hardness, tensile and flexural tests are conducted for the fabricated material. Density and water absorption behavior of the composites are also studied. Conventional hand layup method is used for fabrication of composites. From the results obtained, it is obvious that the flexural strength of hybrid composite is enhanced due to the hybridization with the addition of UHMWPE fabrics on the outermost layers in BFRP composites and augments its suitability in the use of engineering structural applications.

Keywords: UHMWPE, Basalt, Hybridization; Mechanical; Physical properties

1. Introduction

The polymer composites have found its wide applications due to its higher strength to weight ratio, corrosion resistance, lighter weight, applicability, ease of getting shaped and suitability for various engineering applications. Even though the usage of polymer composites has started in the earlier 20_{th} century but the wide use of it have started in the second half of 20th century.

Basalt is a non-polluting material that originate in volcanic rocks from frozen lava, having melting temperature between 1500 ºC and 1700 ºC [1,2]. The density of Basalt fibers ranges between 2.7 g/cm^3 to 2.8 g/cm^3. It is extremely hard and thus has superior abrasion resistance [3]. It is suggested that, the Basalt fiber could be a possible alternative to natural plant fibers [4]. Czigany proved that the Basalt could be considered as natural, because it is formed by the solidification of molten lava, which is bio inert, & having higher mechanical properties. Tensile strength of basalt fiber increased as the drawing temperature is incremented with increased proportions of crystal nuclei of basalt at lower temperatures [5]. Botev et al. [6] reported the effect of basalt fiber that reinforced on commercial grade polypropylene (PP) composites with dynamic mechanical thermal analysis. The result showed that, the storage modulus (E') and also loss modulus (E'') of the produced composite has increased with increased content of basalt fiber. However, the damping value within the region decreased with increase in fiber loading. Fiore et al. [7] have evaluated the influence of uniaxial basalt fabric on the mechanical properties of a glass matrix/epoxy composite that are used for marine applications produced by vacuum bagging technique. They administered three-point bending and tensile tests to gauge the outcome of number of basalt layers and its position on the mechanical behavior of the fabricated composite. The outcome showed the presence of two external layers of basalt resulted in increased mechanical properties of produced laminates in comparison to GFRP laminates.

Ultra-high molecular weight polyethylene (UHMWPE) was first polymerized within the 1950's and the primary UHMWPE fibers were commercialized during 1970. UHMWPE is a Polyolefin composed of long chains of Polyethylene with elevated percentage of parallel orientation and high level of crystallinity. The long polymer chains empower load transfer by strengthening intermolecular interaction.

During early days, UHMWPE fibers were synthesized directly from stirred solutions, possessing a tensile strength of 2.9 GPa and Young's modulus of 101 GPa [8]. UHMWPE fiber have possessed major applications in Ballistic protection, sports, marine and offshore cut-resistant textile, fishing lines, marine cordage, lifting slings, cut resistant gloves, apparel and medical appliances. The tribological characteristics of UHMWPE were extensively investigated because of its exceptional properties that are suitable for tribological studies. UHMWPE possess high chemical stability and compatibility with human tissue, so that the friction and wear behavior of UHMWPE was studied [9]. UHMWPE could be used as substitute for steel , chrome steel and bronze, due to its better anti-chemical behavior, anti-adhesion, high impact resistance, water repellent function, self-lubrication and corrosion resistance [10].

Cao et al. [11] have examined the impact of basalt fiber (BF) in UHMWPE matrix. They showed that, the increased BF content ensured the reduction in toughness and improved strength, creep resistance and hardness properties of the composite material. Raghavendra Rao R et al. [12] have studied the effect of hybridization of UHMWPE & Basalt fibers using L12 epoxy with K6 hardener combination as matrix material. They have concluded that the inclusion of UHMWPE at the outer most layers has increased the flexural strength considerably, whereas the tensile strength reduced minimally. From the existing literature it is found that, not much research articles have reported the effect of addition of UHMWPE fibers on the Basalt fiber-reinforced epoxy composites on their mechanical and physical properties using LY556 epoxy as a matrix.

2. Experimental details

2.1 Materials

Materials used in the present investigation are Basalt plain-woven 2D fabric of 380 GSM; Ultra-high molecular weight polyethylene (UHMWPE) plain woven 2D fabric of 240 GSM is used as reinforcement material. Details of fabrics are illustrated in table 1. Araldite LY556 and HY951 are used as epoxy resin and hardener respectively, both with density of 1.2 g/cc.

Table 1 Represents the basic parameters of Basalt and UHMWPE fabrics

Parameters	Basalt fabrics	UHMWPE fabrics
Density (g/cc)	2.67	0.97
Weight (gsm)	380	240
Thickness (mm)	0.25	0.48
Warp yarns (yarns/m)	600	1000
Weft yarns (yarns/m)	600	1000

2.2 Fabrication of composites

Hand layup method was used in the fabrication of hybrid/non hybrid laminates. The fabrics were prepared to the dimensions of 300 × 300 mm^2 for the fabrication of composite laminates using Aramid cutter. The laminates included seven to eight layers of fabric to maintain approximate thickness of 3 mm for various proportions indicated in table 2. The epoxy resin and hardener were mixed in the ratio of 100:10 by weight. First, wax was applied over the top of a mould and the fibre layers were stacked one after another in the mould and the measured quantity of epoxy was applied on each layer of the fibres. Then the epoxy was uniformly spread using roller. As soon as the last layer of fibre was properly rolled , the granite slab of required size was kept on top of the mould. On the top of the granite, dead weights are placed and sustained in that spot for 24 hours and then withdrawn. The laminates were finally removed from the mould after a period of 24 hours. Then they were cut using water jet machining as per the ASTM standards to conduct several tests.

Weight fraction (%) is calculated by using the Equation (1)[16]

$$W_f = w_f / (w_f + w_m) \text{ and } W_m = w_m / (w_f + w_m) \text{ -------------------------(1)}$$

where W_f is weight fraction of fibre; w_f is weight of the fibre; w_m is weight of the matrix; W_m is weight fraction of matrix.

Table 2. Laminate stacking arrangements.

Laminates	Stacking sequence	Basalt fabric weight fraction in %	UHMWPE fabric weight fraction in %	Matrix Weight Fraction %
L1	B-B-B-B-B-B-B-B	60	-	40
L2	U-B-B-B-B-B-B-U	50	10	40
L3	U-B-B-B-B-B-U	45	15	40
L4	U-B-B-U-B-B-U	40	20	40

3. MECHANICAL CHARACTERIZATION

3.1 Density test

The density is the measure of mass of a material per unit volume. The mass is weighed by weighing machine for a sample of 10 mm x 10 mm and volume is found by Archimedes principle. A 100 ml beaker is taken, and it is filled with distilled water to half a mark. Then the lower meniscus level is noted. Then the samples are dipped one by one to find the final level of water. Final level minus initial level gives the volume of the sample. Then the density of a material is calculated using the formula:

$$\rho = m/V \text{ -------------------------(2)}$$

where, m is the mass; V is the volume.

3.2 Water absorption test

Water absorption test is conducted for 24 hours and 48 hours as per ASTM D570 standard. The sample of 10 x10x3mm^3 is taken and its initial weight is found. Then it is soaked in distilled water for 24 hours. After 24 hours it is withdrawn, and its final weight is found. The same procedure is repeated by dipping the samples for 48 hours.

3.3 Tensile test

Tensile test was carried out on BISS-50KN Universal testing machine (UTM) with data acquisition software. The sample (115 × 19 × 3 mm^3), with gage length of 33 mm and crosshead speed of 1.0 mm/min were selected as per ASTM: D638-IV standard [14] and the test was conducted at a room temperature.

3.4 Flexural test

3- Point bending test is conducted as per ASTM: D790 standard using the same UTM. Testing was conducted with load cell of 10 kN at loading rate of 2.0 mm/min, at room temperature. The dimension of the specimen is 90 x12.5 x 3 mm^3 and the flexural specimens were fixed between two jaws with span length of 60 mm and the load was applied at the center. The specimen is subjected to loading until the gage length increased and the specimen got ruptured. `

3.5 Shore 'D' Hardness test

Shore 'D' hardness test is carried out using the Shore durometer as per ASTM: D2240 standard. Hardness is determined by the penetration of Durometer indenter into the specimen. During the test, the sample was kept on a flat, hard, horizontal surface and held the durometer between both hands over the sample so that the indenter touches it. Then it is pushed down perpendicularly until the presser foot makes firm contact with the sample. Then the reading is taken.

3.6 Surface morphology

Fractured specimens were examined for microstructure using CIRC 20kV scanning electron microscope (SEM). SEM was carried out to examine the dispersion of fibers within the matrix and the adhesion characteristics amongst Basalt and UHMWPE fibers and matrix is analyzed. In order to enhance the conductivity of the samples, the surface of the sample was coated with a thin gold film and the micrographs were captured from the fractured samples.

4. Results and Discussion

4.1 Density test

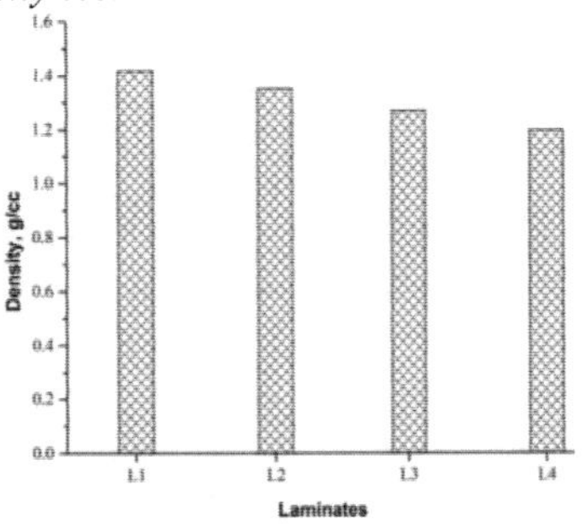

Figure1. Density plot

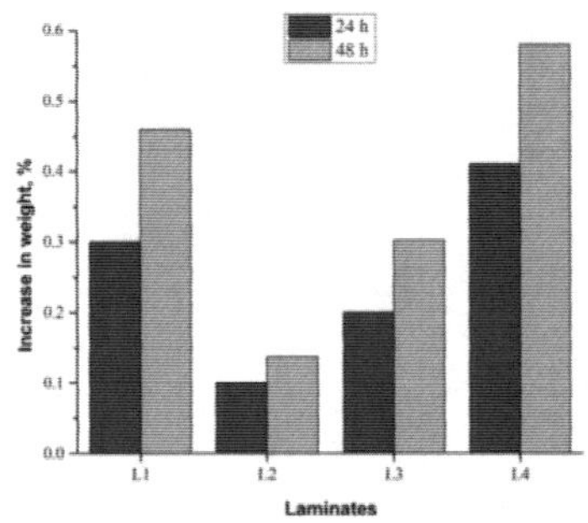

Figure 2. Water absorption comparative plot

The experimental densities of the prepared composite laminates are shown in figure1. As observed in figure, in all the laminate reinforcement percentages, a slight difference in experimental densities were found due to the varying percentages of the two different reinforcement materials used. The non-hybrid laminate L1 exhibited highest density and the L4 laminate exhibited the least density.

4.2 Water absorption test

The water absorption percentages of laminates are shown in figure 2. It is observed from figure that, laminate L4 exhibited highest water absorption percentage & the laminate L2 exhibited least water absorption percentage for both 24 and 48 hours. It is observed that more water absorption is seen in first 24 hours and next 24 hours less rate of water absorption is seen.

4.3 Tensile property evaluation

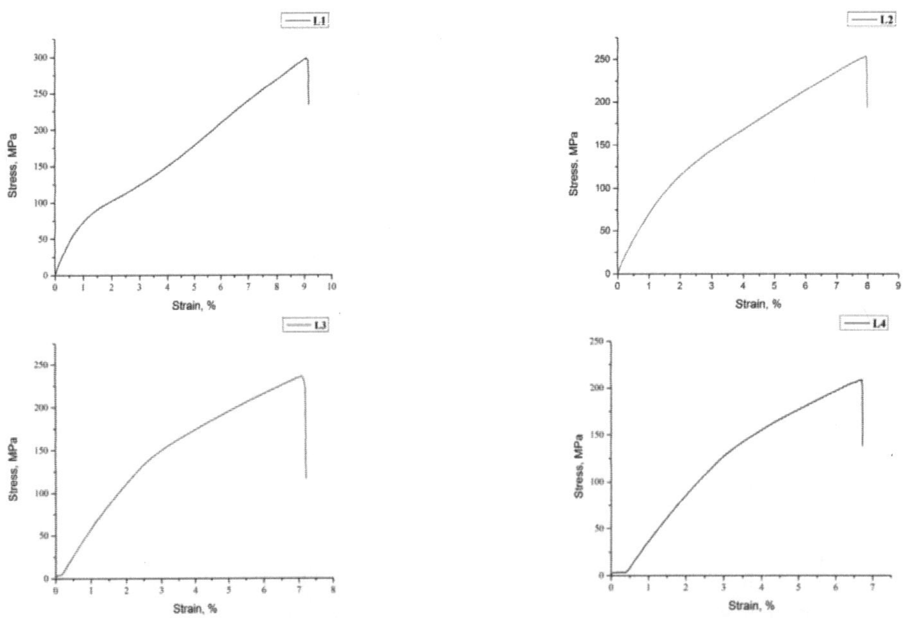

Figure 3. Tensile Strength of laminates

The tensile strength and modulus are shown in the figure 3 and figure 5 respectively. It is seen from the figure that the laminate L1 exhibited highest tensile strength and modulus and laminate L4 exhibited lowest tensile strength & modulus. It is obvious that the addition of UHMWPE fabrics onto the Basalt fiber reinforced composites has substantially reduced the tensile strength and modulus.

4.4 Flexural property evaluation:

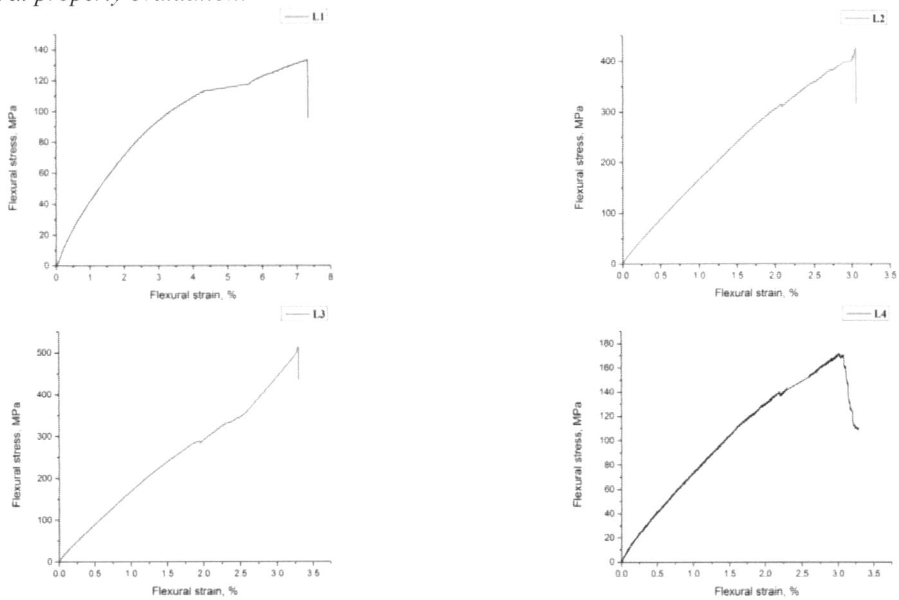

Figure 4. Flexural Strength of laminates

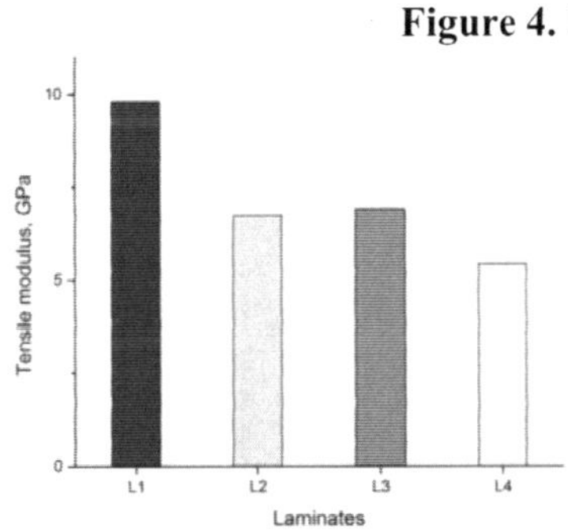

Figure 5. Tensile Modulus of laminates

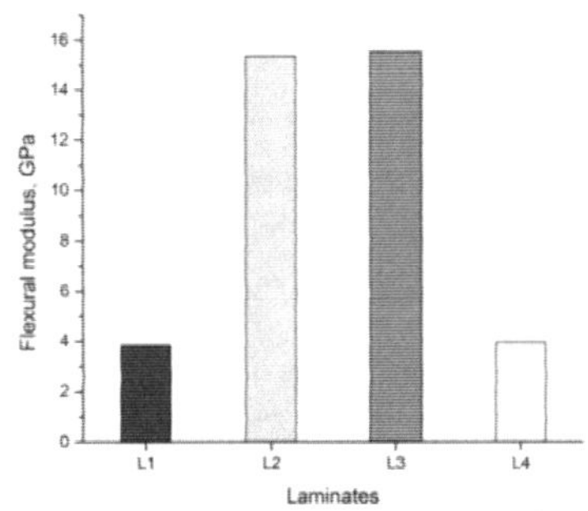

Figure 6. Flexural Modulus of laminates

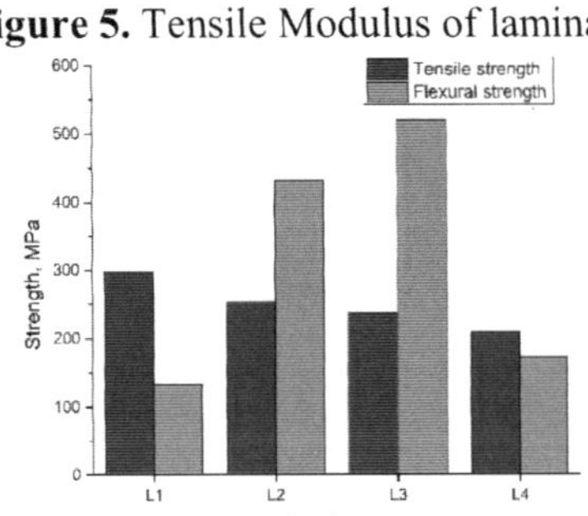

Figure 7. Comparison between tensile and flexural strengths of different laminates

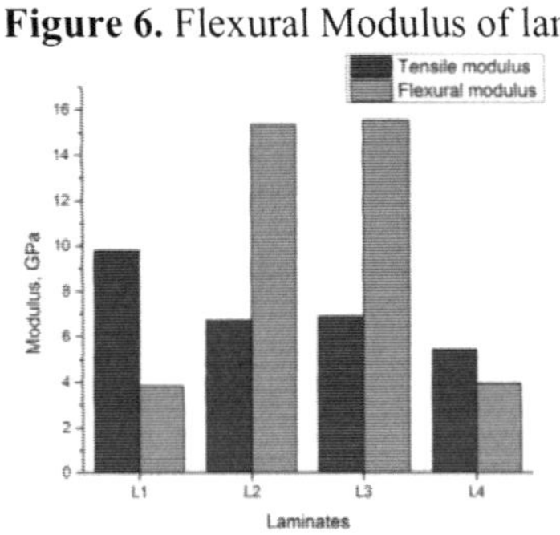

Figure 8. Comparison between tensile and flexural modulus of different laminates

The flexural strength and modulus are shown in the figure 4 and figure 6 respectively. It is seen from that the laminate L3 exhibited highest flexural strength and modulus and laminate L1 exhibited lowest flexural strength and modulus. It is obvious that the addition of UHMWPE fabrics onto outermost layers has increased the flexural strength and modulus of Basalt fiber reinforced composites substantially. The comparison of tensile and flexural strength and modulus are shown in figure 7 and figure 8.

4.5 Hardness test

The Shore D hardness numbers of different laminates are plotted in figure 9. The L1 laminate exhibited the highest hardness of 78 SHN. The laminate L4 exhibited lowest hardness of 63 SHN. Similar trend was observed in tensile strength and appeared interrelated with each other.

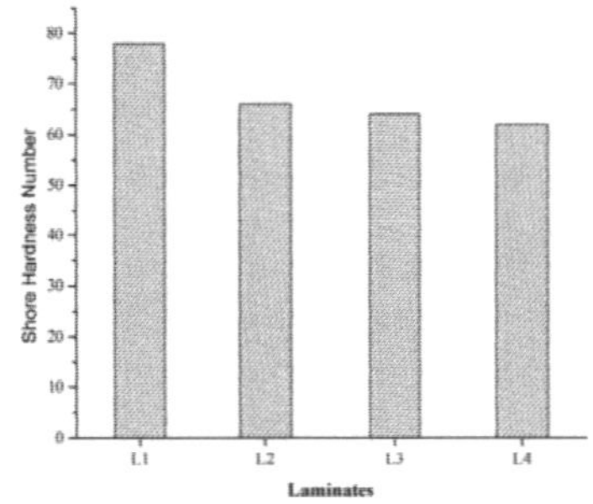

Figure 9. Hardness plot of laminates

4.6 Morphological studies through SEM

Microstructural studies were conducted to visualize the failure surfaces of the fabricated laminates subjected to tensile loading. The void content, fiber matrix adhesion, and pullout properties are analyzed using the SEM images. All the composite specimens were coated with gold before observing them through SEM. The fractured micrographs of various laminate designations L1, L2, L3 and L4 are presented in Figure 10a-d, respectively. It showed fewer amounts of voids in the fractured samples.

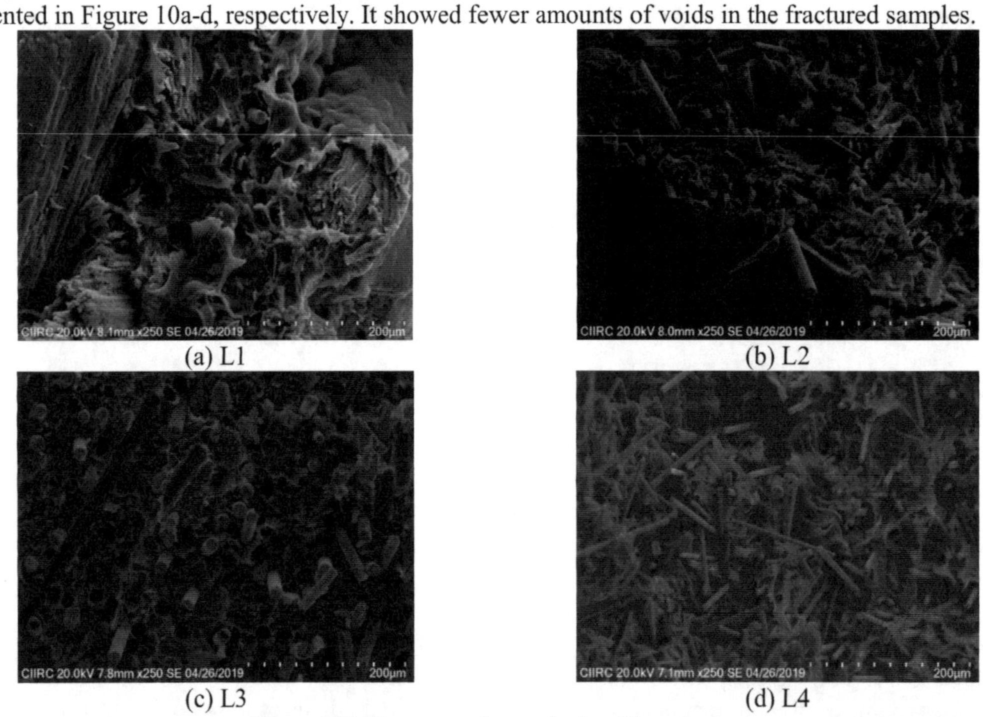

Figure 10. Fractography analysis of Laminates

The detachment of fiber laminates resulted in rupture of fibers and matrix and resulted in phenomenon called fiber pullout. Fiber rupture indicates moderate interfacial bonding between fibers and the matrix.

Above discussion reveled that, lesser voids resulted in better laminate properties proper rupture of fibers and improved adhesion amongst the fibers and matrix.

5. Conclusions

A novel hybrid composites comprising of Basalt & UHMWPE have been developed using hand layup method. The mechanical investigations such as tensile, hardness & flexural tests, physical properties such as density and water absorption behaviour of hybrid/non hybrid composites with four different fibre percentages have been deliberated. These are the conclusions drawn from the research work carried out:

1. UHMWPE fabrics on the outermost layers have resulted in enhancement of flexural properties, whereas the considerable reduction in tensile properties is observed.
2. The hybrid composite laminate L3 indicated highest flexural strength and modulus of 523.6 MPa and 15.4 GPa and laminate L1 indicated lowest flexural strength of 133.488 MPa and modulus of 3.849 GPa.
3. The non-hybrid composite laminate L1 shows highest tensile strength and modulus of 298.36 MPa and 9.80 GPa and L4 laminate shown the least tensile strength and modulus of 208.8 MPa and 5.439 GPa.
4. L1 non hybrid composite laminate shown highest hardness value ranging 77 SHN and laminate L4 shown the least hardness value of 63 SHN.
5. The non-hybrid L1 laminate exhibited highest density of 1.42 g/cc and the L4 laminate exhibited the least density of 1.2 g/cc.
6. Laminate L4 exhibited highest water absorption percentage of 0.44% & 0.56% & the laminate L2 exhibited least water absorption percentage of 0.1% & 0.15% for 24 and 48 hours respectively.
7. SEM analysis confirmed that, the hybrid composite laminates showed lesser voids, better adhesion between fibres and matrix, appropriate fibre breakage and pull-out.

References

[1] J. Militký, V. R. Kovačič, and J. Rubnerova, "*Influence of thermal treatment on tensile failure of basalt fibers*," J. Engineering fracture mechanics, vol. **69**, pp. 1025-1033, 2002.
[2] J. Militky and V. Kovacic, "*Ultimate Mechanical Properties of Basalt Filaments*," J. Textile Research , vol. **66**, pp. 225-229, 1996/04/01 1996.
[3] M. Di Ludovico, A. Prota, and G. Manfredi, "*Structural upgrade using basalt fibers for concrete confinement," J.* composites for construction, vol. **14**, pp. 541-552, 2010.
[4] T. Tabi and J. Kovacs, *"Examination of starch pre-process drying and water absorption on injection moulded poly (lactic acid)/starch blends,"* Polym Eng Sci, vol. **51**, pp. 843-850, 2011.
[5] T. Czigány, "*Basalt fiber reinforced hybrid polymer composites*," J. Materials Science Forum, 2005, pp. 59-66.
[6] M. Botev, H. Betchev, D. Bikiaris, and C. Panayiotou, "*Mechanical properties and viscoelastic behavior of basalt fiber reinforced polypropylene,*" J. Applied Polymer Science, vol. **74**, pp. 523-531, 1999.
[7] V. Fiore, G. Di Bella, and A. Valenza, "*Glass–basalt/epoxy hybrid composites for marine applications*," J. Materials & Design, vol. **32**, pp. 2091-99, 2011.
[8] L. Y. Mwaikambo and M. P. Ansell, "*Chemical modification of hemp, sisal, jute, and kapok fibers by alkalization,"* J. applied polymer science, vol. **84**, pp. 2222-34, 2002.
[9] A. Wang, A. Essner, V. Polineni, C. Stark, and J. Dumbleton, "*Lubrication and wear of ultra-high molecular weight polyethylene in total joint replacements," J. Tribology International,* vol. **31**, pp. 17-33, 1998.
[10] A. Shi and Y. Gong, "*Engineering Plastic: Properties*," *Processing and Applications, Shanghai Science and Technology Press, Shanghai,* 1986.

[11] S. Cao, H. Liu, S. Ge, and G. Wu, "*Mechanical and tribological behaviors of UHMWPE composites filled with basalt fibers,*" J. reinforced plastics and composites, vol. **30**, pp. 347-355, 2011.

[12] Raghavendra Rao R, S. Pradeep, C.K. Yogish, Nikhil R., Geetanjali Patil., "*Studies on Mechanical Properties of Ultra High Molecular Weight Poly Ethylene (UHMWPE)-Basalt Fiber Reinforced Hybrid Polymer Matrix Composite,"*. J. Polymer & Composites. 2019; **7**(3): pp. 8–19

[13] R. M. Jones, *Mechanics of Composite Materials*, Taylor and Francis, 1999.

[14] Standard ASTM. "*Standard test method for tensile properties of polymer matrix composite materials*." ASTM D3039/D M, 039,2008.

Study on Abrasion Wear behaviour of Hybrid Composites using dimensional Analysis and Wear Models

Rudresh B M[1*], Ravikumar B N[2], Madhu D[3]

[1]Department of Mechanical Engineering, Government Engineering College, K R Pet – 571426, Karnataka, India
[2]Department of Mechanical Engineering, Bangalore Institute of Technology, Bangalore -560004, Karnataka, India
[3]Department of Mechanical Engineering, Bangalore Institute of Technology, Bangalore -560004, Karnataka, India

★ **Corresponding author:** Rudresh B M, E-mail: bmrudreshan@gmail.com

Abstract. The prediction of abrasive wear behavior of Glass-Basalt (GB) and Glass –Carbon (GC) hybrid composites in two body wear mode using dimensional analysis has been presented. The blend (PA66/PTFE) in 80/20 wt. % proportion was used as the base material. The individual effect of glass fibers and carbon fibers (10 wt. % each) were also studied and compared. Melt mix method using twin screw extrusion and injection molding process were used for developing hybrid composites. The mathematical model to predict the wear loss has been developed using dimensional analysis for the aforesaid hybrid composites and evaluated through justification using experimentally determined wear loss. Further, the wear behavior due to two body abrasion of hybrid composites is justified using abrasive wear models. It is noticed that developed mathematical model and wear models agrees with the abrasion wear behavior of composites studied. It is noticed that Ratner –Lancaster factor ($\sigma\varepsilon$) was used as one variable for the abrasion wear model. It is proved that specific wear rate is indirectly proportional to the wear resistance. Furthermore, the individual effect of short fibers is compared with GB and GC hybrid composites. It is proved that hybrid composites exhibited the better wear resistance.

Keywords: Dimensional analysis, Abrasion, two body, wear models, glass-Basalt, Hybrid composite

1. Introduction

The abrasion wear behavior of polymer composites must be superior in order to recommend them to the field of structural applications particularly to automobile sectors. There may be two body (2BAW) or three body (3BAW) abrasion effects when the polymer has put in to performance [1]. It is documented from the literature and research that 60% of the total failure of mechanical components was due to 2BAW effect [2]. This can be avoided through polymer modification which led the polymer to acquire resistance against these abrasions. They are modified through blending, adding fillers and fibers and also by copolymerization [3]. The advancement in the development of potential fibers namely glass fibers, carbon fibers, basalt fibers and many more has thrown the light on the development of hybrid composites. Good modulus glass fiber and thermally stable basalt fibers have emerged as the effective combination for hybridization of fibers both for wear and strength [4]. Further, the development of thermoplastics blends such as PA66/PPS, PA66/POM, PA66/PEEK and such related associates has promised to be the best materials system to resist the wear of polymers. The dimensional analysis is one of the mathematical techniques used to establish the equation for a dependent variable using independent variables. It is a method of dimensions. It is used to formulate the mathematical model for a particular research problem using different process variables involved in the process. Srinath and Gnanamoorthy studied the 2BAW characteristics of Nylon clay Nano composites. They reported the impact of grit, load and velocity. They developed the mathematical abrasive wear model using the process parameters involved in the abrasion mechanisms by dimensional analysis. They evaluated the same model using experimentally determined wear values [5]. The effect of fillers and fiber reinforcement on 2BAW performance of polymers is reported by Bijwe et al. [6]. They showed the synergism between fillers and fibers on 2BAW characteristics of thermoplastic composites. They reported the effect of carbon filler, graphite, PTFE and MoS_2 filler on SGF reinforced Polyetherimide (PEI), Polyimide (PI), Nylon 6 and PTFE composites. The results showed that in most of the cases, specific wear rate declined with rise in load. They showed that the reinforcement significantly reduced the ultimate elongation which is the crucial factor in controlling 2BAW of composites. Different wear models were developed to study the abrasive wear behavior of polymer composites. Ratner et al [7] has studied the 2BAW behavior of reinforced polymers. They showed from their model that wear rate is inversely proportional to the product of strain and stress at rupture '$\sigma\varepsilon$' factor. The wear volume of polymer composites varies directly as the product of load and distance which is proposed by Yamaguchi [8]. Horst Czichos [9] proposed the model on abrasive wear volume of polymers stating that the wear volume varied inversely as its rupture stress. Sahin studied the 2BAW characteristics of fiber reinforced PTFE composites through Taguchi technique [10]. He reported that SCF filled PTFE exhibited the better wear resistance and the load has proved to be the most significant factor. Harsha and Tewari [11] studied the abrasion wear characteristics of Polyaryletherketone composites. They showed that specific wear rate is directly proportional to wear volume loss. Further, they showed that the wear rate of composites is a function of $[\sigma]^{-1}$, $[\sigma\varepsilon]^{-1}$ and $[H\sigma\varepsilon]^{-1}$. In the research of polymer composites, the mathematical modeling is must using dimensional analysis for different types of composites. It always gives the effect of one variable on the other in the process.

From the above literature survey, many researchers have showed the relation existing between variables in the wear process. But developing a model for a particular composites using dimensional analysis and evaluating the same using experimented values is not reported. Keeping all these in mind, an attempt has been made to develop a model which incorporates maximum variables affecting the dependent variable (output variable). For this, hybrid composite such as GB and GC composites were used and the response of these composites was compared with the individual fibers to justify the results using wear models.

2. Materials and formulations

The material used for the study and some of their physical data is tabulated in table 1. All the data's are related to room temperature. The formulations of hybrid composites are recorded in the table 2

Table 1. Materials data used for the study

Materials	Mat. ID	Melting point (^{0}C)	Density (g/cc)
Polyamide66	PA66	279	1.75
Polytetrafluroethylene	PTFE	325	2.81
Glass fibers	SGF	---	2450
Carbon fibers	SCF	----	750
Basalt Fibers	SBF	----	2450

Table 2. Materials formulations in weight fraction percentage

Weight fraction Percentage						
Particulars	**Mat.ID**	**PA 66**	**PTFE**	**SGF**	**SBF**	**SCF**
Blend(PA66/ PTFE)	Blend	80	20	---	---	--
Blend/SGF	SGF	80	20	10		
Blend/SBF	SBF	80	20	---	10	---
Blend/SCF	SCF	80	20	----	----	10
Blend/SGF/SBF	GB	80	20	10	10	---
Blend/SGF/SCF	GC	80	20	10	---	10

2.1. Processing and Testing of Composites

The polymer materials for processing were dried in an oven at suitable temperature at about 50 ^{0}C to free them from moisture and plastic effects. The twin screw extruder with five heating zones was used and the composites were processed using melt mix method followed by injection moulding [12]. The processing of composites has been carried out across the five heating zones in the extruder chamber where the temperature of 200, 230, 240, 250 and 255 ^{0}C respectively have maintained [13]. The two body abrasion wear behaviour of hybrid composite was carried out using ASTM G99 method. The abrasion effect was initiated using 180 grit SiC paper at a load of 10 N through 0.25, 0.5, 0.75 and 1 m/s abrading velocity. The interaction span of 100 m has been maintained to study the abrasives effect. The wear volume loss through weight loss has been determined using experimentally determined density [14]. The specific wear rate (Ks) of fibrous composites is determined using the following formula (1):

If D = Abrading distance in m, L = Applied load in N and δV = Wear volume loss in mm^3,

Specific wear rate= $Ks = \delta V / [L*D]$ mm^3/N-m -- (1).

The pin on disc setup of Ducom Bangalore was used as the experimental set up for 2 BAW test and the procedure followed is as per ASTM G99 [14].The experimentation was carried out for a applied load of 10 N, 180 Grit, 100 m abrading distance under the abrading velocity of 0.25, 0.5, 0.75 and 1 m/s in multipass condition. The wear volume loss through multipass has been recorded and the specific wear rate for the applied condition was calculated.

3. Results and Discussion

The development of mathematical model using dimensional analysis has been designed and the same is justified using the experimentally responded values. The table 3 documents the parameters which could effectively control the abrasive wear behaviour.

Table 3. Factors controlling 2BAW behaviour of Hybrid composites

Controlling factors		
Particulars	**GB**	**GC**
Ultimate tensile strength (σ)	88.5	97.22
% elongation (ε)	14	12.5
Factor (σε)	12.39	12.15
Hardness (H)	75	63
1/ σ ε	0.0807	0.0822
1/H σ ε	1.76 x 10^{-3}	1.32x 10^{-3}

3.1 Abrasive Wear Modelling using the Concept of Dimensional Analysis

The Buckingham – Pi (π) theorem is used to establish the relation between variables involved in abrasive wear phenomenon. When the variables involved in the problem are more than four, Buckingham pi theorem method is used for the analysis. This theorem can be applied effectively for the tribological process. In this discussion, 2BAW behaviour of polymer composites was considered. The Wear volume loss (W) of polymer composites due to abrasion depends on abrading load in N (L), abrading velocity in m/s (V), abrading distance in m (D), Ratner- Lancaster factor in mm^2/N (R) (1/σε) and density of composites (ρ) in kg /mm^3.
Therefore, W is a function of L, D, V, R, ρ.
Wear volume is a dependent variable and L, D, V, R and ρ are independent variables. Therefore, W is a function of (L, D, V, R, ρ). From Buckingham- pi theorem method
Therefore, f (W, L, V, D, R, ρ) = 0
Number of fundamental dimensions (Length, Mass and Time) = m = 3
Number of variables in the problem = n = 6

Table 4. Dimensions of variables used in the problem

Variable	Symbol	Units	Dimensions
Volume	W	mm^3	[L^3]
Distance	D	m	[L]
Velocity	V	m/s	[LT^{-1}]
Density	ρ	Kg/mm^3	[ML^{-3}]
Ratner –Lnacaster factor	R	mm^2/N	[$M^{-1}LT^{-2}$]

According to Buckingham Pi- theorem method,
Number of π terms = n - m = 3
Therefore, $f(\pi_1, \pi_2, \pi_3) = 0$ --(2)
According to Buckingham pi theorem,
Number of repeating variables = m = 3
Repeating variables = Geometry variable, flow parameters and fluid property = D,V, ρ
Therefore, $\pi_1 = D^a V^b \rho^c W$, $\pi_2 = D^a V^b \rho^c L$, $\pi_3 = D^a V^b \rho^c R$

Writing the dimensions of each variable involved in the problem.

$\pi_1 = [M^0L^0T^0] = D^a V^b \rho^c W = [L]^a [LT^{-1}]^b [ML-3]^c [L^3]$,

Equating the powers of L M and T on either side of the equation, the term π_1 is obtained.

$\therefore\ \pi_1 = \frac{W}{D^3}$, the arbitrary constants a, b and c are determined using principle of dimensional homogeneity. From the principle of Buckingham pi theorem method, following dimensionless Pi terms are obtained.

$\pi_1 = \frac{W}{D^3}$, $\pi_2 = \frac{L}{\rho V^2 D^2}$ and $\pi_3 = R \rho V^2$

$$\frac{W}{D^3} = f\left[\frac{L}{\rho V^2 D^2}, R \rho V^2\right] \quad \text{--- (3)}$$

Using the concept of dimensional analysis, the following formulations can be written depending on the dependent variable available in the dimensionless Pi term.

$\therefore\ \pi_1 = f(\pi_2, \pi_3)$, $\pi_2 = f(\pi_1, \pi_3)$ $\pi_3 = f(\pi_1, \pi_2)$

$$\text{Hence, } \frac{W}{D^3} = K\left(\frac{L}{\rho V^2 D^2}\right)^a, \left(R\rho V^2\right)^b \quad \text{-- (4)}$$

$$\text{Therefore, } W = K\, L^a\, D^{3\text{-}2a}\, V^{2\,(b\text{-}a)}\, R^b\, \rho^{(b\text{-}a)} \quad \text{-- (5)}$$

Where K is the wear constant. The value of a, b and K are evaluated from experimentation data obtained and are correlated. As the density of composites tested are nearly constant, it is merged with the value of K. Therefore, the final mathematical model obtained for the wear volume loss is:

$$W = K\, L^{\mathbf{0.8}}\, D^{\mathbf{1.32}}\, V^{\mathbf{-0.76}}\, R^{\mathbf{0.46}} \quad \text{--- (6)}$$

Where the value of $K = 4.18 \times 10^{-3}$ and 3.98×10^{-3} for GB and GC hybrid composites respectively. The equation (6) is holds good for the abrasive wear behaviour of studied hybrid composites.

Figure 1 shows the wear volume predicted from both experimentation and wear model. The wear volume loss predicted from the equation obtained from Buckingham Pi theorem method agrees with the experimental data conducted on 180 grit abrasive paper at a load of 10 N. But the large deviation from the experimental data is noticed for higher velocity of 1 m/s. This is due to interfacial temperature during friction. This is not considered for the analysis. The temperature is one of the important factor which can control the wear behaviour of composites. Both GB and GC hybrid composites agreed with the model of Buckingham pi theorem. The wear volume loss for GB and GC hybrid composites for particular conditions obeys the law of dimensional analysis.

IOP Conf. Series: Materials Science and Engineering 1189 (2021) 012026 doi:10.1088/1757-899X/1189/1/012026

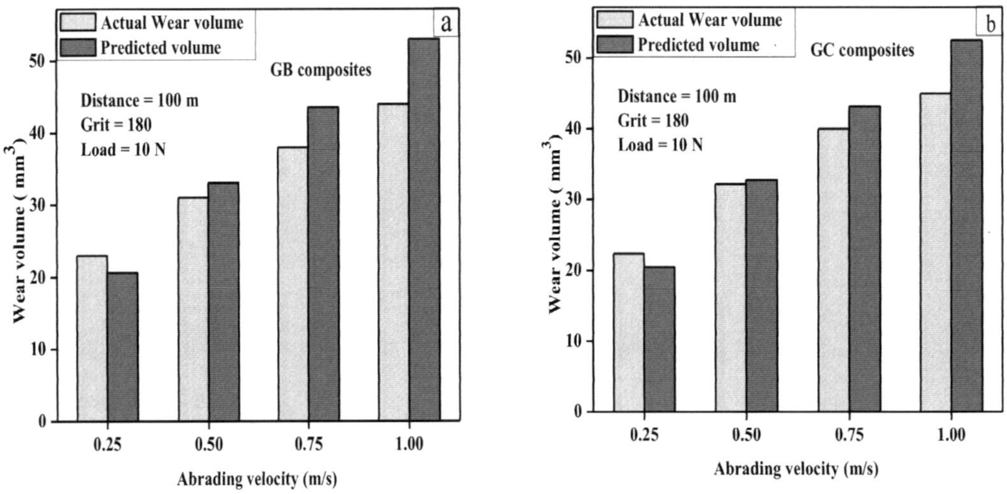

Figure 1. Wear volume prediction by using Buckingham Pi-theorem method for varying velocity at a load of 10 N and 180 grit: a) GB composites and b) GC composites

3.2 Specific Wear Rate as a Function of Mechanical Properties : Abrasive Wear Models

The abrasive wear resistance of polymer composites is greatly influenced by their mechanical properties. These mechanical properties are related to some experimental response of abrasive wear behaviour. The different abrasive wear models have been designed by many researchers. Some of them are given in Table 5. The purpose of reinforcing fibres and fillers into polymer is to enhance the strength by the expense of ductility of a material. As expected, the ductility at yield was decreases with the corresponding increase in strength. According to abrasive wear models (Table 5), the factors 'H', 'σε' and 'Hσε' played the important role in defining the abrasive wear behaviour.

The 'Ks' as a function of some mechanical properties of GB hybrid composites is presented in figure 2(a - c). Figure 2(a) depicts the relation between wear rate and rupture stress. It is found that the influence of hybrid fibres improved the rupture stress and hence the high wear resistance. The wear volume of composites varied inversely as its rupture stress [156]. This is true for the studied composites. Similarly, reinforcing fibers improved the strength of composites with decrease in elongation. The factor 'σε' is high for the composites studied. Hence, GB composites exhibited the good abrasive wear resistance [7, 15]. The wear volume of studied composites is directly proportional to the Yamaguchi factor (PD) which is the product of load and sliding distance [8]. The hybrid GB composites exhibited good 'Hσε' factor which enhanced the abrasion resistance of composites [15]. From all the figures 2 (a - c), it is observed that the specific wear rate is inversely proportional to 'σ', 'σε' and 'Hσε' factors. But wear resistance is directly proportional to all these factors. Hence the composites possessing higher value of these factors exhibit the good abrasion wear resistance. The obtained results agree with the wear models designed for the abrasion behaviour. From the table 5 and the experimental values, it is concluded that the abrasion wear resistance was in the order of GB hybrid >Blend/SGF > Blend/SBF > Blend (PA66/PTFE).

Table 5. Abrasive wear models

Wear models	Authors
$W = \frac{\mu L}{H\sigma\varepsilon}$ Where W = Wear volume/ distance, H = Hardness, L= Load, and $\sigma\varepsilon$ = Product of stress and strain at rupture and μ = friction coefficient	Ratner et al [7]
$W = \beta NPD$ where W = Wear volume , D = Sliding distance P = Normal load, β = abrasive wear factor and N = Scratching efficiency factor ,	Yamaguchi[8]
$W = \frac{K\mu_P}{H\sigma\varepsilon}$ Where μP = Ploughing coefficient, ε = % elongation to failure, and K = Constant , H = Hardness, W = Wear volume per unit distance and σ = tensile strength	Vaziri et al [15]
$W = \frac{F_N S\left[1+4f^2\right]^{0.5}}{\sigma_Y}$ where W = Wear volume, f = friction coefficient, S = Sliding distance, σ_y = Rupture stress of the polymer and F_N = Normal load	Horst Czichos [9]

Similar responses were exhibited by GC hybrid composites. The wear rate of composites is directly proportional to factors $[\sigma]^{-1}$, $[\sigma\varepsilon]^{-1}$ and $[H\sigma\varepsilon]^{-1}$. The variation of 'Ks' of GC hybrid composites as function of aforesaid factors are shown in figure 3(a, b and c). The 2BAW behaviour of GC composites is a function of mechanical properties. The abrasive wear models of composites established the relation between mechanical properties and wear rate of composites. It is well documented through the literature that the wear rate of composites varies inversely as '$\sigma\varepsilon$' factor which is a controlling factor for 2BAW behavior. The wear rate of GC hybrid composites followed the same trend as per the abrasive models [154]. It is well proved through the abrasive models that the 'Ks' of composites is proportional to $[1/\sigma\varepsilon]$ and $[1/H\sigma\varepsilon]$ [7,15,16]. It was also proved that abrasive wear resistance of composites is directly proportional to '$\sigma\varepsilon$' factor. The studied GC hybrid composites exhibited highest '$\sigma\varepsilon$' factor (Table 5) and hence high abrasive wear resistance. All composites studied followed the abrasive wear models. The non-fiber filled composites exhibited the least abrasive wear resistance because of least Ratner - Lancaster factor ($\sigma\varepsilon$). The GC hybrid composites exhibited the better abrasive wear resistance than SGF filled and SGF filled composites. Further, it is observed that the material removal rate of thermoplastic will not improve by reinforcing short fibers if the wear process is highly abrasive [17]. Therefore, it is clear that the short fibers can improve the abrasive wear resistance of thermoplastic composites. In supporting to this, Bijwe et al. [18] reported the abrasive wear behaviour of PEI/PTFE and PEI/SGF composites. The wear volume loss of composites was in the order of PEI/PTFE > PEI/SGF. But the effect of higher weight percentage reinforcement of fibers into polymer increases the abrasivity of surface and hence high wear rate [18]. In the present study, 10 wt. % glass, 10 wt. % basalt and 10 wt.% carbon fibers each enhanced the abrasive wear behaviour of PA66/PTFE blend composites. But 30 wt. % SGF in PSU enhanced the abrasive wear resistance of PSU/SGF composites [19]. This is in good agreement with work of many researchers [18 - 20]

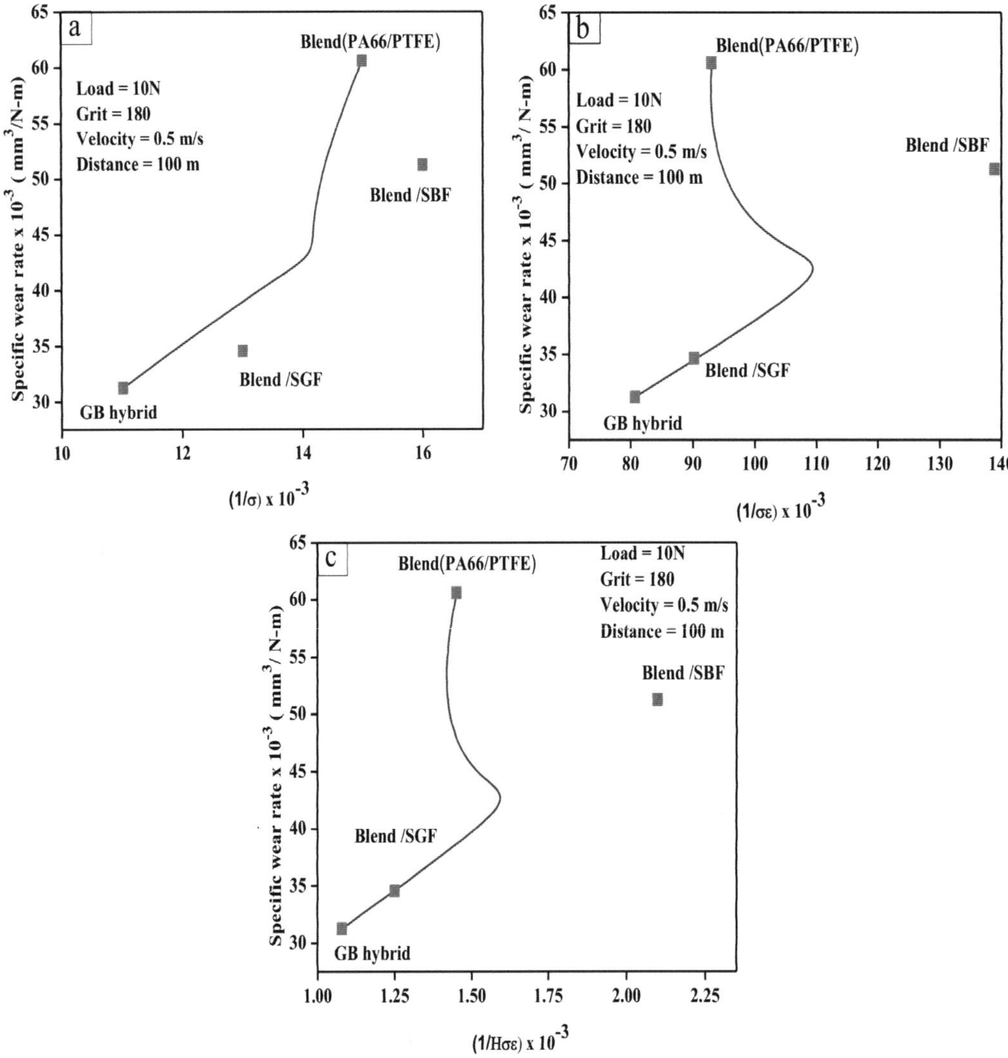

Fig.2. Relation between Specific wear rate and some of mechanical properties of GB hybrid composites: a) $[\sigma]^{-1}$, b) $[\sigma\varepsilon]^{-1}$ and c) $[H\sigma\varepsilon]^{-1}$

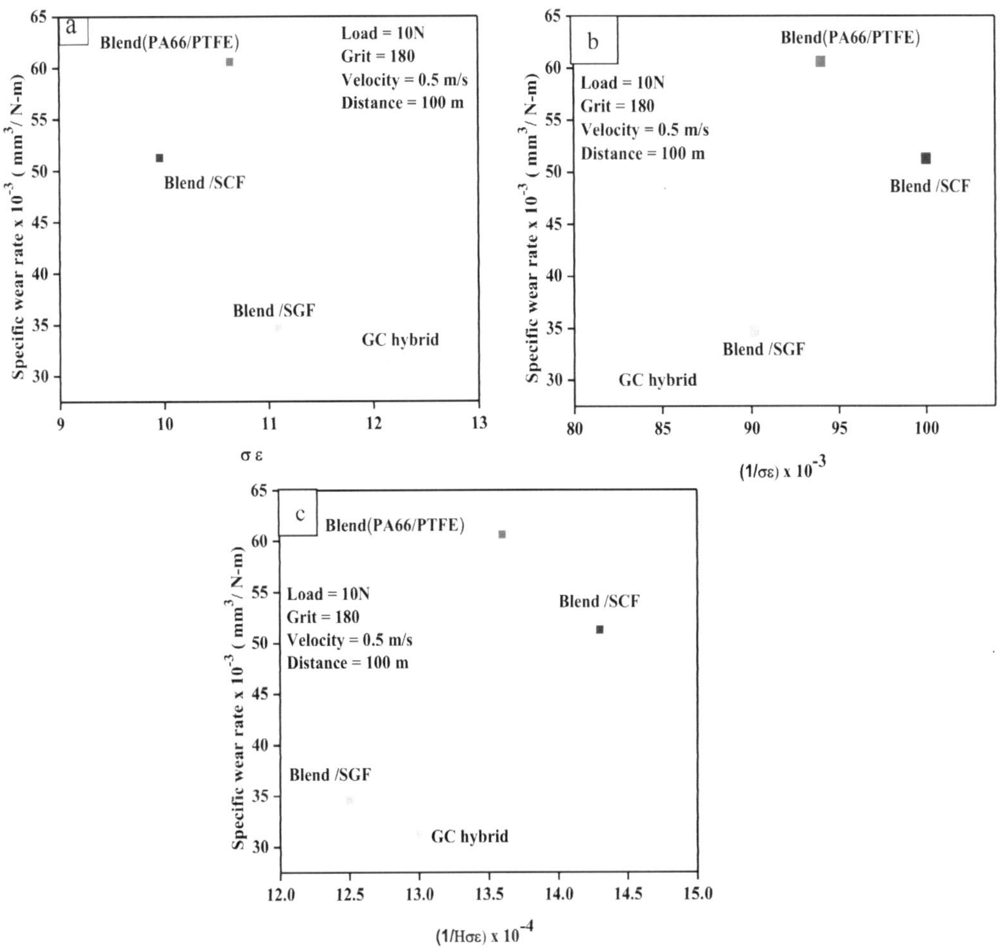

Figure.3. The specific wear rate (Ks) as a function of some mechanical properties under 2-BAW behavior of GC hybrid composites: a) $\sigma\varepsilon$ v/s Ks, b) $[\sigma\varepsilon]^{-1}$ v/s Ks and c) $[H\sigma\varepsilon]^{-1}$ v/s Ks.

4. Conclusion

The following conclusion are made in the investigation of abrasive wear behaviour of GB and GC hybrid composites.

1. The concept of dimensional analysis has been applied effectively for GB and GC hybrid composites
2. The predicted wear volume through designed model agrees with the experimental response for the condition studied
3. The process temperature has not been used as the variable in the analysis which may varied the values to little extent
4. The hybrid composites (GB and GC) have been responded in terms of tribological wear rate which suits the abrasive wear models
5. It is proved from the investigation that the effect of hybrid fiber reinforcement could effectively enhance the abrasion wear resistance
6. Higher percentage of fibre reinforcement transfers the polymer surface to become highly abrasive which is detrimental to abrasive wear resistance

Acknowledgement
I extend all my heartfelt thanks to Mr. Ganesh, CEO, GLS Polymers, Bangalore, India and Dr. Sathisha N.S, Principal, Government Engineering College, K R Pet – 571426, India for their support in preparing this work.

References

[1] K. Friedrich and M. F. Cyffka, 1985 Wear, **103**, 333
[2] J. Bijwe, V. Naidu, N. Bhatnagar and M. Fahim, 2006 Trib.Let. **21(1)** 59
[3] Zhaobin Chen, Tongsheng Li, Yuliang Yang, Xujun Liu, Ringuo Lu, 2004 Wear, **257**, 696
[4] Sultan Ozturk, J.Mater.Sci., 2005 **40(17)** 4585
[5] Srinath and R. Gnanamoorthy, Mater. Sci. Eng: A, 2006 **435**, 181
[6] J. Bijwe, C. M. Logani and U. S. Tewari, 1990 Wear, **138**, 77
[7] S. N. Ratner, I. I. Farberova, O. V. Radynkevich and E. G. Lure, 1964,Soviet Plastics, **7**, 37
[8] Y. Yamaguchi, Tribo. series, 1990 **16**, 125
[9] Horst Czichos, Wear, 1984 **88**, 27
[10] Yusuf Sahin, 2015, Cogent Engineering **2** 1
[11] A.P. Harsha and U.S Tewari, 2002 Ind. J. Engg. Mater. Sci.,2002 **9** 203
[12] B M Rudresh, B N Ravi Kumar, B V Lingesh, 2017 Tran. Ind.Inst. Met.,**70** 2335
[13] Rudresh, B. M. Ravikumar, B. N. Madhu, D, Lingesh B. V, 2019 Tribol. Ind, **41(3),** 416
[14] Rudresh, B. M, Ravikumar, B. N, 2018 Mater. Today: Proce. **5(1) part 3**,2503
[15] M. Vaziri, R. T. Spurr and F. H. Stott,1988 Wear, 122, 329
[16] A. P Harsha, 2011, Wear, **271**, 942,
[17] K. Friedrich, 1986, Friction and wear of polymer composites, 233 Amsterdam: Elsevier
[18] Jayashree Bijwe, J. Indumathi and A.K. Ghosh, 1999, J. Reinf. Plast. Comp, **18 (17)** , 1573
[19] A. P. Harsha and U.S. Tewari, 2003, Polym. Test. **22**, 403
[20] C. Subramanian and S. Senthilvelan, 2010 Ind. Lub. Trib, **62(4)**, 187

IOP Conf. Series: Materials Science and Engineering 1189 (2021) 012027 doi:10.1088/1757-899X/1189/1/012027

Investigation of mechanical properties of ECAP processed AL7068 aluminium alloy

B S Shivashankara[1, 2], K R Gopi[1], S Pradeep[1] and R Raghavendra Rao[1]

[1]Department of Mechanical Engineering, Malnad College of Engineering, Karnataka, Hassan 573202, India
[2]shivnmsv21@gmail.com

Abstract. In this paper, the influence of Equal Channel Angular Pressing (ECAP) on mechanical properties of Al7068 aluminium alloy is investigated. Specimens are subjected to ECAP process by using ECAP split die having an abrupt angle 90° up to 4 ECAP passes using route B_C. The processing temperature was optimized and the ECAP process was conducted at 200°C by route B_C where, the sample is rotated in the same direction between consecutive passes. Unprocessed and ECAP processed samples were subjected to tensile and Vickers microhardness test. ECAP processed samples showed significant improvement in ultimate tensile strength and hardness as compared to unprocessed samples showing improved mechanical properties of the processed aluminium alloy.

Keywords: Equal channel angular pressing, Al7068, Grain refinement, Mechanical properties

1. Introduction

Equal channel angular pressing (ECAP) is one among the important severe plastic deformation (SPD) processes to enhance the mechanical properties of various metals to suit them for different engineering applications by obtaining ultrafine-grained materials. SPD is one of the most promising approach for achieving good mechanical properties by grain refinement and through strain accumulation [1-3]. ECAP process was initially started by Segal with his co-workers during 1970 and 1980 in Minsk in the former Soviet Union [4 and 5]. In ECAP process, the sample is extruded through the sheared formation with dead zone at the outer corner in the die channel [6 and 7].

Z. C. Duan et al. [6], have conducted experiments to assess the capability for producing a high level of homogeneity by ECAP using a split die with a channel angle of 90°. Gopi K. R., et al. [8] worked on magnesium AM90 alloy subjected to ECAP using route B_C. Microstructural characterization revealed the refined grains with an average grain size of 3μm with 4 ECAP passes. Carlo Bruni et al. [9] evaluated the effect of the ECAP on bulk mechanical properties of dual aluminium alloy samples in which one part is cylindrical made of AA6012 aluminium alloy and the other is tubular that is made of AA6026. M. Chegini et al. [10], investigated the impact of grain refinement by ECAP on wear behaviour of Al 7075 material in which, the specimens were processed to four passes using route Bc. Wear resistance of the samples increased substantially because of the development of fine grains by ECAP process. K.R. Cardosoa et al. [11], investigated the effect of ECAP process parameters such as processing route, number of ECAP passes and temperature by pressing AA7050 aluminium alloy through ECAP die. The obtained results revealed that with increase in ECAP passes, refining of microstructure took place by route Bc that achieved more refined microstructure. Cerri E et al. [12] have conducted four ECAP passes via route Bc at room temperature with a die possessing channel of dimension 10 mm in diameter with an internal angle (Φ) of 90° and an outer angle (Ψ) of 35° on Al 6082 alloy. Malek P et al. have conducted experiments to obtain ultra-fine-grained structure in the AA 7075 aluminium alloy through ECAP at pressing temperatures of 120, 170 and 220°C [13]. Zhao X. et al. showed a significant reduction in grain size from 23 μm to 200 nm by pressing the titanium through ECAP die having a channel angle of 120° after 6 or 8 passes at room temperature [14]. Valiev R. Z et al. examined the improvements by using ECAP towards grain refinement included modified conventional ECAP process to improve the process effectiveness and methods for upgrading the technique and to process hard materials [15].

Literature shows limited work on processing of aluminium 7068 (Al7068) metal matrix composite. To the authors knowledge, no work is reported on ECAP process of Al7068. Therefore further investigation is necessary on the above process to enhance the mechanical properties of the material. By secondary processing like ECAP, we can further improve the material properties and its applications can be extended in different areas of manufacturing.

The present work is an effort towards the enhancement of AA7068 Aluminium Alloy material properties by applying a secondary process called ECAP and hence its applications can be extended in different areas of manufacturing. Al 7068 alloys is a heat treatable wrought alloy possessing good fatigue strength, high thermal conductivity and good anodizing response [16 and 17].

2. Material and Experimentation

The elemental composition of Al7068 is shown in table 1. The cast billets of Al7068 were machined to rod form of dimension 15.8 mm in diameter and 100 mm in length. ECAP die of split type was designed and fabricated by hot die steel (HDS) material. Die was designed with two channels of equal cross section that intersects at an abrupt angle of Φ=90° and corner angle of Ψ = 20°. Die was machined to the required dimension and it was heat treated to strengthen the die for ECAP process. The schematic representation of the ECAE process is shown in figure 1. Heating coils are inserted in the holes fabricated within ECAP die for preheating the samples before ECAP process. Four heating coils are used to achieve the required preheat temperature and temperature is monitored using temperature controller.

Table 1. Chemical compositions of Al7068 aluminium alloy in wt.% used in the current work.

Elements	Zn	Mg	Cu	Fe	Zr	Si	Mn	Ti	Cr	Al
Wt.%	8.3	3	2.4	0.15	0.15	0.12	0.1	0.1	0.05	Balance

Experimental optimization was conducted and the preheat temperature was set to 200 °C. Samples are positioned in the ECAP die and the sample is allowed to preheat at 200 °C for 15 minutes. Samples were pressed through ECAP split die at a speed of 50 mm per minute with the application of molybdenum disulfide (MoS2) lubricant to avoid friction between die and sample using route Bc. In route Bc, the sample is rotated in the same direction between consecutive passes through an angle of 90° [2]. ECAP process was repeated up to 4 passes. Grain refinement was inspected by SEM (as per ASTM E-112) [6] to evidence the significant grain refinement achieved through ECAP process. Tensile specimens of ASTM-E8 [8] standards of gauge length 16 mm were prepared and inspected on the BISS Tensile Tester with a cross head speed of 0.24 mm per minute.

Micro hardness test was also done to identify the successive improvement in their hardness with respect to number of passes from 1 to 4 through ECAP die by applying a load of 25g for 15s as per ASTM-E384 [4, 9 and 10].

IOP Conf. Series: Materials Science and Engineering 1189 (2021) 012027 doi:10.1088/1757-899X/1189/1/012027

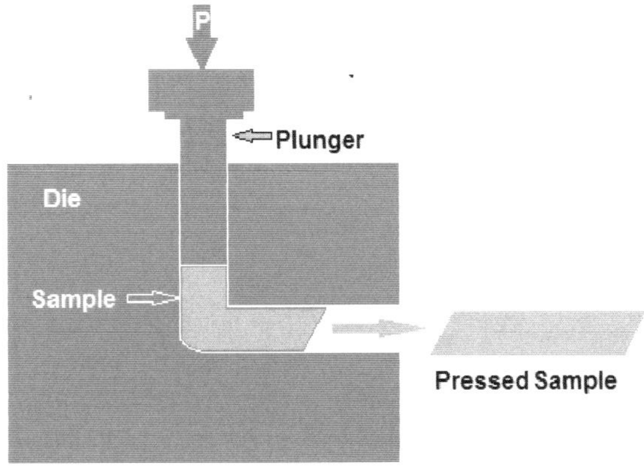

Figure 1.Schematic representation of ECAP process.

3. Results and Discussion

3.1 Microstructural Analysis

Figure 2. illustrates the SEM images of Aluminium Alloy Al7068 before and after ECAP passes from one pass to four passes. Figure 2(a) shows the microstructure of homogenized Al7068 aluminium alloy. The average grain size of Al7086 homogenised aluminium alloy is 80 μm. SEM image evidences the clear visibility of the grain boundaries, which eases the distinguishing between each grain. Further samples were pressed through ECAP die up to 4 passes.

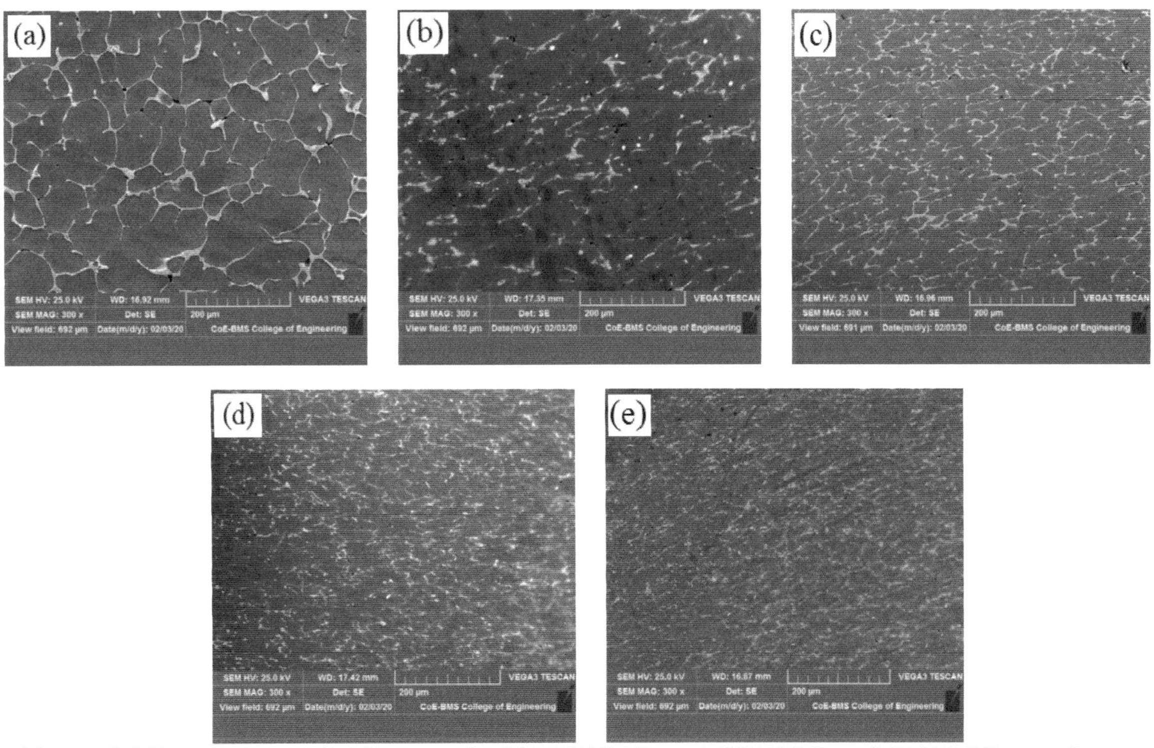

Figure 2: Microstructures of a) Homogenised b) ECAP 1Pass c) ECAP 2 Pass d) ECAP 3 Pass and e) ECAP 4 Pass samples.

Figure 2(b) shows the ECAP 1 pass specimen where initial coarse grains of 80 μm are broken into a average grain size of 60 μm. Figure 2(c), 2(d) and 2(e) represents the further reduction in grain size of 35, 20 and 8 μm during second, third and fourth ECAP passes respectively. From these microstructural images, it is visible that the grains obtained through higher passes are of homogeneous, equiaxed and refined grain size with an influence of continuous dynamic recrystallization occurs during ECAP triggers the formation of UFG material structures [8]. From these SEM images, it is also witnessed that the enhancement in mechanical properties as stated above were achieved due to grain refinement as the Al7068 specimens were processed by severe plastic deformation process through ECAP die.

3.2 Mechanical Properties

Tensile test was performed on the universal testing machine (UTM) with a extreme load capability of 50 KN in research laboratory of Vidyavardaka College of Engineering, Mysore. Specimens were prepared to conduct tensile test according to ASTM E8 standards. All the specimens were tested on UTM in laboratory condition at room temperature by maintaining a constant stroke speed of 0.24 mm per minute. The obtained results were plotted as shown in figure 3. The UTS obtained for homogenized sample is 180 MPa and as the ECAP pass increases from 1 pass to 4 pass the UTS increases to 210, 260, 275 and 350 MPa respectively. It is evident from the graph that ECAP processed specimens resulted with increase in its ultimate tensile strength (UTS) with the ECAP passes from 1 pass to 4 passes as compared to unprocessed Al7068. The increase in UTS for ECAP processed samples is because of grain boundary strengthening mechanism due to reduction in grain size which resulted in increased strength based on Hall Petch equation [18].

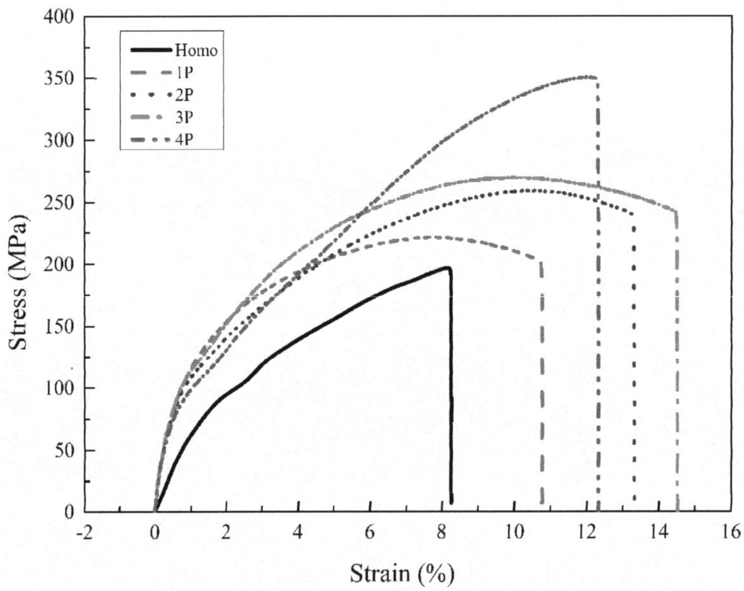

Figure 3: Tensile results of a) Homogenized b) ECAP 1Pass c) ECAP 2Pass d) ECAP 3Pass and e) ECAP 4Pass samples.

Homogenized and ECAP processed samples were exposed to hardness test by applying a load of 0.025 kgf with dwell time for 25 seconds. The test was conducted at room temperature on sophisticated Vickers micro hardness tester and the obtained results are shown in figure 4. ECAP processed samples showed a significant improvement in their hardness values due to grain refinement. Increase in hardness with grain refinement occurs because of strain hardening effect. Enhancement in hardness occurs with increase in ECAP passes due to homogeneous microstructure in the material.

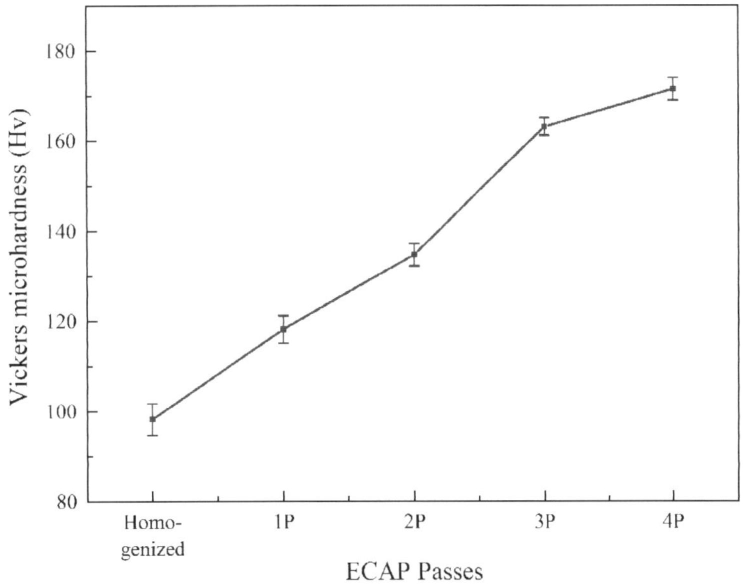

Figure 4: Micro hardness results of a) Homogenized b) ECAP 1Pass c) ECAP 2Pass d) ECAP 3Pass and e) ECAP 4Pass samples.

4. Conclusions

ECAP was performed on aluminium alloy Al7068 and its mechanical properties are investigated. Based on the obtained results, the following inferences are made:

- Aluminium alloy Al7068 was comfortably pressed through ECAP die at 200°C up to ECAP 4 passes using route Bc, which resulted with a better grain refinement of average grain size of 8 µm at the end of 4 pass as compared to homogenized condition.
- Increase in UTS for ECAP processed samples are resulted in 112%, 131%, 136% and 183% more as compared with homogenized sample after 1 pass to 4 pass respectively.
- The results obtained from the microhardness test revealed that successive increase in hardness values for ECAP processed samples due to grain refinement occurred within the material.

References

[1] Torkestani A and Dashtbayazi M R 2018 A new method for severe plastic deformation of the copper sheets Materials *Science & Engineering A,* https://doi.org/10.1016/j.msea.2018.

[2] Liang W, Bian L, Xie G, Zhang W, Wang H and Wang S 2010 Transformation matrix analysis on the shear characteristics in multi-pass ECAP processing and predictive design of new ECAP routes *Materials Scienceand Engineering: A* **Vol 527(21–22)** pp 5557- 64.

[3] Zhao Y H, Liao X Z, Jin Z, Valiev R Z and Zhu Y T 2004 Microstructures and mechanical properties of ultrafine grained 7075 Al alloy processed by ECAP and their evolutions during annealing *Acta Materialia,* Vol **52(15)**, pp 4589–99.

[4] Valiev R Z and Langdon T G 2006 Principles of equal-channel angular pressing as a processing tool for grain refinement*Progress in Materials Science* Vol **51(7)** pp 881–981.

[5] Ravisankar B and Park J K 2008 ECAP of commercially pure titanium: a review *Transactions of the Indian Institute of Metals,* Vol **61(1),** pp 51–62.

[6] Duan Z C and Langdon T G 2011 An experimental evaluation of a special ECAP die containing two equal arcs of curvature *Materials Science and Engineering: A*, Vol **528(12)**, pp 4173–79.

[7] Karoń M and Adamia, M 2018 Characterization of microstructure and mechanical properties of 6060 aluminum alloy processed by ECAP *Solid State Phenomena*, Vol **275**, pp 81–88.
[8] Gopi K R, Nayaka H Sand Sahu S J 2016 Investigation of Microstructure and Mechanical Properties of ECAP-Processed AM Series Magnesium Alloy *Journal of Materials Engineering and Performance*, **Vol 25(9),** pp 3737–45.
[9] Bruni C, Cabibbo M, Ciccarelli D and Paoletti C 2018 Characterization of double aluminium alloy specimens after ECAP *Procedia Manufacturing*, Vol **15**, pp 1517–24.
[10] Cheginia M, Fallahib A and Shaeric M H 2015 Effect of equal channel angular pressing (ECAP) on wear behaviour of Al-7075 alloy *Procedia Materials Science* Vol **11** pp 95–100.
[11] Cardoso, K. R., Travessa, D. N., Botta, W. J., & Jorge, A. M. 2011 *Materials Science and Engineering: A*, Vol **528(18)**, pp 5804–11.
[12] Cerri E, De Marco PP and Leo P 2008A multipass ECAP study of modified aluminium alloys *Materials Science Forum* Vol **604-605** pp 163-70.
[13] Málek P, Turba K, Cieslar M and Harcuba P 2013 Microstructure and high temperature deformation of an ultra-fine grained ECAP AA7075 aluminium alloy International Journal of Materials Research, Vol 104(1) pp 3–10.
[14] Zhao X, Yang X, Liu X, Wang X and Langdon T G 2010The processing of pure titanium through multiple passes of ECAP at room temperature*Materials Science and Engineering: A* Vol **527(23)** pp 6335-39.
[15] Valiev R Z, and Langdon T G 2006 Principles of equal-channel angular pressing as a processing tool for grain refinement *Progress in Materials Science*, Vol **51(7)** pp 881–981.
[16] Azimi A, Shokuhfar A, Nejadseyfi O, Fallahdoost H and Salehi S 2015 Optimizing consolidation behavior of Al 7068–TiC nanocomposites using taguchi statistical analysis Transactions of Nonferrous Metals Society of China, Vol 25(8) pp 2499–2508 doi:10.1016/s1003-6326(15)63868-7.
[17] Osten J, Milkereit B, Reich M, Yang B , Springer A , Nowak K and Kessler O 2020 Development of precipitation hardening parameters for high strength alloy AA 7068 *Materials,*Vol **13(4)** p 918 doi:10.3390/ma13040918.
[18] Nieh T G and Wadsworth J 1991 Hall-petch relation in nanocrystalline solids *Scripta Metallurgica et Materialia,* Vol **25(4)** pp 955–58 doi:10.1016/0956-716x(91)90256-z.

IOP Conf. Series: Materials Science and Engineering 1189 (2021) 012028 doi:10.1088/1757-899X/1189/1/012028

Effect of Shape of Cut-out on Natural Frequency of Square Plate

H K Jayavardhan and Prasanta Kumar Samal

Department of Mechanical Engineering, The National Institute of Engineering, Mysuru, Karnataka, India

Email: jayavardhan457@gmail.com, prasantaku.samal@gmail.com

Abstract. In this paper, the vibration analysis of a square plate with different cut-out shapes with the same area is investigated. Seven square plates of 150 mm x 150 mm size having central cut-out with all outer edges clamped (C-C-C-C) were considered for the study. Different shapes, namely triangle, square, pentagon, hexagon, heptagon, octagon and circle of the same cross-sectional area of 706.86 mm2 were cut out centrally on each square plate. The modal analyses were carried out using finite element package ANSYS Workbench 15.0. The effect of the shape of cut-out on the natural frequency of the plate was studied. It is observed that the values of natural frequencies are increased by increasing the number of edges of the cut-out.

1. Introduction

Plates are one of the most common structural elements in many industries. The Plates are frequently subjected to vibration and controlling the frequency at which a plate vibrates is very important in the field of Engineering. So, it is very important to study the dynamic behaviour of such plate structures subjected to various excitations. The cut-outs are introduced in the plate to provide access, reduce weight, and alter the dynamic response of structures and are often provided to meet some functional or aesthetic requirements. These cut-outs alter the free vibration characteristics of the plates. For plates having multiple cut-outs, the change in frequency may have the variation to a great extent. Hence the study of plates with cut-out is also very important in the field of engineering.

It is important to know the natural frequencies of the component in the analysis of structure subjected to various excitation. Many researchers adopt various Analytical, Numerical and Experimental method for vibration analysis. The free vibration of rectangular plates with an arbitrarily-located hole of different shapes for a rectangular plate with variable thickness is by applying the discrete solution [1]. The ICCM method used for the analysis of a rectangular plate with the change in position of cut-outs of different sizes along its diagonal and major axis line under different aspect ratios of plates [2]. The squared plate of different material with varying boundary condition and the analysis is carried out in both experimental and numerical method to find the effects of different parameters like density, boundary conditions on free vibration characteristics of the plates [3]. The Rayleigh-Ritz [4] and its modified method [5] is adopted for the analysis of plate having different plate and hole domain under varying boundary conditions. The ICCM method for the vibration analysis of a rectangular plate having central cut-out [6]. The Rayleigh-Ritz method is used for analysis of rectangular plates with cut-outs and non-homogeneity [7]. Then the results are compared with both isotropic and orthotropic plates. The method of determining the effects of

IOP Conf. Series: Materials Science and Engineering 1189 (2021) 012028 doi:10.1088/1757-899X/1189/1/012028

openings on the fundamental frequencies in plates with different types of boundary conditions by extending the grid framework model [8] and the results are compared with other methods. The fundamental natural frequency of a thin elastic rectangular, isotropic and orthotropic, plate with clamped corners [9] using Rayleigh's Analytical method. The Green function for the free vibration problem of the plate and it is translated into the eigenvalue problem of the matrix [10]. This method is used for the analysis of free vibration of thin and moderately thick rectangular plates with arbitrary variable thickness.

Literature survey shows that many researchers were adopted different methods for the vibration analysis of rectangular, square and circular plates for the different cut-out shapes of square, rectangle and circle under varying boundary condition. In most of the literature, the study is based on the different shape of the cut-out in the plates, not on the cut-out area of the plates. So, the main objective of the work is to carry out the modal analysis to find the effect of the shape of cut-out on the natural frequency of the square plates having the same cut-out area.

2. Numerical Analysis

The seven homogeneous, isotropic Aluminium plates with size 150mm×150mm×3mm having central cut-out with all outer edges clamped (C-C-C-C) are considered for the study. The properties of the Aluminium specimen are as shown in Table 1.

Table 1. Mechanical properties of Aluminium.

Particulars	Values
Density	2700 Kg/m^3
Young's Modulus	70 GPa
Poisson's Ratio	0.3

The cut-out is made centrally in the plate with different shapes with the same area $706.86mm^2$. Since the different geometrical shapes must be maintained with the same area, the area is taken to the accuracy of two decimal points. Hence the area will be maintained with $706.86 \pm 0.39mm^2$ by considering circular cut-out area as its reference. The cut-out specification is as in Table 2.

Table 2. Cut-out Specifications.

Sl. No.	Cut-out Shape	Dimension	Area
1	Triangle	Side (a) = 40.40 mm	$A = \frac{\sqrt{3}}{4}a^2 = 706.75\ mm^2$
2	Square	Side (a) = 26.59 mm	$A = a^2 = 707.03\ mm^2$
3	Pentagon	Side (a) = 20.27 mm	$A = \frac{1}{4}\sqrt{5(5+2\sqrt{5})} \times a^2 = 706.90\ mm^2$
4	Hexagon	Side (a) = 16.49 mm	$A = \frac{3\sqrt{3}}{2} \times a^2 = 706.47\ mm^2$
5	Heptagon	Side (a) = 13.95 mm	$A = \frac{7}{4} \times \frac{a^2}{\tan\left(\frac{\pi}{7}\right)} = 707.17\ mm^2$
6	Octagon	Side (a) = 12.10 mm	$A = 2(1+\sqrt{2})a^2 = 706.93\ mm^2$
7	Circular	Radius (r) = 15 mm	$A = \pi r^2 = 706.86\ mm^2$

The Modal analysis was conducted in the Finite analysis package ANSYS Workbench 15.0. In this the Plate is modelled with the size of 150×150×3mm and a cut-out has been made centrally for required dimensions. The properties of the material are entered in the Engineering data section. For meshing hexahedral element with size 10 is considered. Then the plate is fixed at all the outer edges. Finally the analysis is carried for the required amount of modes. The results obtained will be the Natural frequency and its corresponding Mode shape.

3. Results and Discussion

The Modal analysis is used to find the Natural frequencies and its corresponding Mode shapes. This helps us to find the dynamic behaviour of components or structures. For the above mentioned seven plates, it is observed that the values of natural frequencies are very less in plate having triangular cut-out, it is increased little more for the square and then it goes on increasing in the order of plate having cut-out shape of pentagon, hexagon, heptagon and octagon. Finally, it has the maximum natural frequency for the plate with circular cut-out and it is considered to have an infinite number of sides. The Frequency results are shown in Table 3.

Table 3. Frequencies (Hz) of plate with different cut-out shape.

Frequency	Cut-out Shape						
	Triangle	**Square**	**Pentagon**	**Hexagon**	**Heptagon**	**Octagon**	**Circular**
1st	1186.4	1197.7	1202.9	1202.9	1203.4	1203.9	1205.8
2nd	2289.0	2323.4	2336.3	2338.0	2338.6	2339.2	2346.9
3rd	2297.0	2323.7	2337.2	2338.3	2339.5	2339.8	2348.7
4th	3449.4	3442.9	3459.5	3460.1	3460.2	3459.5	3464.1
5th	4163.3	4188.8	4178.3	4175.1	4175.7	4177.8	4185.1
6th	4469.1	4525.5	4538.8	4542.5	4544.9	4545.4	4552.7

The results show that the natural frequency of the plate with cut-outs is increased by increasing the number of edges of the cut-out and has the maximum frequency for circular cut-out plate which is considered to have an infinite number of edges.

The mode shape results for the corresponding frequency results show similar mode shapes in each mode for different cut-out plates. Even though it has a similar mode shape, the orientation of its pattern may change according to the shape of the cut-out. This may be due to a very small variation in the frequency. Table 4 shows the mode shapes results of the first three modes. Similarly, for the rest of the modes, the mode shape follows the same trend.

Table 4. Mode shapes of plates having different shape of cut-outs.

Shape of cut-out	1st Mode Shape	2nd Mode shape	3rd Mode shape
Triangle	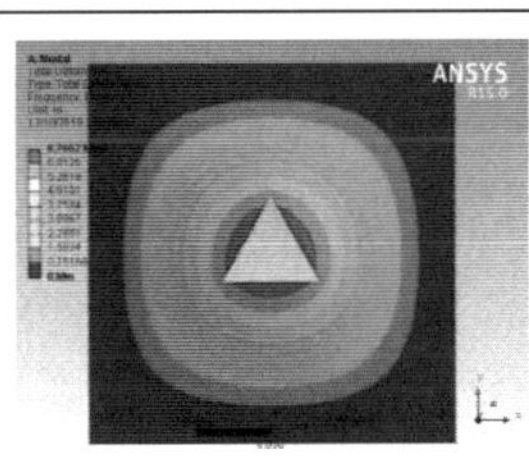	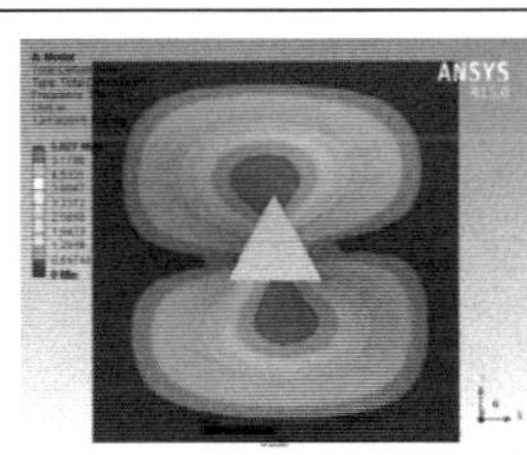	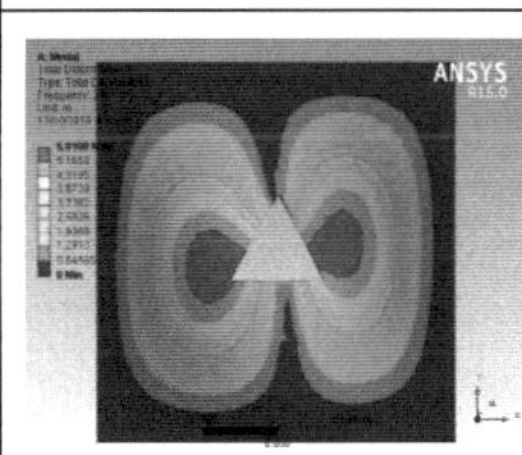

IOP Conf. Series: Materials Science and Engineering 1189 (2021) 012028 doi:10.1088/1757-899X/1189/1/012028

Square	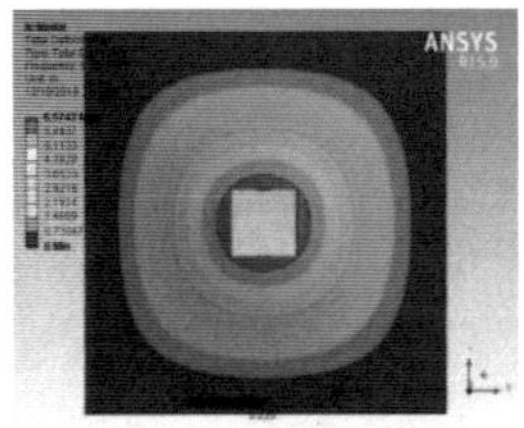	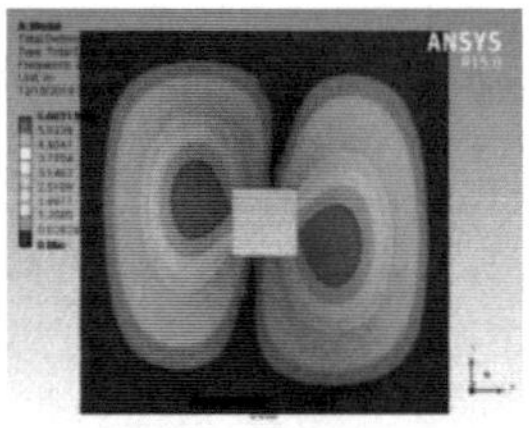	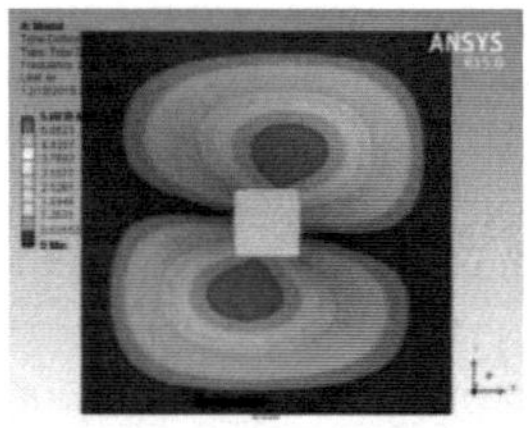
Pentagon	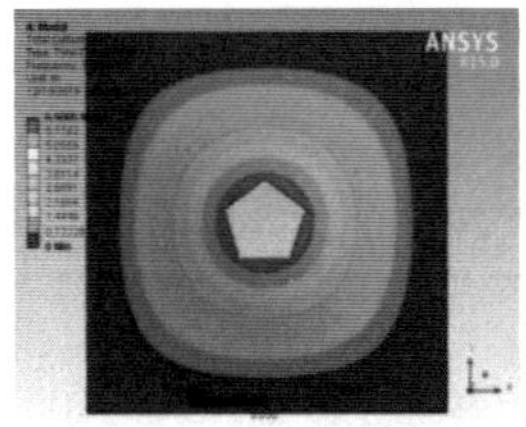	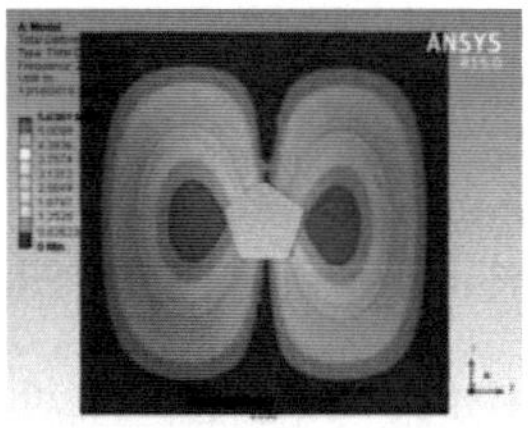	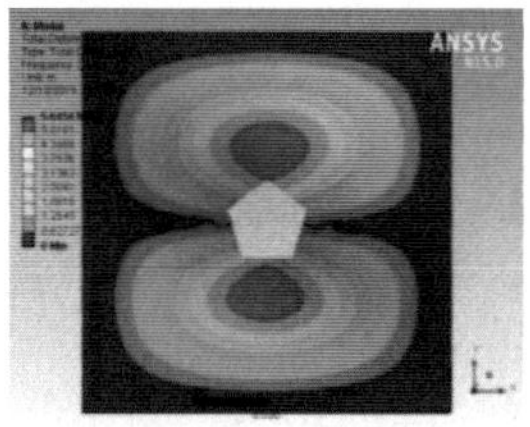
Hexagon	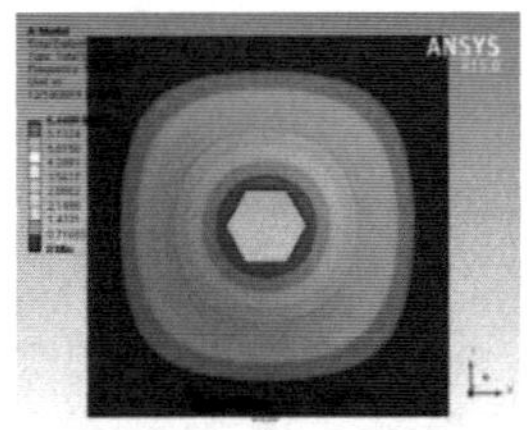	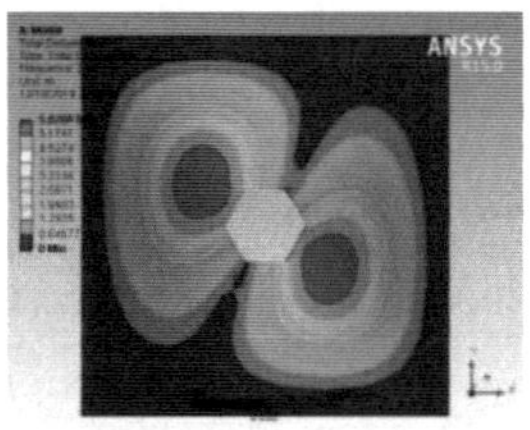	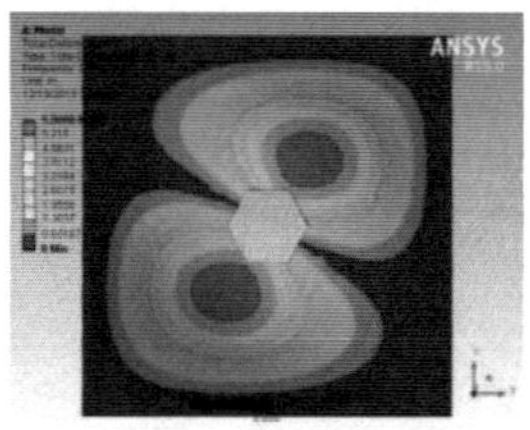
Heptagon	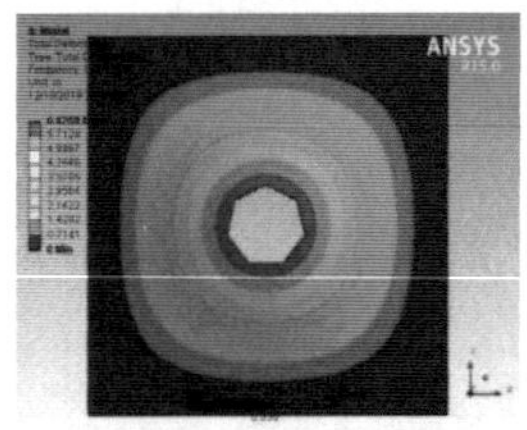	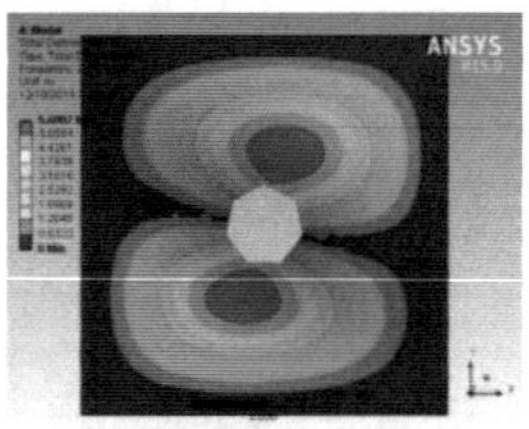	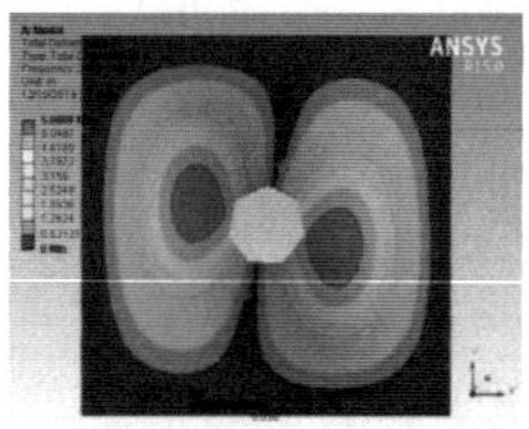
Octagon	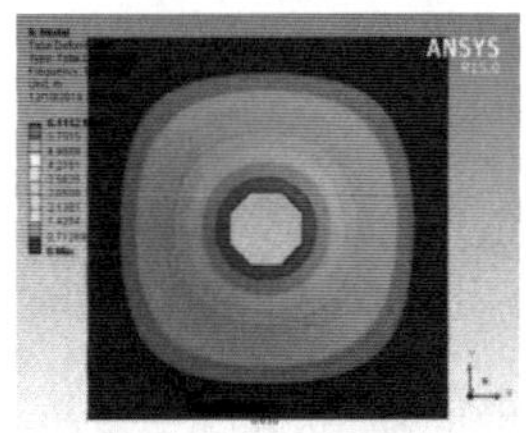	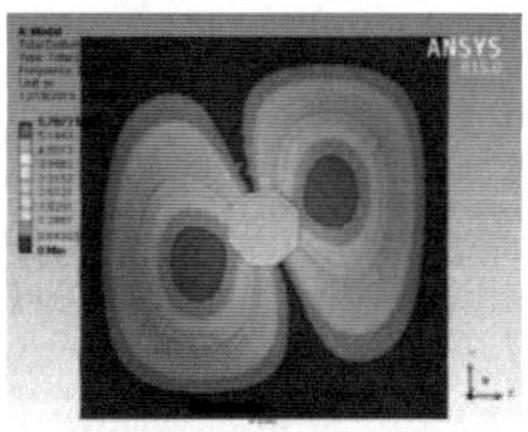	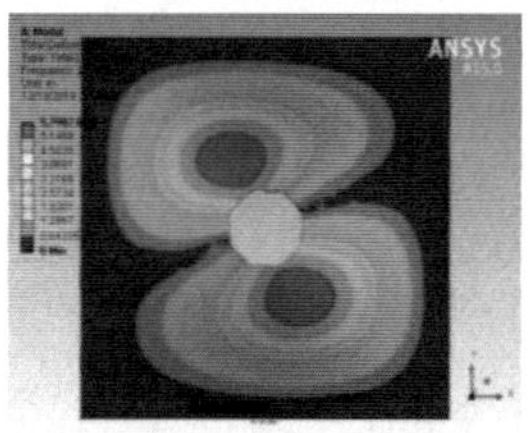
Circle	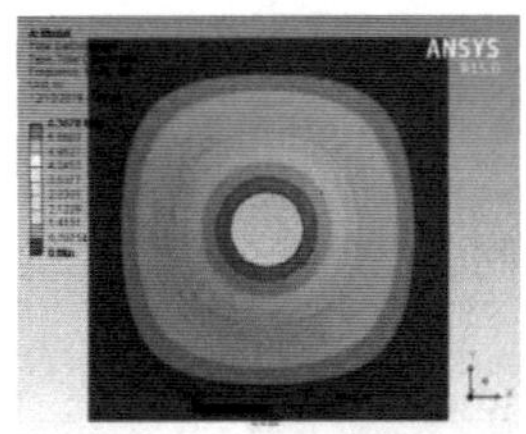	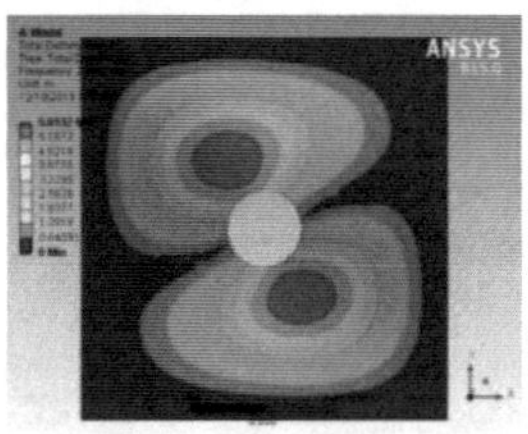	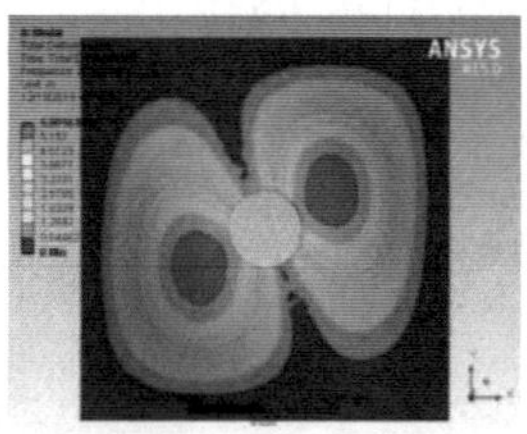

Figures 1, 2 & 3 shows the graph of frequency versus different cut-out shape of the plate in first, second and third modes respectively. Where it can be seen that the frequencies of plates in each mode are increased as the number of edges of the cut-out is increased. Similarly, for the rest of modes, the graph of frequency versus the shape cut-out follows the same trend.

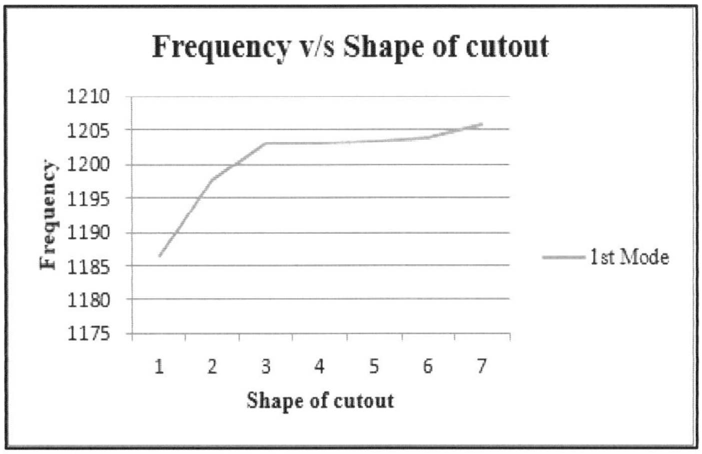

Figure 1. Frequency v/s the shape of cut-out in First mode.

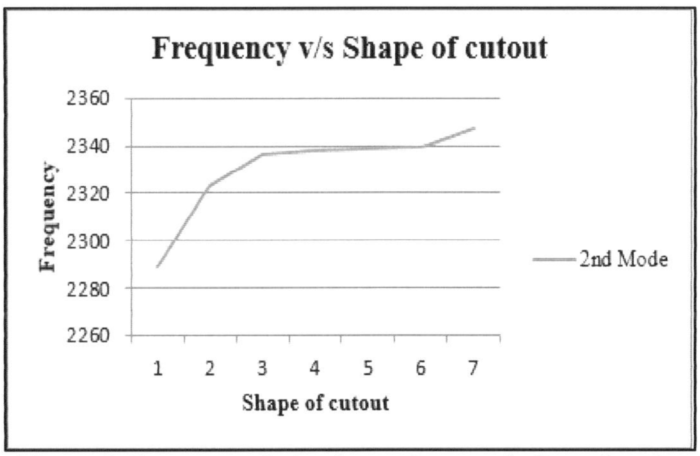

Figure 2. Frequency v/s the shape of cut-out in Second mode.

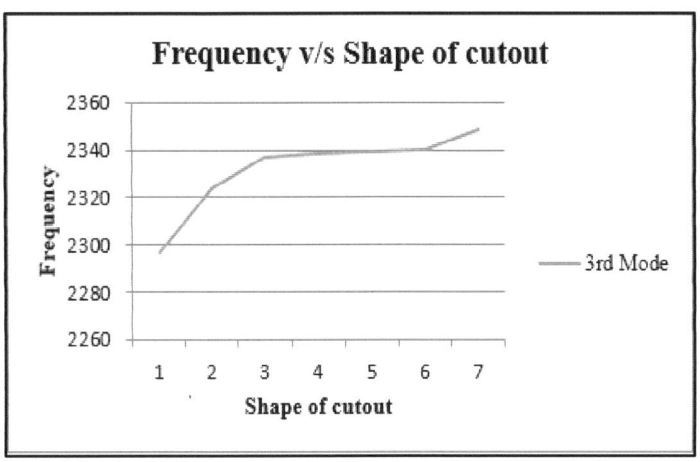

Figure 3. Frequency v/s the shape of cut-out in Third mode.

Note: In Figures 1 to 3 the number 1 to 7 represents the cut-out shape of the plate in order Triangle, Square, Pentagon, Hexagon, Heptagon, Octagon and Circle respectively.

4. Conclusion

In this work, the effect of the shape of cut-out on the natural frequency of the plate is analysed. The seven homogeneous, isotropic aluminium square plates with 150×150×3 mm size having central cut-out of different shape with the same area of 706.86±0.39 mm^2 with all four edges clamped are considered. The modal analysis was carried out using a finite element analysis package ANSYS Workbench 15.0. The result shows that the plate having a triangular cut-out has very less frequency, further the frequency will go on increasing for the plates having a square, pentagon, hexagon, heptagon, octagon cut-out shapes and finally the frequency will be maximum for circular cut-out plate. So, it is observed that the natural frequency of the plate is increased by increasing the number of edges of the cut-outs. The circular cut-out plate shows the maximum frequency because it is considered to be having an infinite number of edges.

The plate with multiple cut-outs may have a very large variation in the frequency. If all the cut-outs are necessary, then the proper designing of the plate structure is done with the help of changing its cut-out shape according to its functionality. The present work may help in choosing the correct shape of the cut-out for the plate structures.

References

[1] Huang M and Sakiyama T 1999. Free vibration analysis of Rectangular plates with variously-shaped holes. *Journal of Sound and Vibration*, **226**(4), 769–786.

[2] Merneedi A, RaoNalluri M & Rao V V S 2017. Free vibration analysis of a thin rectangular plate with multiple circular and rectangular cut-outs. *Journal of Mechanical Science and Technology*, **31**(11), 5185–5202.

[3] H Kemparaju and P Samal, Experimental Investigations on Free Vibration of Plates, *Journal of Testing and Evaluation* **47**, no. 4 (2019): 2750-64.

[4] K Torabi, A R Azadi 2014. Vibration Analysis for Rectangular Plate Having a Circular Central Hole with Point Support by Rayleigh-Ritz Method. *Journal of Solid Mechanics* Vol. **6**, No. 1 (2014) pp. 28-42.

[5] Hussain M and Naeem M N 2018. Effect of Various Edge Conditions on Free-Vibration Characteristics of Isotropic Square and Rectangular Plates. *Advanced Engineering Testing*.

[6] Kwak M K and Han S 2007. Free vibration analysis of rectangular plate with a hole by means of independent coordinate coupling method. *Journal of Sound and Vibration*, **306**(1-2), 12–30.

[7] Lam K Y, Hung K C and Chow S T 1989. Vibration analysis of plates with cut-outs by the modified Rayleigh-Ritz method. *Applied Acoustics*, **28**(1), 49–60.

[8] Paramasivam P 1973. Free vibration of square plates with square openings. *Journal of Sound and Vibration*, **30**(2), 173–178.

[9] Gharaibeh M A and Obeidat A M 2018. Vibrations Analysis of Rectangular Plates with Clamped Corners. *Open Engineering*, **8**(1), 275–283.

[10] Sakiyama T and Huang M 1998. Free vibration analysis of Rectangular plates with variable thickness. *Journal of Sound and Vibration*, **216**(3), 379–397.

Experimental determination of mode shapes of a plate using speaker as excitation device

K Gurukiran[1*], **Prasanta Kumar Samal**[2]

[1*, 2] Department of Mechanical Engineering, The National Institute of Engineering, Mysuru, Karnataka, INDIA.
E-mail:gurukirank9@gmail.com, [2]prasantaku.samal@gmail.com

Abstract. Experimental modal analysis is usually carried out in two methods, namely impact hammer method and shaker method. The limitation of both the methods is determination of mode shapes. Of course, the mode shapes are obtained using FFT analyser which is costly software. In this paper, however an attempt is made to determine the mode shapes of a plate experimentally using a speaker as excitation device. An experimental setup was prepared using a speaker to excite a plate at its central portion. An aluminium plate was mounted at its central portion and excited. Dust particles were sprinkled over the plate and when the excitation frequency matches the natural frequencies, the corresponding mode shape of the plate was obtained. The numerical modal analysis was carried out for the plate with proper boundary conditions using finite element analysis package ANSYS Workbench. The mode shapes obtained numerically and experimentally were compared and observed to be in good agreement.

1. Introduction

Vibrations occur in numerous mechanical and structural systems. In the event when vibration is uncontrolled, it can prompt to catastrophic circumstances. When the mechanical structure is forced to resonance vibration, at one of the natural frequency standing wave pattern are obtained. In eighteenth century, German researcher Ernst Chladni found that when a specific size of the board is at a specific frequency of vibration, because of wave reflection at the edge of a plate, it will deliver standing waves, forming nodes of no vibration and antinodes with huge vibration. As it is the vibration of a plane, the nodes are arranged in sections on the curves, forms Chladni pattern.Not all structures can be impact tested, however. For instance, structure with delicate surfaces cannot be impact tested. When impact testing cannot be used, FRF measurements must be made by providing artificial excitation with shaker. Shaker such as Electrodynamics shaker is used in the determination of mode shapes which cost more. So the speaker is used as an excitation device to determine the mode shapes which reduces the cost of the excitation device.

Xun Ma et al [1]. Experimental modal analysis and modal extraction experiment for the plate and it has the same results as of finite element analysis. H.R.Kemparaju et al [2]. The development of experimental setup to study the free vibration characteristics of the plate gives good results. A. E. Ikpe et al [3]. The Chladni patterns were obtained on the thin plate by acoustic excitation. The plate was analyzed using CATIA, ANSYS, and HYPERMESH, and the mode shapes and natural frequencies are obtained. P. H. Tuan et al [4] chladni nodal pattern for a thin plate was obtained using a mechanical oscillator. The resonant frequency was also measured experimentally by attaining the effective impedance of exciter. Owunna et al [5] The mode shapes and natural frequencies of the plate were determined using electromagnetic shaker.

IOP Conf. Series: Materials Science and Engineering 1189 (2021) 012029 doi:10.1088/1757-899X/1189/1/012029

Frequencies acquired by experimental analysis, the resonant frequency are obtained by varying impedance of the oscillator with the thin plate. Shridhar [6] the friction and adhesion present in the plate effects the formation of chladni patterns. This came to know by spatial autocorrelation analysis in investigating the effect of frictional forces and adhesion on the rate of pattern formation. Luo, Yan & Feng [7] introduces a simple approach to determine chladni patterns on an elastic plate by the optical lever method. In this method, mode shapes are acquired by integrating the slope distribution. Kaczmarek A et al [8] Mode of vibration for square plate which has built-in corners and inducted at the centre does not correspond to vibration modes for plates with free edges. Van Gerner et al [9] the inverse Chladni patterns are obtained on the plate when little particles are dispersed on the resonating plate. Xavier Escaler et al [10] utilized the Chladni procedure to obtain the mode shapes of the plate both in air and completely submerged in water.

The literature survey shows the researcher adopted different methods to determine the mode shapes of the plate. Some of the methods were time-consuming. Some researchers used electrodynamic shaker for mode shape determination which is costly. So, the main objective of the project is to determine the mode shapes of the plate using a speaker as an excitation device.

2. Methodology

2.1. Specimen details

An Aluminium square plate of size 305 x 305 mm and 1 mm thickness is taken for the modal analysis.

Table 1: Properties of the specimen

Young's modulus	70 GPA
Poisson's ratio	0.33
Density	2720 Kg/m^3

2.2. Experimental modal analysis

An experimental setup is prepared so that plate can be excited at the centre by the speaker is as shown below in fig 3. Frequency of vibration is varied by the function generator. An aluminium square plate was fixed on the mounting plate of the excitation device using nut and bolts. The sound signals obtained from the function generator were supplied through an amplifier. When the plate starts vibrating due to excitation from the speaker the silica powder sprinkled periodically will start vibrating. At the place of the node, there is no vibration and silica powder settle there and at the place of antinodes where vibration is very high, the silica powder moves away from it towards nodes and settles there.

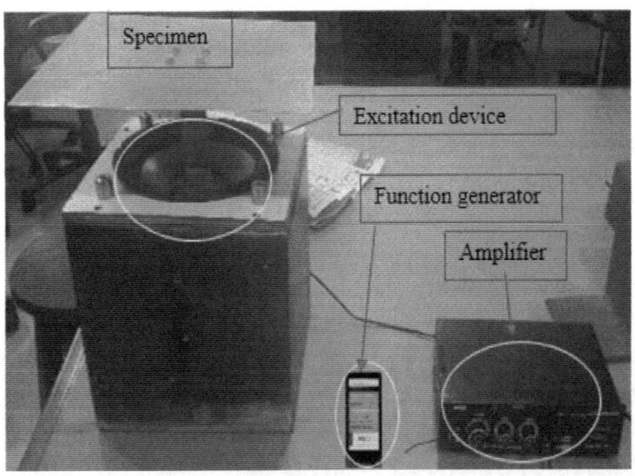

Fig 1: Experimental setup

This leads to the generation of nodal patterns which are the mode shapes of the plate. For the particular natural frequencies of vibration, corresponding mode shapes were thus obtained. These mode shapes obtained at different natural frequencies are recorded.

2.3. Numerical Modal Analysis

Numerical modal analysis of the aluminium square plate was done using ANSYS Workbench. The meshing of the plate was done using hexahedral elements. The plate was fixed at the centre and edges are free as shown in figure 3.7. The first eight mode shapes of the plate are extracted by carrying out numerical analysis utilizing ANSYS Workbench. The results obtained the numerical modal analysis are tabulated.

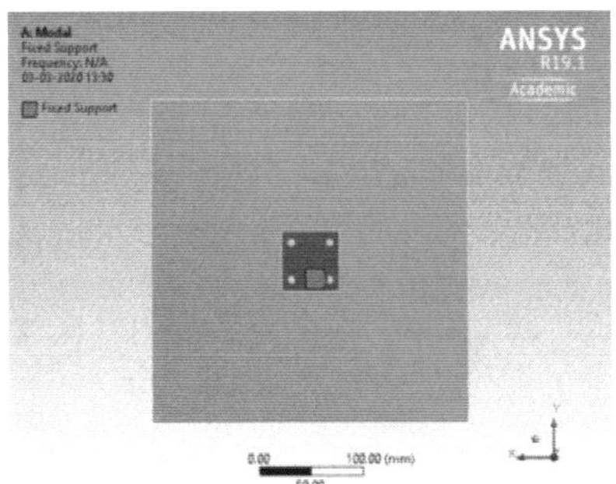

Figure 2: Boundary Condition

3. Results and discussion

Table 2 shows the comparison of the mode shapes of the plate. The first column of table shows the mode number. The second column in table shows the mode shape of the plate obtained from the numerical modal analysis. The third column of table shows mode shapes of the plate obtained from experimental modal analysis using the speaker as an excitation device.

Table 2: Comparison of the mode shapes

Mode Number	Numerical Modal Analysis	Experimental Modal Analysis
1	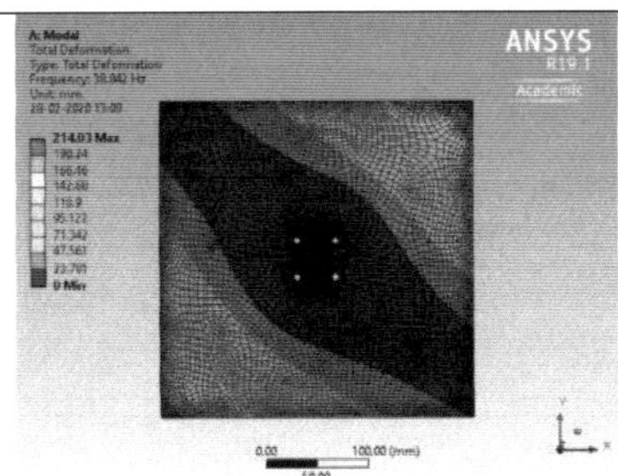	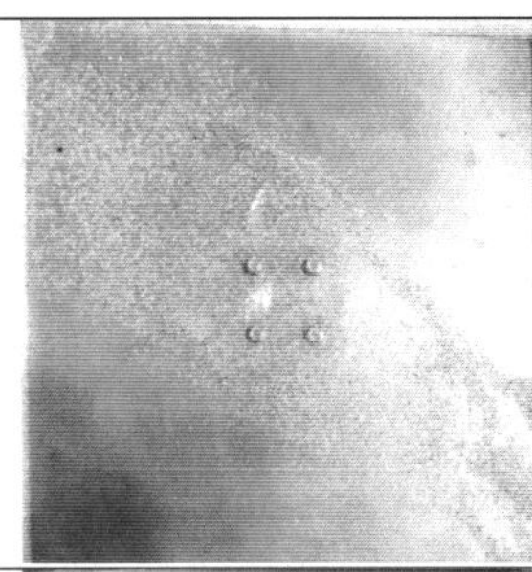
2	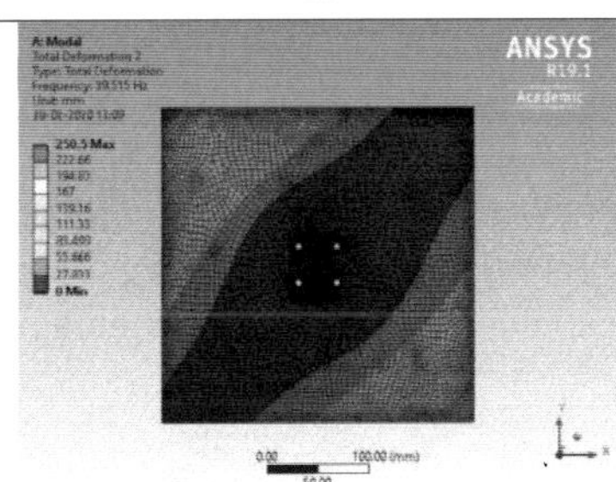	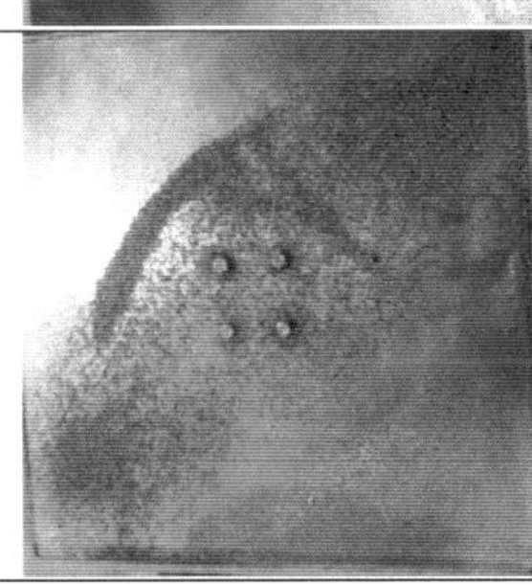

IOP Conf. Series: Materials Science and Engineering 1189 (2021) 012029 doi:10.1088/1757-899X/1189/1/012029

3	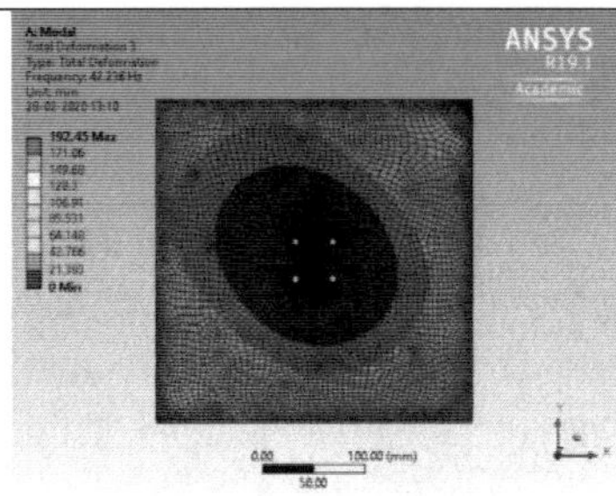	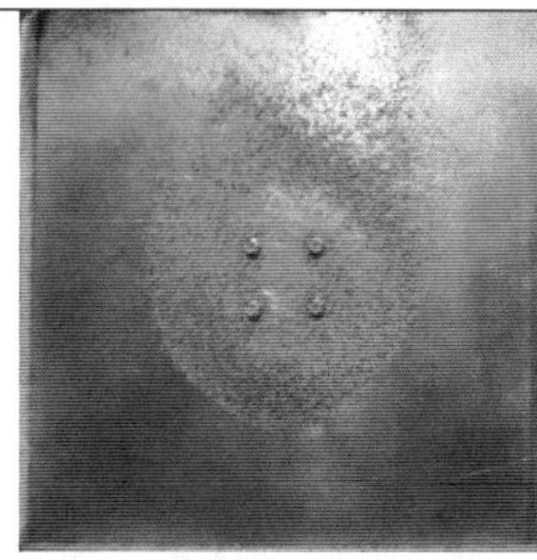
4	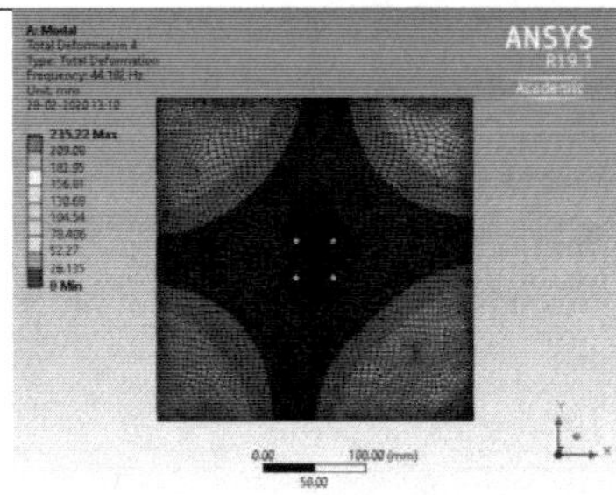	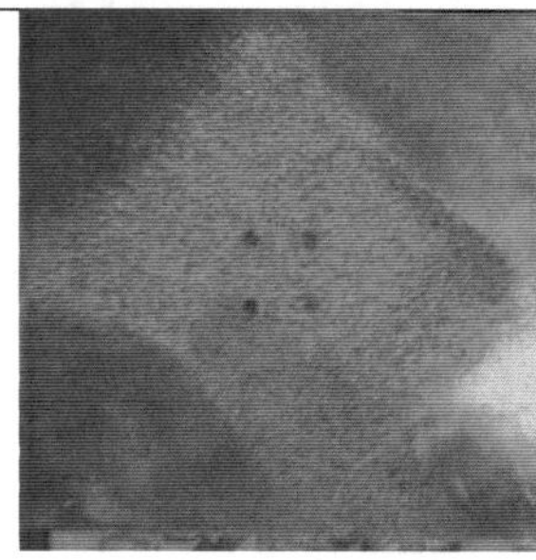
5	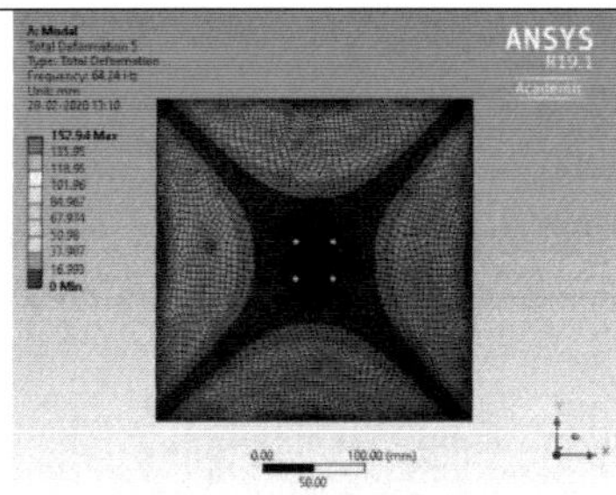	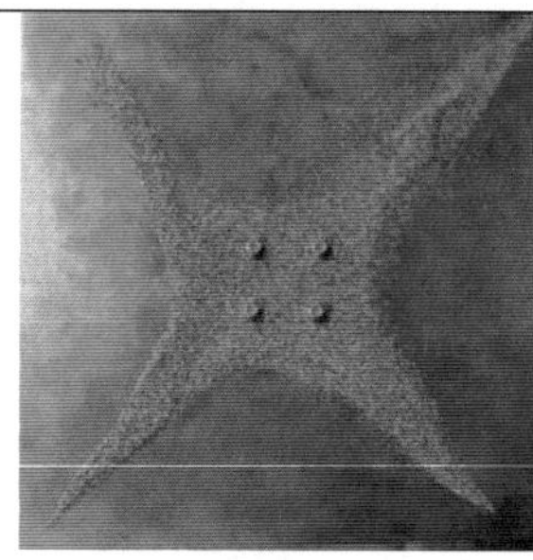
6	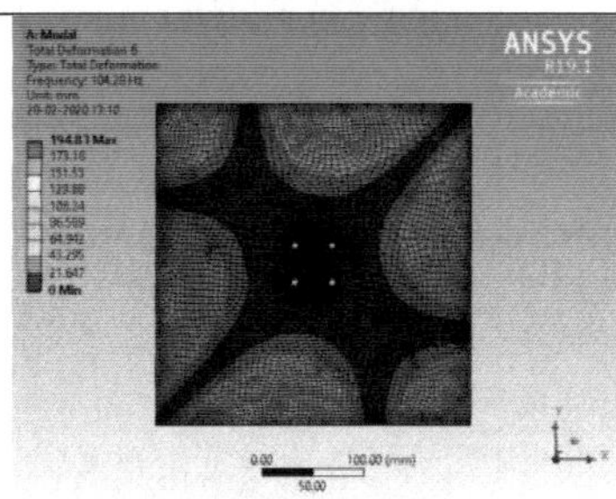	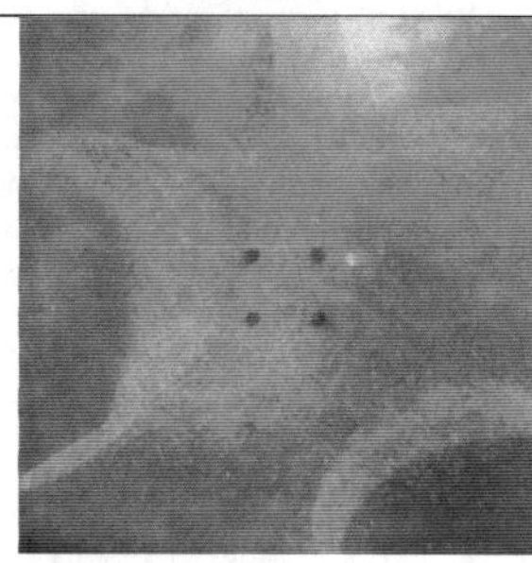
7	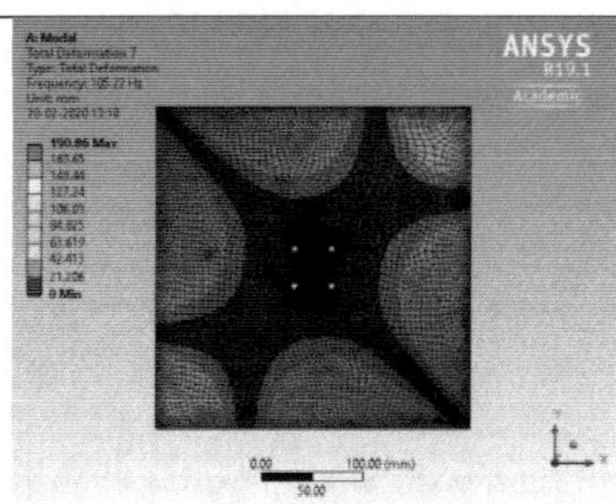	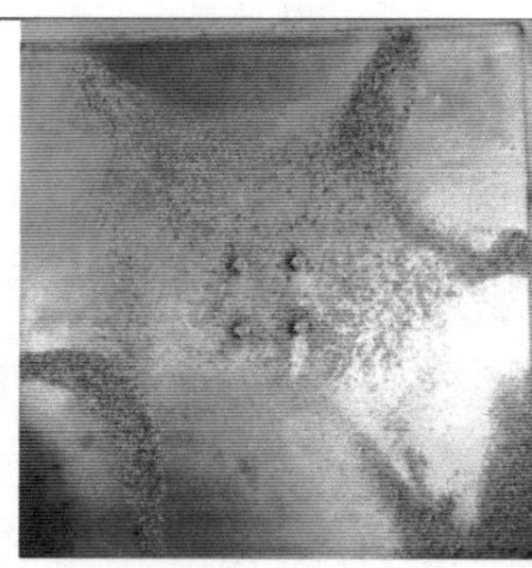

8	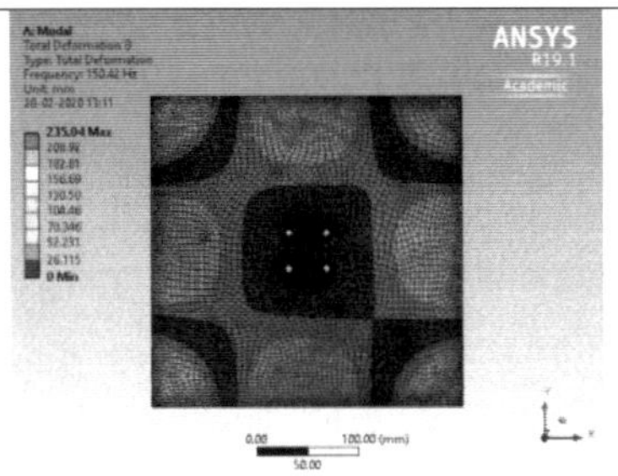	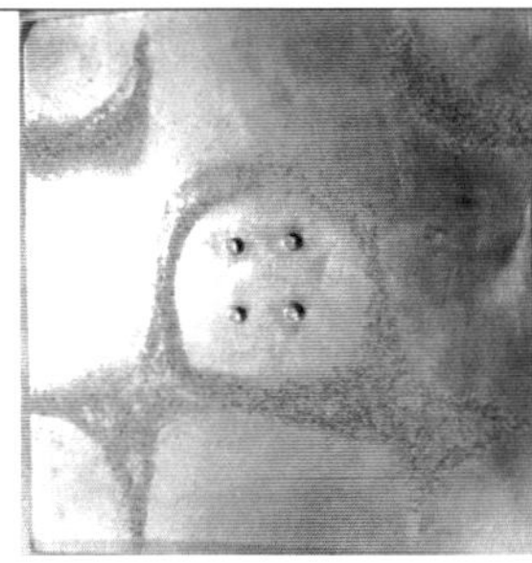

Table 2 shows the mode shapes of the plate obtained from numerical and experimental modal analysis. The mode shape of the plate acquire from experimental modal analysis were nearer to the mode shapes acquired from numerical modal analysis. Thus, by using the speaker as an excitation device the mode shapes of the plate of different boundary conditions can be determined, as mode shapes of the cantilever plate.

Table 2.1: Comparison of the Natural Frequencies

Mode Number	Numerical Modal Analysis[Hz]	Experimental Modal Analysis[Hz]	% of Difference
1	38.8	40.8	4.90
2	39.5	42	5.95
3	42.2	45.9	8.06
4	44.1	48	8.12
5	64.2	70	8.28
6	104.2	116	10.17
7	105.2	117.9	10.77
8	150.4	163.3	7.89

Table 2.2 shows the comparison of natural frequencies obtained from numerical and experimental modal analysis. Comparisons of results give good agreement. Although results are closer there are some deviations. It was found that the deviation of the results is less than 11% as shown in table. These deviations between experimental and numerical modal analysis results may be due to density which was considered homogeneous during the numerical analysis of the plate. Using the speaker as an excitation device the cost of the excitation device can be considerably reduced.

4. Conclusion

In this paper, experimental modal analysis of the plate was conducted by fixing the plate at centre using the speaker as an excitation device. Numerical modal analysis of the plate was conducted by applying the boundary conditions using ANSYS Workbench. The comparison of results obtained from experimental and numerical modal analysis gives good agreement. The results obtained from experimental modal analysis were closer to the results obtained from numerical modal analysis. Thus, by using the speaker as an excitation device the mode shape of the plate with different boundary conditions can be determined, such as modes shape of the cantilever plate. By using speaker as an excitation device cost of the excitation device can be considerably reduced.

5. Reference

[1] Xun Ma, Jiwei Zhang, Shiming Yan (2012) *Experimental Modal Analysis and Modal Reproduce Experiment Research of a Chladini Plate*, Applied Mechanics and Materials, *Vols **152-154** pp 1401-1405.*

[2] H. R. Kemparaju and P. K. Samal (2019), *Experimental Investigations on Free Vibration of Plates*, Journal of Testing and Evaluation, Vol. **47**, No. 4.

[3] A. E. Ikpe1, A. E. Ndon and E. M. Etuk (2019), *Response variation of chladni patterns on vibrating elastic plate under electro-mechanical oscillation,* Nigerian Journal of Technology, Vol. 38, No. 3, pp. 540 – 548

[4] P. H. Tuan, C. P. Wen, P. Y. Chiang, and Y. T. Yu, *Exploring the resonant vibration of thin plates: Reconstruction of Chladni patterns and determination of resonant wave numbers.* J. Acoust. Soc. Am., Vol. **137**, No. 4.

[5] Owunna, I., Ikpe, A., Satope, P. and Ikpe, E (2016), *Experimental Modal Analysis of a Flat Plate SubjectedTo Vibration,* AmericanJournalof Engineering Research, vol.**5**,No. 30-37.

[6] Mohit Shridhar (2011*), Manipulating Standard and Inverse Chladni Patterns by Modulating Adhesive, Frictional, and Damping Forces*, International Journal of Scientific and Engineering Research, International Journal of Scientific & Engineering Research, Volume 3, Issue 5, May-2012 1 ISSN 2229-5518

[7] Yan Luo, Rong Feng, Xiaodong Li and Donghuan Liu (2019), *A simple approach to determine the mode shapes of Chladni plates based on optical lever method*, Eur. J. Phys. **40** 065001,pp 12.

[8] Kaczmarek A,Javorek L, Orłowski K (2014) , *Mode vibrations of plates – experimental analysis, Forestry and Wood Technology*, vol.**88**, pp 97-101

[9] Van Gerner, Henk & Hoef, Martin & Meer, Devaraj & Weele, Ko. (2010*). Inversion of Chladni patterns by tuning the vibrational acceleration.* Physical review. E, Statistical, nonlinear, and soft matter physics.

[10] Xavier Escaler, Oscar De La Torre (2018), *Axisymmetric vibrations of a circular Chladni plate in air and fully submerged in water*, Journal of Fluids and Structures, IOP Conf. Series: Journal of Physics: Conf. Series **1106** (2018) 012031.

Theoretical Study of influence of MHD in an infinitely long rough Porous Journal Bearing

G K Kalavathi[1],Yuvaraja B K [2], Dinesh P A[3] ,Vasundhara M G [4]
[1] Department of Mathematics, Malnad College of Engineering, Hassan, Karnataka, India
[2]Department of Mathematics, B G S Institute of Technology, Mandya, Karnataka, India
[3]Department of Mathematics, M. S. Ramaiah Institute of Technology, Bangalore, Karnataka, India
[4]Department of Mechanical Engineering, Malnad College of Engineering, Hassan Karnataka, India

Abstract. Analytical study of porous rough journal bearing with magneto hydrodynamic fluid in the presence of longitudinal and transverse roughness is investigated. Surface roughness effect is studied by using Christensen stochastic theory. The generalized Reynolds equations for both types of roughness patterns are derived. Reynolds boundary conditions are used to study the characteristics of infinitely long journal bearing. Cavitation angle is obtained by using bisection method. The expressions for pressure, load carrying capacity and attitude angle are obtained. The results are shown in the graph for longitudinal roughness pattern and concluded that, load carrying capacity increases for the increasing values of Hartmann number.

Keywords Porous, Long Journal bearing, Christensen Stochastic roughness theory, Load carrying capacity.

1. Introduction

An analytical work is made on infinitely long journal bearing adopting an electrically conducting lubricant with magnetic field applied to the surfaces of bearing. M I Anwar and C M Rodkiewicz [1] investigated non uniform magnetic field effects in MHD slider bearing, a result indicate that increase of Hartmann number decreases inertia contribution but increases the load carrying capacity. The roughness effects in porous journal bearing studied by Gururajan K and Prakash [2-6] and it was found that the roughness structure increases the bearing performance. Numerical study of the MHD Reynolds equation for squeeze film lubrication between two parallel surfaces was investigated by Ramesh et al. [7] author shows that the lubricating characteristics of bearing surfaces such as pressure, load carrying capacity increased by Hartmann number. Kalavathi et al. [8-11] studied the influence of roughness on narrow, long and finite journal bearing surface with slip/no-slip and also studied the combined effect of MHD on narrow porous journal bearing surface with a heterogeneous slip/no slip and author concluded that the combined effect of roughness and MHD increases the load carrying capacity. Qiyin et al. [12] investigated the transient behavior of journal bearing with surface structure based on fluid structure interaction approach and found depending on the position of circumferential direction, surface structure increase or decrease the performance of journal bearing. Recently, Yuvaraj et al. [13] have studied the effect of MHD on narrow porous journal bearing of longitudinal roughness pattern and found that introduction of MHD increases the load carrying capacity.

In this paper, influence of MHD on long porous rough journal bearing is presented. Differential equations for two different roughness types like longitudinal and transverse types are derived and the roughness effects are studied by considering Christensen stochastic theory. Reynolds boundary conditions are used to obtain the expressions for bearing characteristics. Cavitation angle is determined by using bisection method, and expectations values are obtained by using Simpsons $1/3^{rd}$ rule. Load carrying capacity is evaluated by using Gaussian Quadrature 16 point formula. In this paper the characteristic of infinitely long porous journal bearing like load carrying capacity for longitudinal roughness type is studied theoretically.

2. Analytical approach

The physical configuration of the rough porous journal bearing is as shown in the figure. 1. Consider the journal bearing and shaft rotating with velocityU, eccentricity ratio ε^*, $H_0{}^*$ is the wall thickness of the porous media with a lubricant of constant viscosity.

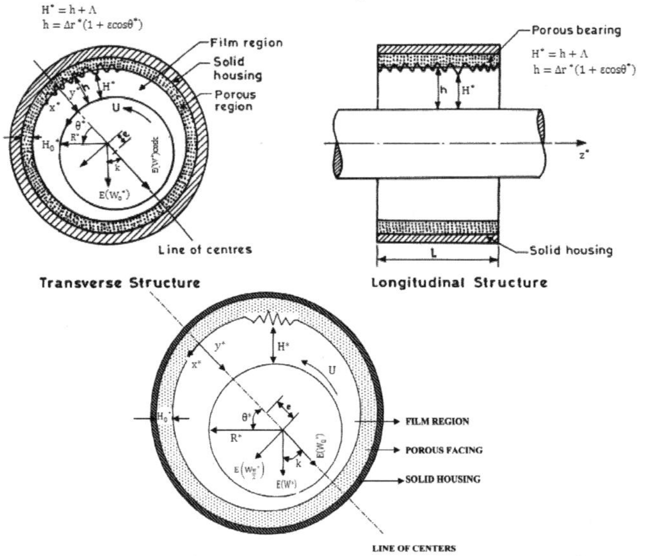

Figure. 1 Bearing Geometry and Journal bearing configuration

The governing equations for MHD fluid are given by [1]

$$\frac{\partial^2 u^*}{\partial y^{*2}} - \frac{\left(B_0{}^{ll}\right)^2}{\left(h_m{}^{ll}\right)^2} u^* = \frac{1}{\gamma^*}\frac{\partial p^{ll}}{\partial x^*}. \tag{1}$$

$$\frac{\partial^2 w^*}{\partial y^{*2}} - \frac{\left(B_0{}^{ll}\right)^2}{\left(h_m{}^{ll}\right)^2} w^* = \frac{1}{\gamma^*}\frac{\partial p^{ll}}{\partial z^*}. \tag{2}$$

$$\frac{\partial p^{ll}}{\partial y^*} = 0. \tag{3}$$

where $B_0{}^{ll} = \frac{M_0 h_m{}^{ll}}{\sqrt{\sigma/\gamma^*}}$ is Hartmann number, M_0 is applied magnetic field, γ^* is the viscosity of the fluid, σ is electrical conductivity of fluid, $h_m{}^{ll}$ denotes the minimum film thickness. u^*, v^* and w^* are velocity components along x^*, y^* and z *direction, p^{ll} is pressure in the porous matrix.

The equation of continuity is given by

$$\frac{\partial Q_{u^*}}{\partial x^*} + \frac{\partial Q_{v^*}}{\partial y^*} + \frac{\partial Q_{w^*}}{\partial z^*} = 0 \text{ or } \frac{\partial Q_{u^*}}{\partial x^*} + \frac{\partial Q_{w^*}}{\partial z^*} = -\frac{\partial Q_{v^*}}{\partial y^*} \tag{4}$$

where, $Q_{u^*} = \int_0^{H^*} u^*\, dy^*$

Solving Equations (1) and (2) for u^* and w^* we obtain

$$u^* = A_1 \cosh\left(\frac{B_0{}^{ll}}{h_m{}^{ll}} y^*\right) + A_2 \sinh\left(\frac{B_0{}^{ll}}{h_m{}^{ll}} y^*\right) - \frac{1}{\gamma^*}\frac{\left(h_m^{ll}\right)^2}{\left(B_0^{ll}\right)^2}\frac{\partial p^*}{\partial x^*}. \tag{5}$$

IOP Conf. Series: Materials Science and Engineering 1189 (2021) 012030 doi:10.1088/1757-899X/1189/1/012030

$$w^* = B_1\cosh\left(\frac{B_0{}^{II}}{h_m{}^{II}}y^*\right) + B_2\sinh\left(\frac{B_0{}^{II}}{h_m{}^{II}}y^*\right) - \frac{1}{\gamma^*}\frac{(h_m^{II})^2}{(B_0^{II})^2}\frac{\partial p^*}{\partial z^*}. \quad (6)$$

The boundary conditions for u^*, v^* and w^* are

$$u^*(x^*,0,z^*) = 0,\ v^*(x^*,0,z^*) = -V_0^*,\ w^*(x^*,0,z^*) = 0. \quad (7)$$

$$u^*(x^*,H^*,z^*) = U, v^*(x^*,H^*,z^*) = 0,\ w^*(x^*,H^*,z^*) = 0. \quad (8)$$

where H^* is the film thickness, V_0^* is velocity component in porous media, and is given by [13]

$$V_0^* = -\frac{K}{\gamma^*}H_0^*\left[\frac{\partial^2 p^{II}}{\partial x^{*2}} + \frac{\partial^2 p^{II}}{\partial z^{*2}}\right]. \quad (9)$$

where K is the permeability.

Applying boundary conditions Equations (7) and (8) in to Equations (5) and (6) we obtain

$$u^* = \frac{1}{\gamma^{*2}}\frac{(h_m^{II})^2}{(B_0^{II})^2}\frac{\partial p^{II}}{\partial X^*}\left\{\cosh\left(\frac{B_0{}^{II}}{h_m{}^{II}}y^*\right) + \left(\frac{1-\cosh\left(\frac{B_0{}^{II}}{h_m{}^{II}}H^*\right)}{\sinh\left(\frac{B_0{}^{II}}{h_m{}^{II}}H^*\right)}\right)\sinh\left(\frac{B_0{}^{II}}{h_m{}^{II}}y^*\right) - 1\right\} + U\frac{\sinh\left(\frac{B_0{}^{II}}{h_m{}^{II}}y^*\right)}{\sinh\left(\frac{B_0{}^{II}}{h_m{}^{II}}H^*\right)}. \quad (10)$$

$$w^* = \frac{1}{\gamma^{*2}}\frac{(h_m^{II})^2}{(B_0^{II})^2}\frac{\partial P^{II}}{\partial z^*}\left\{\cosh\left(\frac{B_0{}^{II}}{h_m{}^{II}}y^*\right) + \left(\frac{1-\cosh\left(\frac{B_0{}^{II}}{h_m{}^{II}}H^*\right)}{\sinh\left(\frac{B_0{}^{II}}{h_m{}^{II}}H^*\right)}\right)\sinh\left(\frac{B_0{}^{II}}{h_m{}^{II}}y^*\right) - 1\right\}. \quad (11)$$

Integrating Equations (10) and (11) across the fluid film thickness H^*, substituting in to continuity Equation (4) we obtain

$$\frac{\partial}{\partial x^*}\left\{\left[2\left(\coth\left(\frac{B_0{}^{II}}{h_m{}^{II}}H^*\right) - \operatorname{cosech}\left(\frac{B_0{}^{II}}{h_m{}^{II}}H^*\right)\right) - \frac{B_0{}^{II}}{h_m{}^{II}}H^*\right]\frac{\partial p^{II}}{\partial x^*}\right\} + \frac{\partial}{\partial z^*}\left\{\left[2\left(\coth\left(\frac{B_0{}^{II}}{h_m{}^{II}}H^*\right) - \operatorname{cosech}\left(\frac{B_0{}^{II}}{h_m{}^{II}}H^*\right)\right) - \frac{B_0{}^{II}}{h_m{}^{II}}H^*\right]\frac{\partial p^{II}}{\partial z^*}\right\} = -V_0{}^*\frac{B_0{}^{II^3}}{h_m{}^{II^3}} - \gamma^* U\frac{B_0{}^{II^2}}{h_m{}^{II^2}}\frac{\partial}{\partial x^*}\left\{\left[\coth\left(\frac{B_0{}^{II}}{h_m{}^{II}}H^*\right) - \operatorname{cosech}\left(\frac{B_0{}^{II}}{h_m{}^{II}}H^*\right)\right]\right\}. \quad (12)$$

3. Generalized stochastic Reynolds's equation

The geometry of the lubricant film thickness is made up of two parts

$$H^* = h_m{}^{II} + \Lambda$$

$$H^* = \Delta r^*(1 + \varepsilon^*\cos\theta^*) + \Lambda(\theta^*, z^*, \xi^*). \quad (13)$$

where $h_m{}^{II}$ denotes the nominal smooth part of the film thickness, and Δr^* is the radial clearance, ε^* is the eccentricity ratio, Λ represents the random part which is regarded as randomly varying quantity with zero mean, θ^* is the angular coordinate and ξ^* is the random variable .

Taking the expected value of Equation (12) and non-dimensionalising, we obtain the modified Reynolds equation as

$$\frac{\partial}{\partial x^*}\{E[f(N^*,\bar{H}^*) + \Theta N^{*3}]\}\frac{\partial E(p^{II})}{\partial x^*} + \frac{\partial}{\partial z^*}\{E[f(N^*,\bar{H}^*) + \Theta N^{*3}]\}\frac{\partial E(p^{II})}{\partial z^*} = \frac{\gamma^* U}{(\Delta r^*)^2}N^{*2}\frac{\partial}{\partial x^*}E[g(N^*,\bar{H}^*)]. \quad (14)$$

IOP Conf. Series: Materials Science and Engineering 1189 (2021) 012030 doi:10.1088/1757-899X/1189/1/012030

where $f(N^*, \bar{H}^*) = 2[\coth(N^*\bar{H}^*) - \mathrm{cosech}(N^*\bar{H}^*)] - N^*\bar{H}^*$ (15)

$$g(N^*, \bar{H}^*) = -[\coth(N^*\bar{H}^*) - \mathrm{cosech}(N^*\bar{H}^*)] \tag{16}$$

$$N^* = \frac{B_0{}^{ll}}{h_0{}^{ll}}, h_0{}^{ll} = \frac{h_m{}^{ll}}{\Delta r^*} \bar{H}^* = \frac{H^*}{\Delta r^*} \Theta = \frac{K H_0{}^*}{(\Delta r^*)^3}$$

where Θ is the non-dimensionalised permeability parameter
Here $E(\bullet)$ is the expectancy operator defined by

$$E(\bullet) = \int_{-\infty}^{\infty} (\bullet) f(\Lambda) d\Lambda . \tag{17}$$

where $f(\Lambda)$ is the probability distribution function for the stochastic variable Λ. Since many of the engineering surfaces shows a roughness height distribution which is Gaussian in nature, the polynomial form of the Gaussian distribution probability density function is [15]

$$f(\Lambda) = \frac{35}{32a^7}\left(a^2 - \Lambda^2\right)^3, -a < \Lambda < a.$$

$$= 0 \qquad \text{otherwise} \tag{18}$$

where a is the total ridge of the random film thickness variable and the function terminates at $a = \pm 3\sigma$ and σ is the standard deviation.
The film thickness of one dimensional longitudinal roughness pattern is assumed to have the form

$$H^* = h_m{}^{ll} + \Lambda(z^*, \xi^*) \tag{19}$$

$$\frac{\partial}{\partial x^*}\{E[f(N^*, \bar{H}^*) + \Theta N^{*3}]\}\frac{\partial E(p^{ll})}{\partial x^*} + \frac{\partial}{\partial z^*}\left\{\frac{1}{E\left[\frac{1}{f(N^*,\bar{H}^*)}\right]} + \Theta N^{*3}\right\}\frac{\partial E(p^{ll})}{\partial z^*} = \frac{\gamma^* U N^{*2}}{(\Delta r^*)^2}\frac{\partial}{\partial x^*}E[g(N^*, \bar{H}^*)]. \tag{20}$$

For transverse roughness film thickness is given by $H^* = h_m{}^{ll} + \Lambda(\theta^*, \xi^*)$ (21)

$$\frac{\partial}{\partial x^*}\left\{\left[\frac{1}{E\left(\frac{1}{f(N^*,\bar{H}^*)}\right)} + \Theta N^{*3}\right]\frac{\partial E(p^{ll})}{\partial x^*}\right\} + \frac{\partial}{\partial z^*}\{E[f(N^*, \bar{H}^*) + \Theta N^{*3}]\}\frac{\partial E(p^{ll})}{\partial z^*} =$$

$$\frac{\gamma^* U N^{*2}}{(\Delta r^*)^2}\frac{\partial}{\partial x^*}E\left[\frac{g(N^*,\bar{H}^*)/f(N^*,\bar{H}^*)}{1/f(N^*,\bar{H}^*)}\right]. \tag{22}$$

4. Infinitely long porous journal bearing

It is assumed that the bearing is infinitely long in the axial direction, hence no variation of pressure in z^*-direction is found, therefore $\frac{\partial p^{ll}}{\partial z^*}$ term can be neglected.
For longitudinal pattern Equation (20) becomes

$$\frac{d}{d\theta^*}\left\{\left[E\left(f(N^*, \bar{H}^*)\right) + \Theta N^{*3}\right]\frac{dE(p^{ll})}{d\theta^*}\right\} = \frac{\gamma^* U N^{*2}}{(\Delta r^*)^2}\frac{d}{d\theta^*}(E[g(N^*, \bar{H}^*)]). \tag{23}$$

For transverse roughness pattern Equation (22) becomes

$$\frac{d}{d\theta^*}\left\{\left[\frac{1}{E\left[\frac{1}{f(N^*,\bar{H}^*)}\right]} + \Theta N^{*3}\right]\frac{dE(p^{ll})}{d\theta^*}\right\} = \frac{\gamma^* U N^{*2}}{(\Delta r^*)^2}\frac{d}{d\theta^*}E\left[\frac{g(N^*,\bar{H}^*)/f(N^*,\bar{H}^*)}{1/f(N^*,\bar{H}^*)}\right]. \tag{24}$$

where $x^* = R^*\theta^*$, R^* is the radius of the journal (25)

Reynolds boundary conditions are given by

$$E\left(p^{ll}\right) = 0 \qquad \text{at} \qquad \theta^* = 0. \tag{26}$$

$$\frac{d[E(p^{ll})]}{d\theta^*} = 0 \qquad \text{at} \qquad \theta^* = \theta_c. \tag{27}$$

$$E\left(p^{ll}\right) = 0 \qquad \text{at} \qquad \theta^* = \theta_c. \tag{28}$$

where θ_c is angular extent of the film

5. Longitudinal roughness

Integrating Equation (23) and using the boundary conditions Equations (26) and (27)
The dimensionless mean pressure in the integral form is obtained as

$$p^{ll^*} = \frac{E\left(p^{ll^*}\right)(\Delta r^*)^2}{\gamma^* UR^*} = N^{*2} \int_0^{\theta^*} \frac{E[g(N^*,\bar{H}^*)] - \{E[g(N^*,\bar{H}^*)]\}_{\theta^*=\theta_c}}{E\{f(N^*,\bar{H}^*) + \Theta N^{*3}\}} d\theta^*. \tag{29}$$

where $E[g(N^*, \bar{H}^*)] = \frac{35}{32.a^7} \int_{-a}^{a} g(N^*, \bar{H}^*)\left(a^2 - \Lambda^2\right)^2 d\Lambda$

The cavitation angle θ_c is calculated by using the boundary condition Equation (28). Therefore, from Equation (29) we have

$$-E\{[g(N^*, \bar{H}^*)]\}_{\theta^*=\theta_c} \int_0^{\theta_c} \frac{d\theta^*}{E[f(N^*,\bar{H}^*) + \Theta N^{*3}]} + \int_0^{\theta_c} \frac{E[g(N^*,\bar{H}^*)]}{E[f(N^*,\bar{H}^*) + \Theta N^{*3}]} d\theta^* = 0. \tag{30}$$

Above integral is evaluated to obtain θ_c using bisection method
The average load along the line of center $\bar{W}_0^*$ is given by

$$E(W_0^*) = E(W^*)\cos\varphi = -LR^* \int_0^{\theta_c} E\left(p^{ll^*}\right)\cos\theta^* \, d\theta^*. \tag{31}$$

The non-dimensional form of Equation (31) is

$$\bar{W}_0^* = \frac{E(W_0^*)(\Delta r)^2}{\gamma^* \, ULR^{*2}} = -\int_0^{\theta_c} p^{ll^*} \cos\theta^* \, d\theta^*. \tag{32}$$

The average load capacity along normal to the line of centre $\bar{W}_{\frac{\pi}{2}}^*$ is

$$E\left(W_{\frac{\pi}{2}}^*\right) = E(W^*)\sin\varphi = LR^* \int_0^{\theta_c} E\left(p^{ll^*}\right)\sin\theta^* d\theta^*. \tag{33}$$

The non-dimensional form of Equation (33) is

$$\bar{W}_{\frac{\pi}{2}}^* = \frac{E\left(W_{\frac{\pi}{2}}^*\right)(\Delta r^*)^2}{\gamma^* ULR^{*2}} = \int_0^{\theta_c} p^{ll^*} \sin\theta^* \, d\theta^*. \tag{34}$$

Integral involved in Equations (32) and (34) are evaluated using Gauss quadrature16-point formula
The total dimensionless load carrying capacity is given by

$$\bar{W}^* = \sqrt{\bar{W}_0^{*2} + \bar{W}_{\frac{\pi}{2}}^{*2}}. \tag{35}$$

Attitude angle φ is given by

$$\varphi = \tan^{-1}\left(\frac{\overline{W}_{\frac{\pi}{2}}^{\ *}}{\overline{W}_0^{\ *}}\right). \tag{36}$$

6. Transverse roughness

Integrating Equation (24) with respect θ^* using the boundary conditions Equation (26) and Equation (27)
The mean pressure in dimensionless form is obtained as

$$\bar{p}^* = \frac{E(p^*)(\Delta r^*)^2}{\gamma^* UR^*} = N^{*2} \int_0^{\theta^*} \frac{E\left[\frac{g(N^*,\bar{H}^*)/f(N^*,\bar{H}^*)}{1/f(N^*,\bar{H}^*)}\right] - E\left[\frac{g(N^*,\bar{H}^*)/f(N^*,\bar{H}^*)}{1/f(N^*,\bar{H}^*)}\right]_{\theta^*=\theta_c}}{\frac{1}{E[1/f(N^*,\bar{H}^*)]} + \Theta\, N^{*3}} d\theta^*. \tag{37}$$

The angle of cavitation θ_c is calculated by using the boundary condition Equation (28)

$$-E\left[\frac{g(N^*,\bar{H}^*)/f(N^*,\bar{H}^*)}{1/f(N^*,\bar{H}^*)}\right]_{\theta^*=\theta_c} \int_0^{\theta_c} \frac{d\theta^*}{\frac{1}{E[1/f(N^*,\bar{H}^*)]} + \Theta\, N^{*3}} + \int_0^{\theta_c} \frac{E\left[\frac{g(N^*,\bar{H}^*)/f(N^*,\bar{H}^*)}{1/f(N^*,\bar{H}^*)}\right]}{\frac{1}{E[1/f(N^*,\bar{H}^*)]} + \Theta\, N^{*3}} d\theta^* = 0. \tag{38}$$

The non dimensional mean load carrying capacity $\overline{W}_0^{\ *}$ and $\overline{W}_{\frac{\pi}{2}}^{\ *}$ are given by Equations (32) and (34). The total dimensionless load carrying capacity and attitude angle are given by Equations (35) and (36).

7. Theoretical case study

In this paper, the bearing characteristics are the functions of permeability parameter $\Theta\left[=\frac{KH_0^{\ *}}{(\Delta r^*)^3}\right]$, ε^* eccentricity ratio, Hartmann number $N^*\left[=\frac{B_0^{\ ll}}{h_0^{\ ll}}\right]$ and the roughness height parameter $\chi\left[=\frac{a}{\Delta r^*}\right]$. It may be noted that definite integrals involved in various parts of analysis are evaluated numerically by using Simpson's 1/3rd rule and Gaussian quadrature formulae. Cavitation angle θ_c is obtained by using Bisection method. The results are compared with some of Gururajan and Prakash [12], as $N^*\to 0$ the results of non-magnetic fluid are recovered.

8. Angle of cavitation

Figure 2, explains the angular extent of the film θ_c for different values of eccentricity ratio. The roughness parameter is fixed at $\chi = 0.2$. It is observed that longitudinal roughness decreases cavitation angle, θ_c as compared with smooth. The increasing value of Θ increases the θ_c.

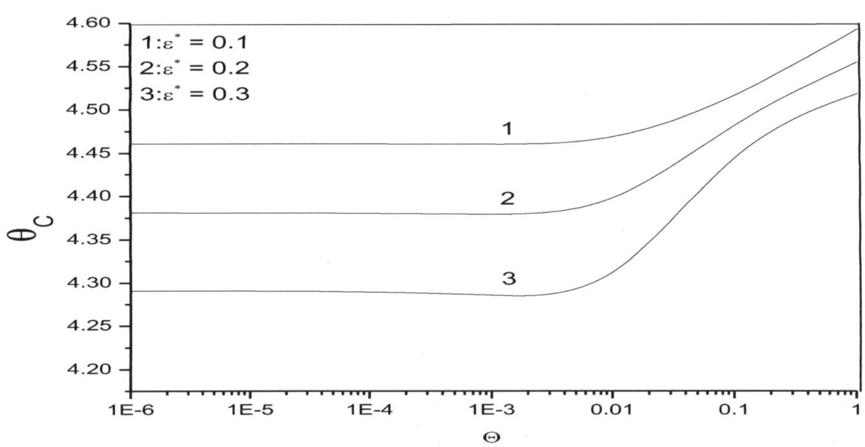

Figure 2 Angle of cavitation θ_c versus permeability parameter Θ for various values of ε^*, $\chi = 0.2$ and $N^* = 1$

9. Load Carrying Capacity:

The Variation of load carrying capacity as a function of eccentricity ratio (ε^*) for different values of Hartmann number N^* by keeping $\chi = 0.49$ and $\Theta = 0.01$ is shown in the figure 3 The Hartmann number and eccentrity ratio (ε^*) increases the load carrying capacityas compared to the smooth case. figure 4 shows that the variation of load capacity with respect to the $N^*$for different values of eccentricity ratio, with fixed value of roughness and permeability parameters. It is seen that the load increases slowly with increasing value of N^* from0 to 3, and after that it decreases. The variation of dimensionless load carrying capacity versus the roughness parameter χ for the various values of the N^* is shown in figure 5 it was observed that load carrying capacity increases with increasing N^* and also it decreases for increasing the value of χ. Interpretation of the variation of load carrying capacity as a function of Θ by fixing the other parameters constant is shown in figure 6 The Hartmann number is to increases the load capacity.It is observed that load capacity decreases with increasing value of Θ, this is because permeable surface becomes the main path to pass a fluid through a porous region, this decreases the load carrying capacity and is significantly high for the particular value $\Theta = 0.01$.

10. Attitude angle:

Attitude angle φ for various values of $\varepsilon^*$and Θ is shown in the table 1 by considering $\chi = 0.2$ and $N^* = 3$. It is observed that attitude angle varies insignificantly with increase of Θ and ε^*.

Table 1: Attitude Angle

ε^*	$\Theta = 0.001$	$\Theta = 0.01$	$\Theta = 0.1$	$\Theta =1.0$
0.100000	68.935165	68.762904	70.937906	69.956387
0.200000	53.571599	54.075043	53.574444	53.163393
0.300000	38.355185	40.095019	36.373348	36.840805
0.400000	22.921001	23.455840	23.214575	23.855422
0.500000	78.896767	33.779022	13.987312	14.431891
0.600000	23.495498	18.601278	7.306093	7.558808
0.700000	15.763951	10.265674	1.944821	2.317761
0.800000	19.718913	8.353334	10.096708	-1.962381

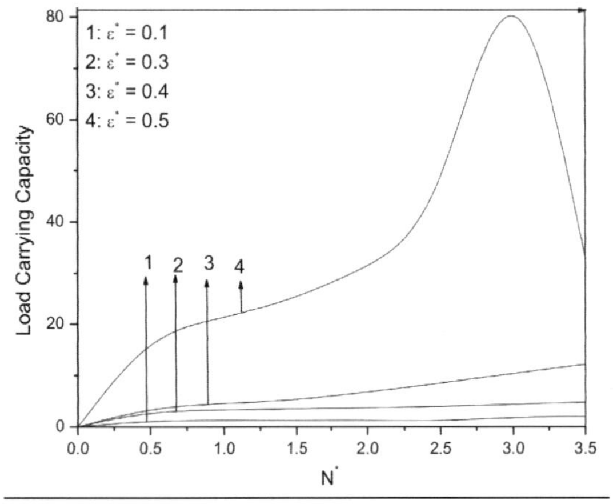

Figure 3 Load carrying capacity versus e for various values of Hartmann Number N* with c = 0.49 and Q = 0.01

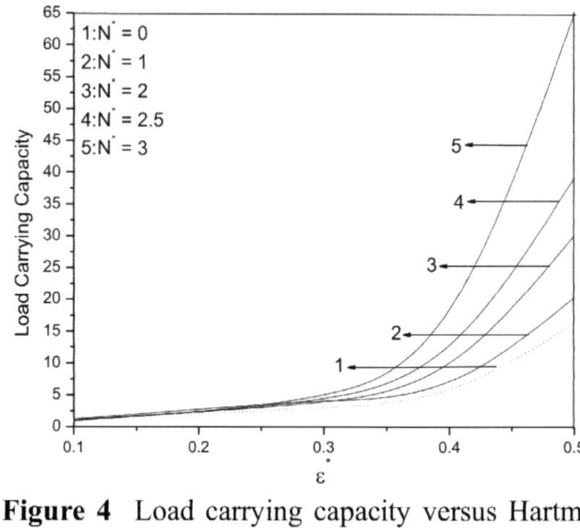

Figure 4 Load carrying capacity versus Hartmann Number N* for Various values of Eccentricity ε* with χ = 0.49 and Θ = 0.01

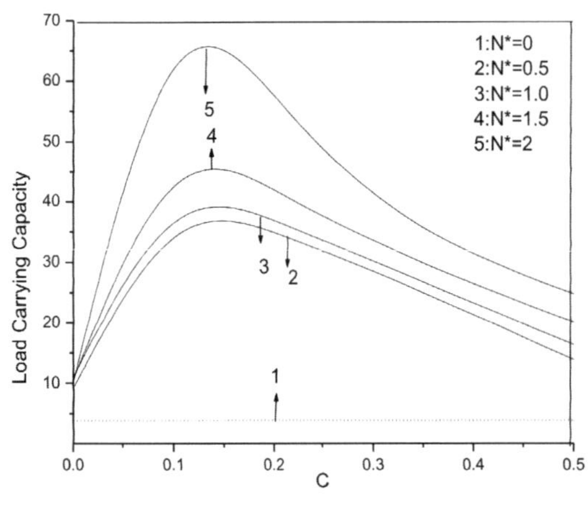

Figure 5 Load carrying capacity versus χ for various values of Hartman number N* with ε*=0.5 and Θ=0.01

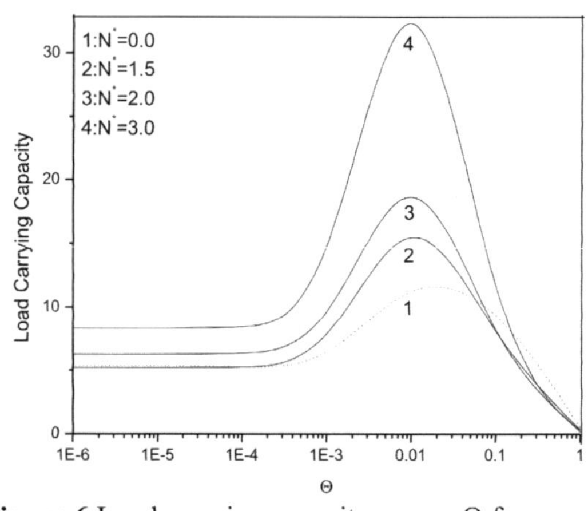

Figure 6 Load carrying capacity versus Θ for various values of Hartmann number N* with χ = 0.5 and ε*= 0.49

IOP Conf. Series: Materials Science and Engineering 1189 (2021) 012030 doi:10.1088/1757-899X/1189/1/012030

11. Conclusion

In this paper, using Christensen stochastic method, a theory was developed to study the influence of MHD fluids in a hydrodynamically lubricated rough porous journal bearing. As an illustration, the case of an infinitely long journal bearing for longitudinal roughness pattern is analyzed. It is observed that the effect of magnetic field is to increase the load carrying capacity. Load carrying capacity is highly significant for the values of $N^* = 3$, χ between 0.1 to 0.2 and $\Theta = 0.01$. Finally, it is concluded that addition of MHD considerably influences the porous bearing performance.

References

[1] Anwar M I, and C. M Rodkiewicz C M, 1972 Non uniform Magnetic Field Effects in MHD Slider Bearing Journal *of Lubrication Tech. ASME*, **94** 101-105.

[2] Gururajan K, and Prakash J, 1999 Surface roughness effects in infinitely long porous journal bearing Journal *Bearing. Tribology Transactions,* **121** 139-147.

[3] Prakash J, and Gururajan K, 1999 Effect of Velocity Slip in an Infinitely Long Rough Porous Journal Bearing, *Tribology Transactions*, **42** 661-667.

[4] Gururajan K, and Prakash J, 2000 Effect of surface roughness in a narrow porous journal bearing, *ASME Journal of Tribology*, **122** 472-475.

[5] Gururajan K, and Prakash J, 2002 Roughness effects in a narrow porous journal bearing with arbitrary porous wall thickness, *International Journal of Mechanical Sciences*, **44** 1003-1016.

[6] Gururajan K, and Prakash J, 2003 Effect of velocity slip on a narrow porous journal bearing *In proceedings of institution of mechanical Engineers, Part J: Journal of Eng. Tribology*, **217** 59-70.

[7] Ramesh B Kudenetti, and Basti D P, 2012 Numerical solution of the MHD Reynolds equation for squeeze film lubrication between two parallel surfaces *Applied Mathematics and Computation,* **218** 9372-9382.

[8] Kalavathi G K, Gururajan K, Dinesh P A, Gurubasavaraj G, 2014 Effect of surface roughness in a Narrow porous journal bearing with a heterogeneous slip/no-slip surface, *International journal of Scientific and Innovative Mathematical Research*, **2** 944-959.

[9] Kalavathi G K, Dinesh P A, Gururajan K, 2016 Influence of Roughness on Porous Finite Journal Bearing with Heterogeneous slip / no-slip surface, *Tribology International*, **102** 74-181.

[10] Kalavathi G K, Dinesh P A, Gururajan K, 2016 Numerical Study of Effect of Roughness on Porous Long Journal Bearing with Heterogeneous Surface, *Journal of Nigerian Mathematical Society*. **35** 468 – 487.

[11] Kalavathi G K, Dinesh P A, Gururajan K, 2017 Numerical Study of Magnetic Field on Rough Porous Narrow Journal Bearing with Heterogeneous Surface, *Materialstoday: Proceedings* **4** 10539–10543.

[12] Qiyin Lin, Qingkang Bao, Kejia Li, M M, Khonsari, Hong Zhao, 2018 An investigation into the transient behavior of journal bearing with surface structure based on fluid- structure interaction approach", *Tribology international*, **118** 246-255.

[13] Yuvaraja B K, and Kalavathi G K, Dinesh P A, Gururajan K, 2018 Effects of Magneto Hydrodynamic Fluids in a narrow Porous rough Journal Bearing, *IJISM*, **6** 7-15.

[14] Prakash J, and Vij S K, 1973 Load capacity and time height relations for squeeze films between porous plates, *Wear,* **24** 309-12.

[15] Christensen H 1969 Stochastic model for hydrodynamic lubrication of rough surfaces, *Part 1. Proceedings of the Institution of Mechanical Engineers,* **184** 1013-1026.

IOP Conf. Series: Materials Science and Engineering 1189 (2021) 012031 doi:10.1088/1757-899X/1189/1/012031

Application of data science in automotive design optimization for faster convergence

Rajeshwar S Kadadevaramath[1] and Mohan Sangli[2]

[1]Professor *& Head, Dept. of Industrial Engineering and Management, SIT,Tumkur*
[2]Research *Scholar, Dept. of Industrial Engineering and Management, SIT,Tumkur*
corresponding *authors:* msangli@intuceo.com, dr.rskmutt@gmail.com

Abstract:

The objectives of simulations during design process is to meet certain objectives, be it minimization of weight, maximization of strength, or having certain response values within some range. The simulations in Automotive design takes several phases and iterations consuming both cost and money. Though the optimization process based on blackbox surrogate models of finite element analysis as validation function has been around for some time, the adoption has been slower than what it should be due to lower confidence and also not having sufficient tools and technology. This paper provides a framework on optimization process using data science based surrogate models. The process involves identifying a baseline, design of experiments to generate initial samples, traditional finite analysis-based simulations, building surrogate models and the using genetic algorithms for optimization of stated objectives of response variables.

Keywords:Design of Experiments (DoE), Response Surface Modeling(RSM), Meta-Model, Surrogate models, Dimensionality Reduction, Multi Disciplinary Optimization (MDO)

1. Introduction

Computer aided engineering (CAE) has been a boon to automotive industry for decades now. Over the period, several computer applications have been developed for every stage in Automotive design. Numerous tools are available for CAD/modeling, pre-processing for solvers, to conduct simulations using Finite element analysis or computational fluid dynamics and then post-processors to visualize and understand the design outputs. These tools require special skills and it takes several hours for each simulation cycle. The automotive design involves thousands of simulations across various disciplines of design for each automobileand the typical concept to final design takes about 18 months or more.

While design process is streamlined over the years, the market conditions have changed enormously. Faster go-to-market on one hand and need for innovative design on the otherhand are placing stress on the design cycles. Product development is becoming increasingly expensive because of regulatory compliances, testing and validation. These factors call for the review of current practices and innovations in design optimization process. The data science based surrogate models, also called as meta-model or response surface models (RSM) are used as validation functions to predict the response values for set of derived design parameter values during the optimization process.

2. Literature survey

The process for design optimization and in particular meta-model based optimization involves understanding steps involved such as Design of experiments (DoE), Finite element analysis, dimensionality reduction, Response surface model (RSM) and genetic algorithm (or other) based optimization techniques. For body-in-white (BIW) development as part of traditional automotive design requires several simulations for each discipline, such as Noise-Vibration-Harshness (NVH), Front crash, Rear crash, Dynamic stiffness etc.[1] There are several approaches available for DoE and the popular ones are Random, Latin Hyper Cube, Full factorial, adaptive design samples, Box-Behnken, Optimal design methods for RSM. While DoE is a first step in selecting the set of experiments to be carried out, any free or commercial Finite Element Analysis (FEA) tools can be used for simulations and obtain the corresponding values for one or multiple responses in each experiment.[2] During a panel discussions on "approximation methods", many researchers observed that there is an important distinction between physical experiments or simulations.[3] Meta-model, being function of design parameter set evaluate the response value much faster. As an example one of the experimentshowed that meta-model took 1 second to produce response value where the traditional FEA took 30 minutes to produce results for same input parameters[4] These meta-models and associated qualitative or sensitivity analysis allow designers to gain insight into the characteristics of the product being designed or analysed[5]. In automotive design, several parameters are involved. The existence of parameters that are degenerative, redundant or even noisy could significantly hamper the performance and interpretability of models built for data containing such features.[6] Moreover dimensionality reduction, i.e. reducing the number of parameters reduces the number of samples as the design space to be covered stands reduced. Certain algorithms like Random Forest, XG Boost are commonly used to analyze the importance of parameters for a given response and have been found to be fairly successful as evidenced by researchers.[7][8] Some practices also involve the use of linear regression model coefficients.[9] The coefficients are obtained by iteratively solving the system of linear equations. However, since the nature of parameters and their influence vary by the problems to be solved, we suggest that parameter importance be obtained by multiple algorithms and then normalize importance scores. Robust rank aggregation method[10] proposed by R. Kolde et al. is one such method found useful in normalizing importance score. By placing a multi-disciplinary design optimization (MDO) within the development cycle, not only results in efficient BIW structure be developed, but a mass reduction otherwise unobtainable by traditional means may be achieved.[11]

3. Research problem and objectives

In most engineering simulations, the objective is to arrive at optimal parameter values to achieve the desired results – meaning, for example, arriving at minimum thickness (weight), material or shape of each component in the system under consideration. Unlike most research papers that focus on limited dimensions (design variables), the reality is that the variables involved are in large numbers. Each simulation takes several hours and is expensive.

The objectives of improvements in design process to meet the new demands should reduce cycle time and provide platform for automation. The current design systems for pre-processing, solvers and post-processing are all work in silos whereas most disciplines are interconnected. The data preparation and exchange of the intermediate results to next in line system is cumbersome. There are tools to automate some part of this process but still cycle times are longer than desired.

Traditional multidisciplinary optimization involving, for example, NVH, Durability, Stress or Crash or other disciplines involves several iterations. The vehicle parts involved in BIW analysis

IOP Conf. Series: Materials Science and Engineering 1189 (2021) 012031 doi:10.1088/1757-899X/1189/1/012031

are common to many disciplines and most common optimization done is for sheet thickness which will reduce weight of vehicle and there by costs.

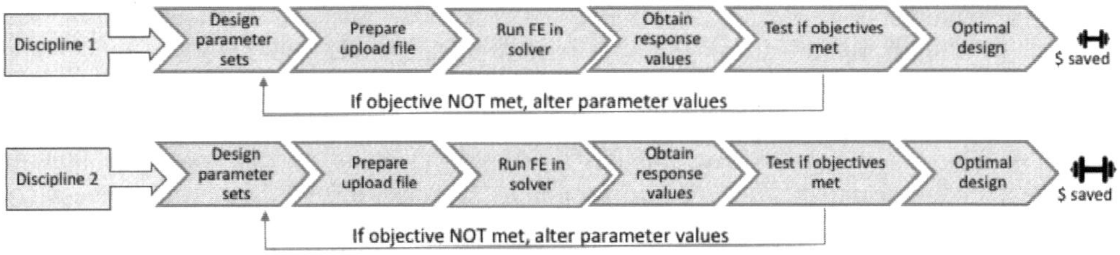

Figure 1. Design optimization

Assuming *W1* and *W2* are savings in each optimization process, as the same design parameters should meet both optimized designs, the actual savings would be limited by retaining the higher of the thickness for each common part. Thus the effective saving *W*, can only be subset of the two savings i.e.

$$(W1 + W2) - \sum (\Delta Wn)$$

where *Wn* is delta weight of parts 1…n due to selection of higher of the 2 optimized thickness.

Objective of this is to put a framework around Design Optimization using the alternative concepts in the process. This shall limit the number of traditional simulations needed in automotive design process through

- scientific selection experiments (DoE),
- identification of sensitive parameters to reduce dimensions,
- developing surrogates to FEA analysis and
- using algorithmsfor optimization of design parameter combinations instead of multiple simulations

4. Methodology

Framework

Meta-model and optimization tools have been gaining importance. The elastic processing power of the cloud and advancementsin data sciences have provided the opportunity to augment the design process through meta-models for responses and then optimizing the designs using these models. Wherever large number of simulations are involved, using the initial few sample design combinations and their responses, the meta-models can be built, and these models can then take the role of solvers reducing significant number of simulations. The new process would change to process as shown in figure 2:

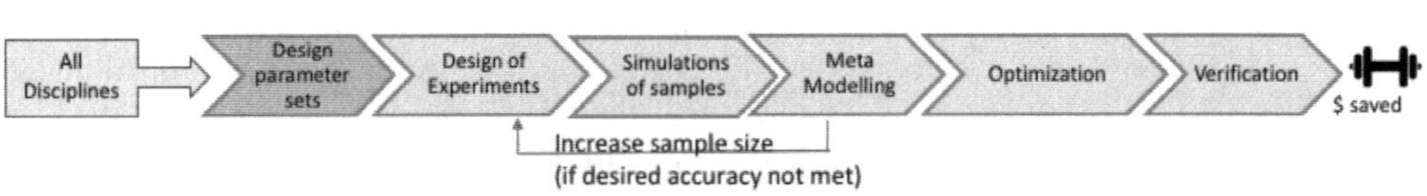

Figure 2. Process change

The detail process looks like what is shown in figure 3 below:

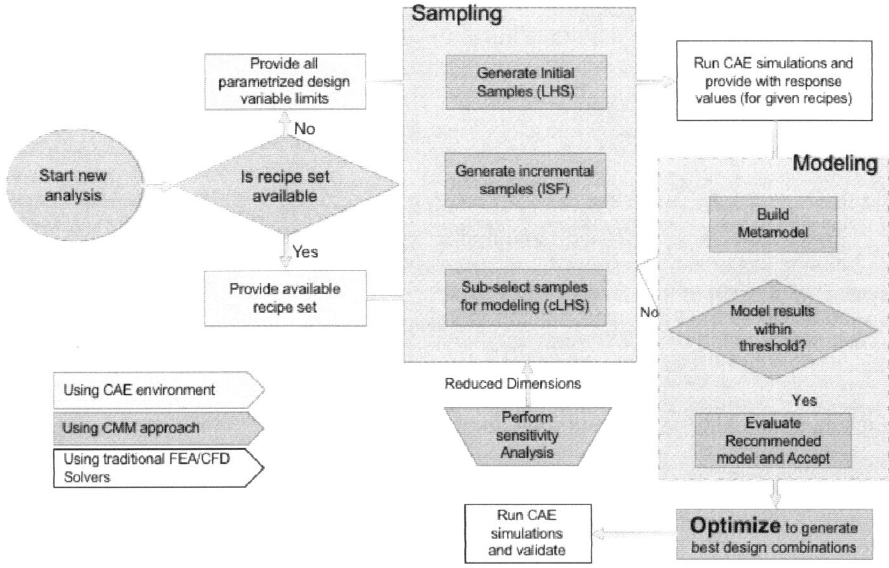

Figure 3. Sampling and Modeling

Optimized designs are of good quality when the models from machine learning tools are built with least error. Larger the size of DoE,lesser is the model error but each simulation in solver is expensive and time consuming. Hence the challenge is to get good design space coverage with least number of samples to produce models with highest accuracy.

Design of Experiment – DoE is a selection of multiple design combinations wherein the design variable valuestake different values within the allowed values. These sample designs are then passed through traditional solvers along with their geometric parameters to get the response values for each design combination. Each simulation in solvers is computationally intensive and expensive. Advanced algorithms are used to achieve minimum samples requirement.

Sensitivity Analysis and Dimensionality reduction - As simulations are expensive, it is suggested to perform DoE in couple of iterations – first one to do sensitivity analysis for dimensionality reduction and then second DoE with reduced dimensions. Various quality insights similar to what is shown in figure 4 below can easily be obtained from underlying analytics.

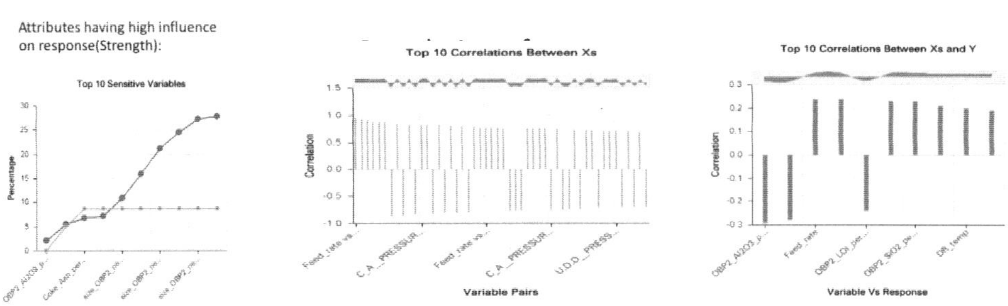

Figure 4. Attributes

Meta-modeling – Several advanced algorithms and machine learning tools are now available to build fairly accurate surrogate models for linear or non-linear behavior of responses with respect to design variables. Dimensionally reduced models some time provide best results by reducing the noise. Metamodel concepts eliminates the designer bias and model parameter tuning requires adept skills in data science.

Optimization–It is a CPU intensive quick checks of combinations of design variable values that provide nearest designs meeting the objectives and constraints setup for design. Choice of algorithms is key to ensure best designs are uncovered by traversing through all possible zones in design space in a manner that converges faster to meet the objectives. Reducing the design variables to only those that are sensitive to set objectives is a big plus. When multi objectives are involved, the desired design can be selected from the designs on Pareto front (figure 5).

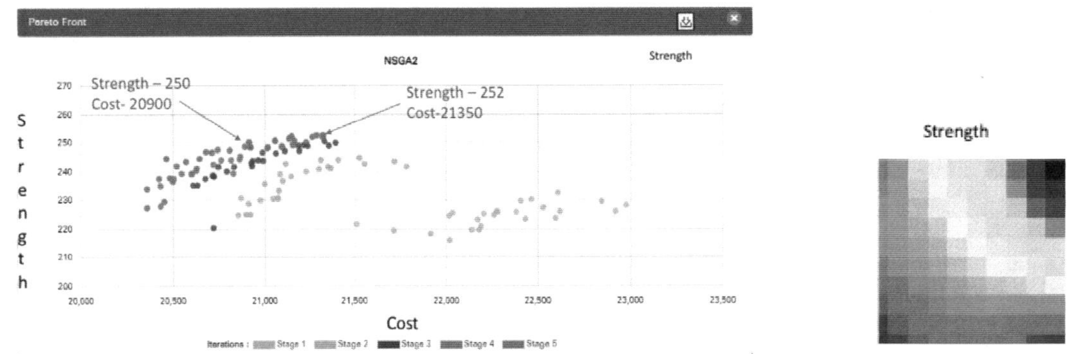

Figure 5. Pareto font image

Unlike the output of solvers in traditional computer aided engineering (CAE), which is large data, this data science-based approach has an opportunity to store the information of all the current and past analysis. Statistical data analysis on this data in addition to insights from previous models will provide good inputs to designers. Augmented analytics now allows to merge human intelligence with machine learning to get most of the data produced in the process.

A simple sample experiment done following the above framework is presented below:

Objective:

1. Minimize the number of Spotwelds while retaining the strength at different locations within the given minimum constraints.
2. Reduce the number of recipes (samples/simulations) that are required to build surrogate model within 10% error for strength predictionsand optimized designs.

Method adopted:

Latin Hypercube based DoE was used to get Spot pitch for 24 lines, spot counts at 25 clustersfrom within the constraints set for each.

- Using DoE and the range of values for spot pitch, we generated 50 samples.
- Traditional analysis methods were used to derivecorresponding response values Lateral, Bending, Torsion and Spot weld counts.
- Sensitivity analysis conducted to analyze the critical parameters and correlations between the weld locations.

- Meta model were built using various algorithm and the best one selected; the error observed was sub 3%.
- For the constraints set on parameter ranges, objectives of maximization of bending, lateral and torsion strengths and minimization of spot weld counts were set.
- Genetic algorithm used for optimization runs and provided the optimized designs.
- Results: In addition to effort and time saved for convergence, the 4% reduction in spot welds count per vehicle could mean savings in millions over product line lifetime (Table 1).

Table 1. Convergence table.

Parameter	Traditional Optimization	Metamodel based MDO
Turnaround Timeline (days)	530	510
Efforts (man-hours)	8	3
# of Simulations runs on HPC Computer	80	50
Easy to change the target responses (Global modes, global stiffness etc)	No	Yes

There are now tools and utilities available to streamline all the steps. For DoE, sensitivity analysis, meta-modeling and optimization, we used Intuceo-Ex, a metamodel based optimization application.

The benefits of such framework are multitude:

- Faster convergence of optimized design and hence faster time to market
- Qualitative data analysis is possible throughout the optimization process
- Unlike in traditional simulations, the intermediate data is small, all results can be saved for continuous learning
- Multi-discipline multi-objective optimization is made easy
- Multiple functions of design process can be streamlined into one simplified design optimization process

5. Conclusion

The innovations in both methods and tools have been substantial to augment the traditional style of design with the meta-model, multi-objective and multi-discipline optimization methods. The ever-reducing design cycle time for automotive design to meet the time to market demands at reduced overall costs can be partially achievedby aligning the skill sets to the newer methods. As the MDO is based on meta-model, the framework can be used in multiple engineering problems where the simulations are involved or the observed behaviors of responses for changes in system parameters is available. Authors believe that further work can be carried out to tailor the presented framework and technologies to cover material and shapevariations and new disciplines in optimization process.

6. References

[1] Salehi M, Noordermeer JWM, Reuvekamp LAEM and Blume A. 2020 *Materials and Design.* 2020 Sep;194. 108879

[2] Husslage B G M, Rennen G., van Dam E R and Hertog Dick den 2011 *OptimEng* **12**, 611–30 Space-filling latin hypercube designs for computer experiments.

[3] Simpson TW, Booker AJ, Ghosh D, Giunta AA, Koch PN and Yang R-J 2004 *Struct MultidiscipOptim* **27(5)** 302–13 Approximation methods in multidisciplinary analysis and optimization: a panel discussion.

[4] Liang L, Liu M, Martin C and Sun W 2018 *J. R. Soc. Interface* **15: 20170844** A deep learning approach to estimate stress distribution: a fast and accurate surrogate of finite-element analysis.

[5] Sobieszczanski-Sobieski J and Haftka RT 1997 *Struct MultidiscipOptim* **14(1)** 1–23 Multidisciplinary aerospace design optimization: survey of recent developments.

[6] Guyon, Isabelle and André Elisseeff 2003 *Journal of machine learning research* **3**. 1157-82 An introduction to variable and feature selection

[7] Díaz-Uriarte, Ramón, and Sara Alvarez De Andres 2006 *BMC bioinformatics* **7.1** Gene selection and classification of microarray data using random forest

[8] Bjoern H Menze, B Michael Kelm, Ralf Masuch, Uwe Himmelreich, Peter Bachert, Wolfgang Petrich and Fred A Hamprecht 2009 *BMC bioinformatics* **10.1** A comparison of random forest and its Gini importance with standard chemometric methods for the feature selection and classification of spectral data.

[9] Xing, Eric P, Michael I Jordan and Richard M Karp 2001 *ICML* **1** Feature selection for high-dimensional genomic microarray data

[10] Kolde Raivo, Laur Sven, Adler Priit and Vilo Jaak 2012 *Bioinformatics* **28.4** 573-80 Robust rank aggregation for gene list integration and meta-analysis

[11] Sheldon A, Helwig E and Cho Y 2011 *8th European LS-DYNA Users Conf.,* **8** Investigation and Application of Multi-Disciplinary Optimization for Automotive Body-in-White Development

Design and Development of interactive and low cost Autonomous Ergonomic Weed Eliminator

Saurabh Suman[1], Vijay Kamate' S Mohan Kumar[3], Umakant P Kulkarni[4]

[1]Computer Science & Engineering, GEC Munger, Munger-811213, Bihar, India
[2]Mechanical Engineering, SDMCET, Dharwad- 580 002, Karnataka, India
[3]Mechanical Engineering, Malnad College of Engineering, Hassan-573201, Karnataka, India
[4]Computer Science & Engineering, SDMCET, Dharwad- 580 002, Karnataka, India

E-mail: [1]saurabhsuman2406@gmail.com,[2]vijay.kamate@gmail.com, [3]dr.s.m.kumar@gmail.com,[4]upkulkarni@yahoo.com

Abstract. Recently it has been observed that there is a wide gap between agricultural income and non agricultural income. This has compelled huge population to look for job in other sectors where income is high. Thus Indian farmer needs accessible, affordable, user friendly farm automation system which encompasses true cost cutting and improving yield. It is a big challenge to have unified approach to design operator neutral platform, conforming to uniform application standards, localized and easy to use agriculture information system that can eliminate distance, labour and time with low cultivation cost. The aim of this paper is to structure, functionality, ergonomically usage with illustration of user friendly collaborative, interactive and low cost autonomous weed eliminator. The discussed weed eliminator can detect weeds using camera vision and eliminates weed through actuation of cutters using smart phone as a client. Further it has been improvised to accept the Voice Command for its operation. A low cost futuristic autonomous weed eliminator has been designed and developed by the authors. The objective of this weed eliminator equipment is to eliminate weeds from the field ergonomically.

1. Introduction

Any unwanted plant in agricultural fields restricting the growth of the crops and thereby reducing the agricultural yield can be termed as a weed. It reduces the agricultural products value, either by reducing the plant growth or by destroying the agro based plants itself. The yield reduction is mainly due to consuming water, light, soil nutrients and agricultural space by the weed [1].

`Impact of Weeds on Crops:` Weeds have high growth rates as compared with agro based crops, and if not controlled properly, they may dominate in the cultivating area. Apart from reducing the yield, the weed also hosts for many plant diseases by providing protection to various deadly worms, insects and microorganisms.

`Classification:` There are around thirty thousand types of weeds in all around the world, out of which about eighteen thousand species are known to cause serious losses for plants, animals and humans. As per agricultural scenario is concerned, they can be broadly categorized into in-row or intra row weeds (IR weed) and inter row or between row weeds (BR weed).

IOP Conf. Series: Materials Science and Engineering 1189 (2021) 012032 doi:10.1088/1757-899X/1189/1/012032

`Economical and Labor Issues Related to Weed Management in Agriculture:`It is estimated by agricultural researchers that around 30% of oil seeds, 50% of food grains and pulses cultivated per annually are lost due to weeds alone, which accounts around 1,05,000 crores of rupees every year [2] forecasted that, if not treated at the proper time, the weeds may results in losses of around 65% of crops. Most of the farmers are still following traditional method of manual weed removal methods like use of (Khurpi), hoe (Salaki), animal driven and simple manual weeders to remove weeds from the fields. Drudgery involved in such operations is more and continuous use of traditional methods may affect the physical and mental well being of the farmers. A weeding process of this kind without considering human capabilities may cause work related injuries to the farmers thereby reducing working efficiency and comfort [3]. In these manual weed elimination stages, most of the farmers follow an unhealthy and awkward working posture which leads to pain in different body parts and further results into work related musculoskeletal disorders (WMSDs).
In-row weed removal is one of the crucial farm operations in crop production and protection system. This kind of weed developed between the crops is a main problem for both dry and wet land farming process, causing a significant loss of agricultural yield. Weed elimination from the crop rows is most tedious and difficult, which needs to be accomplished without damaging the main crops. Farmer in traditional in-row weeding approach removes the weed either by hand plucking or using simple tool like sickle. Thus, a farmer engaged in in-row weed removal process occupies squatting or forward bend postures for prolong period, which is harmful and further leads to work related musculoskeletal disorders (WMSDs) like lower back pain, knee pain, shoulder and neck pain. For in-row weeding many sophisticated weeders are available but these are not affordable by the marginal farmers [4]. The main objective of this work is to design and develop next generation collaborative, interactive and low cost autonomous ergonomic weeder to reduce the risk of musculoskeletal disorders (WMSDs)[5][6].

2. Improvised In-Row Weed Eliminator based on Ergonomics

The initial version of In-row weed Eliminator designed and developed by the authors is shown in figures from Figure 1 to Figure 3. In this equipment camera high resolution camera fixed towards the crop side and display unit is fixed nearer to the farmer end. While moving inside the row, the camera captures the images of both crop and weed. The image thus captured is displayed on the display screen. The farmer after looking into the screen identifies or differentiates between crop and weed in a comfortable standing posture. A cutting tool and its actuation lever are attached to the main chassis to perform the weed removal task. A DC motor is fixed for the operation/rotation of cutting tool using a simple switch. After identifying the weed and its exact location in the display screen, the farmer triggers the cutter with the help of a switch and removes the weed by simply moving the actuating lever comfortably.
The initial version of In-Row weed eliminator has been designed based on the ergonomic guidelines to address these issues of discomfort. The equipment helps in achieving the weeding operation in comfortable working posture with reduced drudgery and work related musculoskeletal disorders. In this approach of in-row weed removal method though achieved its intended goal but still cutter has to be actuated manually by the user. In-Row weed eliminator has been further refined and made much sophisticated to handle by adopting a touch screen based mobile application interface to actuate the cutter by simply touching the weeds on the display screen or giving voice command. This allows the cutter to go to the weed location automatically and destroys it. This equipment can further sense real time farm data for further analytics and visualization for scheduling of other agricultural process. Different profiles of cutting tools and their actuating mechanisms can be designed as an add-on to the existing system to suit the different weed types. The conceptual model of the improvised In-row weed eliminator is shown in Figure 4.

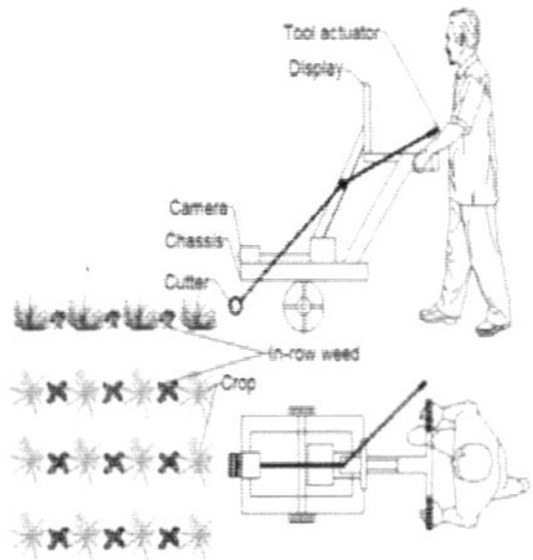

Figure 1. Conceptual depictions of In-row weed removal method

Figure 2. 3D CAD Model of In-row weed eliminator

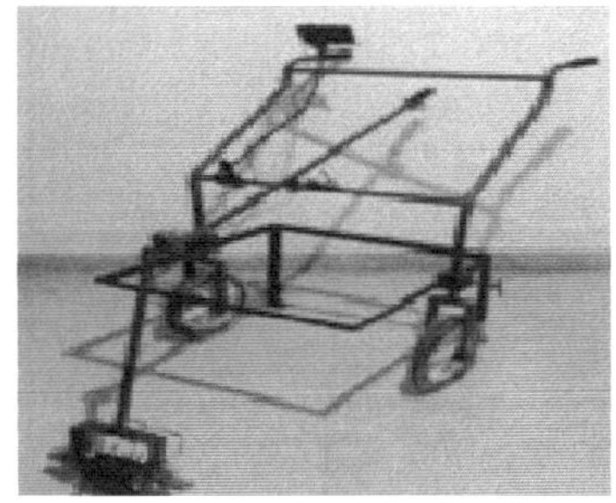

Figure 3. Fabricated In-row weed eliminator

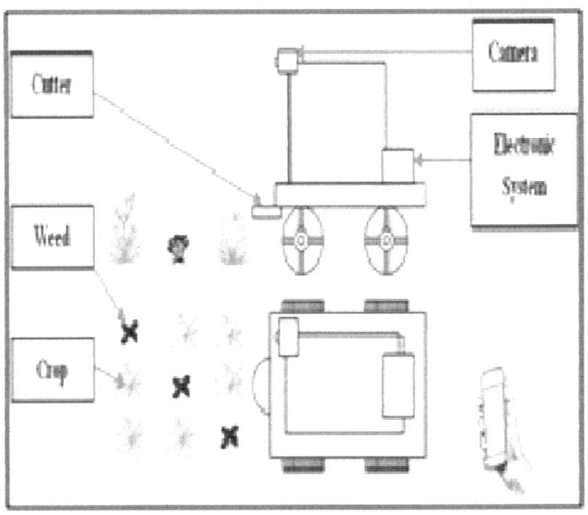

Figure 4. Schematic diagram of conceptual model of In-Row Weed Eliminator

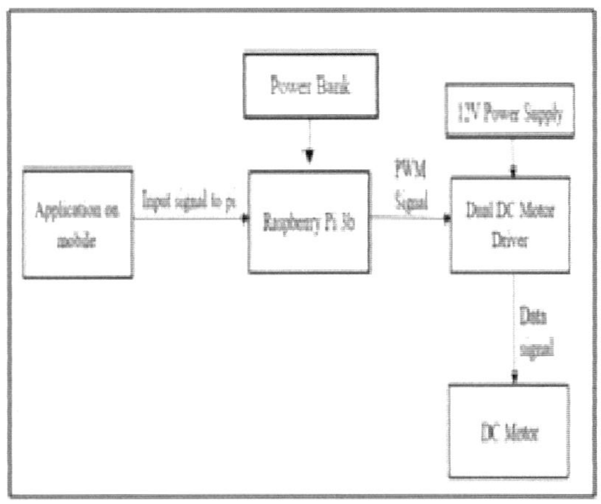

Figure 5. Block diagram of working of electronic circuit

2.1. Details, Specifications and Fabrication

Based on the crop spacing data available from different sources, the chassis length and width is taken as the average value of 600 mm. The chassis height is made adjustable to suit the crop growth at various phases of its existence. The height from the camera lens to the ground was set to 1m in order to allow the solution to target in-fallow weed regrowth for a variety of weed species. The ground clearance underneath the robotic vehicle is fair to adopt Indian farm land. No external shading or lighting has been used for the optical system not to limit vehicle's maneuverability. A high-end image capturing camera captures the images of the cultivated crop and weeds. The operator accordingly then positions the equipment at an appropriate location to remove the weed using the direction control buttons provided in user interface of the mobile client application. The cutter is actuated both in clockwise and anticlockwise by using the buttons provided in user interface of the mobile client application. After successful removal of weed, the operator moves the equipment to next location where weed is present through mobile application.

The `cutter profile` is prepared based on the soil interaction while operating the in-row weed removal process. To a cylindrical drum, various sharp nails are welded all along its surface for better weed removal. Based on the torque generated by the motor, the cutter is tested in the field to ensure its intended performance of removing weeds and loosening of soil for better aeration and moisture absorbing capability. The CAD image of cutter profile and the fabricated cutter are shown in the following Figure 6. In order to actuate the mechanical systems of the current equipment an interface with electronic system is necessary. The user operates the equipment and the cutter using mobile application. When a button is pressed, a signal is sent to the Raspberry Pi board. The Raspberry Pi receives the unique input signal from the mobile application and sends high or low voltage pulse i.e., Pulse Width Modulation (PWM) signal to the dual DC motor driver. Power bank is used to supply power to the Raspberry Pi board. The motor driver supplies power to the motor from the 12V power supply i.e., battery based on the PWM signal from Raspberry Pi board and sends data signal to the DC motor which then starts operating accordingly.

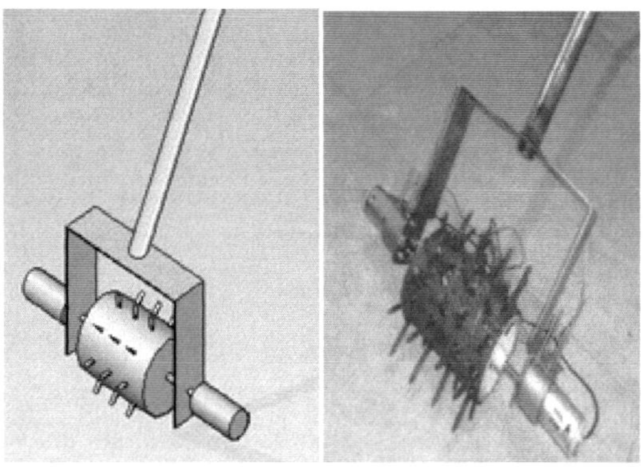

Figure 6. In-row weed cutter profile

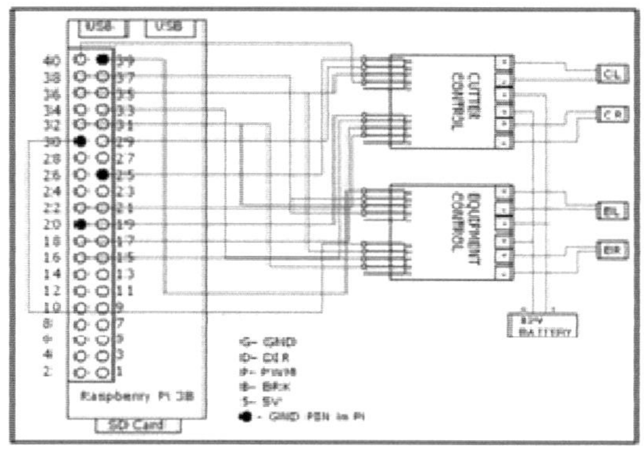

Figure 7. Pinout configuration for Raspberry Pi B3 to motor control driver

2.2. Motor control using Python Programming

Python code has been developed for controlling the vehicle movement by controlling the DC motors using the Raspberry Pi. The code contains of four modules:

- `Pin configuration specification:` It is necessary to specify the pins used in the Raspberry Pi to be connected to the motor driver so as to send signal from the board to the particular pin of the motor driver.
- `Camera feed:` The camera connected to the Raspberry Pi board is programmed to send the feed directly to the mobile application through wireless transmission in real-time situations.
- `Signal receiving and transmitting:` Each button in the mobile application when operated sends a unique signal in the form of alphabets to the Raspberry Pi board which is programmed accordingly to receive, decode and instruct the driver to operate the required motor.
- `Voice Control:` It is used for providing voice based command Like Move Forward, Move Backward, Turn Left, Turn Right, Turn Soft Left, Turn Soft Right, Stop and equipment speed control command.

3. Results and Discussion

The Improvised In-Row Weed Eliminator was physically built and tested for its intended purpose. The salient features of the physical prototype are mentioned below:

- The body/chassis of the equipment was built with mild steel angle bars and plates.

- Three varieties of cutters were fabricated for different weed variants commonly observed in agricultural fields.
- Sufficient clearance was ensured between the crop and chassis structure so as to move the equipment in the agricultural fields without damaging the main crops.
- Electronic components, power source, battery and its related items were mounted on the top of the equipment.
- DC motors are assembled with wheels and cutter shafts, which further receives signals from the operator to function according to the instructions given.
- High definition camera was mounted on to the chassis frame at an appropriate location so as to capture the cutter and its positional area clearly. This live video is noticeable on the mobile screen in real time.
- A separate camera is mounted to capture images of different weeds to collect the image data set of weeds. The prepared image data set of weeds can be used to train the system

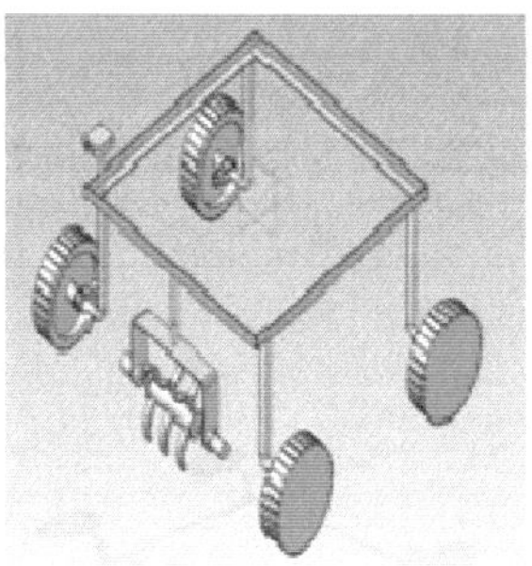

Figure 8. CAD Model

Figure 9. Fabricated equipment

Figure 10. Electronic Component

Figure 11. Different cutters

Figure 12. Different cutters

Figure 13. Weed removed by cutter

The CAD model and the fabricated functional working prototype of the mobile app controlled agricultural weeding equipment are shown in the following figures from figures from Figure 8 to Figure 15.

4. Conclusion

Field test of the developed weed eliminator prototype has been conducted and the found that the equipment is comfortable in its operation. A comfortable working condition for farmer is

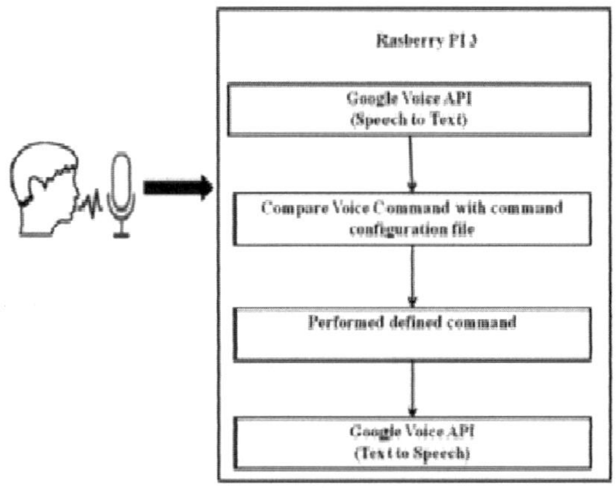

Figure 14. Voice Command to Control Movement of weed eliminator

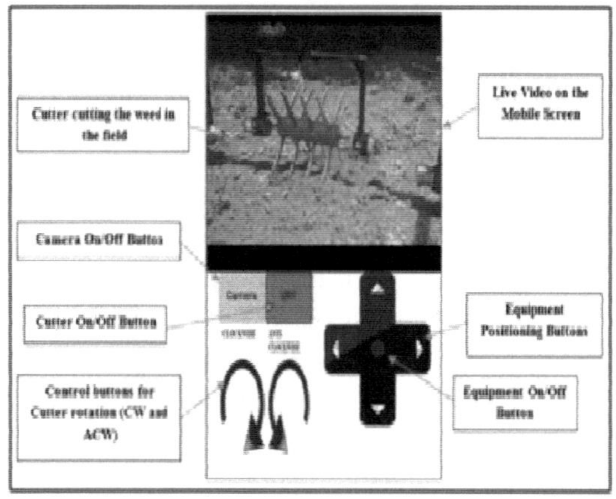

Figure 15. Mobile app screen for equipment and weed control

achieved as there was no need for physical presence in the field during weed removal. Mobile application program can be easily installed to any android operating mobile smart phone and was user friendly in controlling the equipment movement and cutter actuation. Different cutters can be installed to ensure proper weed removal for variety of weeds and in different soil conditions. This method of weed elimination technique avoided traditional way of harmful weeding positions like squatting postures, forward bend postures thus reducing work related injuries in farming community. This In-row weed eliminator can be further improved by implementing deep learning for more accuracy in identification of weed and actuation.

References

[1] Slaughter, D., Giles D., and Downey D. (2008), "Autonomous robotic weed control systems: A review, **Computers and Electronics in Agriculture**, 61 (1), 63-78.

[2] Yaduraju NT., Prasad Babu MBB. and Poonam Chandla. (2006), "Herbi-cide Use. In: Agriculture and Environment", Swaminathan, M.S.and Chadha, K.L. (Eds.) **Malhotra Publishing House, New Delhi, India**, 192-210.

[3] Villarejo D. & Baron S.L. (1999), "The occupational health status of hired farm workers". **ccupational Medicine: State of the Art Reviews**, 114(3), 613-635.

[4] Kumar Satish, Kumar Ashok and Kumar Sanoj (2017), "Performance Evaluation of Developed Manually Operated Rotary Weeder for Vegetable Crops", **Performance Evaluation of Developed Manually Operated Rotary Weeder for Vegetable Crops**, 6(11), 4012-4019.

[5] Lynn McAtamney and E Nigel Corlett (1993), "RULA: a survey method for the investigation of work-related upper limb disorders".**Applied Ergonomics**,24(2), 91-99.

[6] Mohan Kumar S and Vijay Kamate (2019) Anthropometric Data of Farm Workers of North-Karnataka Region of India and its Practice for Better Design of Agricultural Tools and Implements**J Robot Mech Eng Resr 3(2): 8-21**,doi: https://doi. org/10.24218/jrmer.2019.31. Copyright: © 2019 Mohan Kumar.

IOP Conf. Series: Materials Science and Engineering 1189 (2021) 012033 doi:10.1088/1757-899X/1189/1/012033

Optimization of Profile Shift Co-efficient for Highest Contact Ratio in Non-standard Gearing

Dr. A R Rajesh[1,a], Dr. Joseph Gonsalvis[2,b] and Dr. K A Venugopal[3,c]
[1] Selection Grade Lecturer, Department of Mechanical engineering, Government Polytechnic, Kushalnagar, India.
[2] Professor, Department of Mechanical engineering, RNS Institute of Technology, Bangalore, India.
[3] Professor, Department of Mechanical engineering, Malnad College of Engineering, Hassan, India.

E-mails: [a]raajhassan@gmail.com, [b]drjgonsalves@gmail.com, [c]venumce@gmail.com

Abstract. The power transmission in mechanical drives is undoubtedly handled by gears. Tooth profile of a gear characterizes the power transmission. Standard gears mostly employ involute tooth for reasons of simplicity, while non-standard gears employs involute tooth profile shifting either positive or negative to modify the tooth geometry. Profile shifting is practiced in two ways, first known as S-gearing, either S_0 or S± without change in tooth-sum while the second known as Altered tooth-sum gearing, aptly called Z± gearing with change in tooth-sum Z_s, tooth-sum being the sum of teeth on pinion and gear. In S± gearing the center distance will change while in S_0 gearing and Z± gearing the center distance remains unchanged. Altered tooth-sum gearing is a novel method of increasing (Z+) or decreasing (Z−) the tooth-sum of standard gearing either by adding or removing one or more teeth ($\pm Z_e$) in either pinion or gear or both and eventually distributing the resulting total profile shift co-efficient among the gear pair, thus maintaining the same center distance. In designing Non-standard gearing the amount of profile shift is expressed as 'Xm', where 'X' is known as Profile shift co-efficient and 'm' is module of the gear. This has a significant influence on the gearing in terms of load carrying capacity, contact ratio (CR), load sharing etc. In this study contact ratio of altered tooth-sum gearing is focused, as it affects the tooth load sharing with influence on vibration and noise levels as well. Optimization of the profile shift co-efficient will be meaningful in designing gears to give highest contact ratio. Load sharing is observed to be beneficial in Altered tooth-sum gearing using increased tooth-sum (Z+ gearing), besides this method offering unparallel benefits to gear engineering.

Key words: Profile shift, Altered tooth-sum gearing, High contact ratio, Vibration, Noise, Load sharing.

1.Introduction

Gears used in power transmission mostly have involute tooth profile. Most of the involute gears have their tooth profile shifted constituting Non-standard gears. The amount of profile shift is expressed in terms 'Xm', 'X' being the profile shift co-efficient and 'm' is module of the gear. Profile shift is common in gear design to modify the tooth geometry in order to meet certain requirements that are not

IOP Conf. Series: Materials Science and Engineering 1189 (2021) 012033 doi:10.1088/1757-899X/1189/1/012033

fulfilled by standard gears. Such Non-standard gears perform better in terms of tooth strength, contact ratio and many other parameters. Profile shifting in S-gearing may be S_0 gearing or S± gearing, S_0 gearing involves shifting of the tooth profiles in opposite direction such that the center distance remains unaltered, while S± gearing involves shifting of the tooth profiles independently such that their center distance alters. Non-standard gearing can also be obtained using a novel and unique way by altering the tooth-sum of a standard gearing without change in center distance. In this, tooth-sum Z_s of a considered standard gearing is the sum of the number of teeth Z_1+Z_2 of the gear pair operating between a standard centre distance a_o. Altered tooth-sum gearing is different to S± gearing with the possibility of conforming different tooth-sums Z_s' (or $Z_1'+Z_2'$) involving eventual profile shifting without change in center distance, this necessitates the gearing to work on a different pressure angle known as operating (working) pressure angle, α_e. The tooth-sum Z_s of a standard gear pair is altered by adding (Z+ gearing) or removing (Z− gearing) one or more teeth $\pm Z_e$ to give altered tooth-sum Z_s' with eventual negative or positive profile shift respectively, while still working on the same centre distance for a given module and pressure angle. The required amount of total profile shift co-efficient X_e is computed for each number of teeth altered which must be distributed among the meshing gears. Further tooth topping has also to be done to ensure backlash free contact.

In total, altered tooth-sum gearing will undergo the following geometrical changes that are computed using the equations given in table 1:

i. The number of teeth either on pinion or gear or both changes, this alters the size of base circles defining a new common tangent that introduces operating (working) pressure angle α_e.
ii. Eventual profile shifting such that $X_e=X_1+X_2$, here X_1 is decided by designer and $X_2=X_e-X_1$.
iii. Equal amount of tooth topping 'ym' on both the meshing gears.

Table 1. Equation of the parameters in Z± gearing.

Parameters	**Operating (Working) pressure angle (deg)**	**Sum of profile shift co-efficients**	**Tooth topping**
Equations	$\alpha_e = \cos^{-1}\left(\frac{Z_s' \cos(\alpha)}{Z_s}\right)$	$X_e = \frac{Z_s'(\mathrm{inv}\alpha_e - \mathrm{inv}\alpha)}{2\tan\alpha}$	$ym = a_o + (X_e)m - a$

From the computed value of total profile shift X_e, X_1 is decided by the designer, while X_2 is obtained using $X_e=X_1+X_2$ (i.e., $X_2=X_e-X_1$). As same center distance can be specified, altered tooth-sum gears will be subjected to tooth topping 'ym' ensuring backlash-free contact. For the purpose of visualization of altered tooth-sum gearing Figures 1.1 (a, b and c) show the images of complete gear belonging to GR=1:1, m=4mm, $\alpha=20^o$ deg, $Z_s=100$ (50x50), altered with $Z_e=\pm2$, i.e., $Z_e=-2$ so that $Z_s'=48$(24x24) constituting Z− gearing, and $Z_e=+2$ so that $Z_s'=52$(26x26) constituting Z+ gearing.

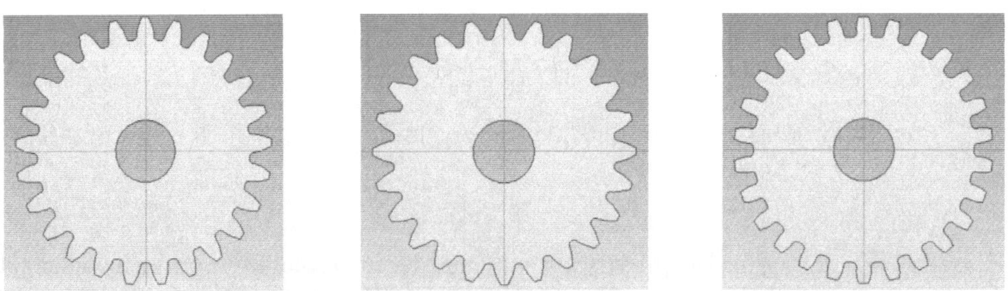

(a) Standard gear (25 teeth) (b) Z− gear (24 teeth) (c) Z+ gear (26 teeth)

Figures 1.1 (a), (b) and (c) CAD models of complete gear.

Contact ratio (CR) refers to the number of pairs of teeth in mesh during the contact period. It is a very important parameter to be considered while designing gears because it decides the smoothness of engagement and also its noise behaviour. A CR between 1 and 2 is referred as normal contact ratio while that between 2 and 3 is referred as High Contact Ratio (HCR). In gear design, practically a minimum contact ratio of 1.4 is specified for satisfactory performance of gears involved in power transmission in order to take care of any misalignments.

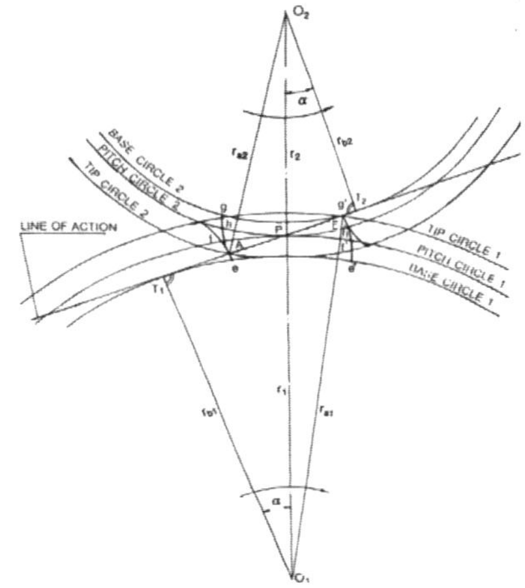

Figure 1.2

A geometrical representation of a gear pair in mesh is shown in Figure 1.2 in which a pinion is driving a gear. Referring to geometry of the figure, CR which is denoted by ε is given by equation (1).

$$\varepsilon = \frac{\sqrt{(r_{a1}^{2}-r_{b1}^{2})}+\sqrt{(r_{a2}^{2}-r_{b1}^{2})}-a_0^{2}\sin\alpha}{Pb} \qquad (1)$$

It can be seen that the contact ratio is dependent on addendum radius r_a, base radius r_b, centre distance a_o, pressure angle α and base pitch p_b, all these parameters in turn are influenced by the profile shift. Hence profile shifting has a significant effect on contact ratio which is discussed in sections ahead.

2.Literature review

Considerable amount of research has been carried in regard to standard gears and Non-standard gears. The amount of profile shift required, its calculation and use are discussed by **Niemann.G** [1]. **Merritt H.E** [2] has mentioned that S± gearing can be used to confirm different tooth-sums to a specified centre distance. **Maitra** [3] has suggested altering the tooth-sum for correcting small deviations in gear ratio. **Joseph Gonsalvis and Sachidananda H.K** [4] have shown using large number of monographs that altered tooth-sum design can tailor the gear drive either for normal or high contact ratio. **Dr.Gonzalo Gonzalez Rey** [5] has opined that gear noise is an indication of a flawed design or inferior product. It is reported that gear noise in standard gears can be reduced by increasing the contact ratio by adding one or two teeth subjected to negative profile shift. **Moldovean Gheorghe et al.** [6] have shown that CR of a gearing can be increased either by increasing the number of teeth, lowering the pressure angle or by increasing the addendum factor. These studies reveal that no attempts are made to optimize the profile shift co-efficient for achieving highest contact ratio.

3.Objectives

In the light of the available literature the following objectives are identified:

i. To understand profile shifting in non-standard gearing by altered tooth-sum design.
ii. To optimize the profile shift co-efficient in order to achieve highest contact ratio.
iii. To identify the high contact ratio cases in non-standard gearing and its useful discussion.

4.Contact ratio in Non-standard gearing

Contact ratio is the average number of pairs of teeth in mesh at any given instance. If CR is between 1 and 2 it is normal CR whereas if it is between 2 and 3 it is HCR. For smooth and continuous meshing the CR must be high. A minimum value of contact ratio specified is 1.4 for practically satisfactory performance. Equation (2) is the modified form of equation (1) that suits non-standard gearing.

$$\varepsilon = \frac{\sqrt{(r_{a1}^{2}-r_{b1}^{2})}+\sqrt{(r_{a2}^{2}-r_{b1}^{2})}-a^{2}\sin\alpha_{e}}{p_b} \tag{2}$$

Low noise behavior in standard industrial gear units is becoming an important selection criterion, it is a factor indicating gear quality to the customer. In several gear tests and practical research the gear contact ratio has been reported to have large effect on noise level, especially in spur gear applications. Lower noise levels are generally associated with gears having higher contact ratio. For this reason gear design leading to high contact ratio is an important key for reduced noise levels. In this study, guidelines to design non-standard gears generated with standard tools of 20° pressure angle and tooth modification using profile shift to have high contact ratio are presented.

5.Optimization of Profile shift co-efficient

In gears contact ratio is undoubtedly an important issue. Standard gearing can fetch a highest CR of 1.94 when meshed with a rack, but still it is less than 2. High Contact Ratio (HCR) is desirable in applications that are characterized by low load carrying capacity, quieter operation and vibration free running. This is possible by profile shift, because the CR increases with increase in profile shift upto certain limit beyond which it decreases. If the GR is 1:1, highest CR can be obtained by equal distribution of resulting total profile shift between the mating gears ($X_1=X_2$), but for other GRs, this does not holds good. Hence there is a need to optimize the profile shift co-efficient in order to achieve highest CR. In profiles shifted gears the geometrical equations involve the terms X_1 and X_2, besides the operating pressure angle α_e. Optimization of profile shift co-efficient X_1 for highest CR is done using $\frac{d\varepsilon}{dX_1} = 0$, the optimized value of X_1 is given by equation (3).

$$X_1 = \frac{(a+m)(r_{b1}-r_{b2})+m(X_e)r_{b2}-r_1r_{b1}+r_2r_{b2})}{m(r_{b1}+r_{b2})} \tag{3}$$

Where, a = specified centre distance; r_{b1}, r_{b2} = base circle radii of pinion, gear
m = module; r_1, r_2 = pitch circle radii of pinion, gear
X_e = total profile shift co-efficient

6.Results and discussion

The CRs obtained using profile shift co-efficient X_1 from equation (3) for different altered tooth-sums and GRs are tabulated in table 2. For illustration Z_s=100 (50x50), m=2 mm α =20°, GR 1:1 and 1:2 are considered. It shows Z± gearing offering a family of altered tooth-sums for a given standard tooth-sum, the amount of profile shift X_1 to be allowed on the pinion that gives highest contact ratio for each tooth altered is determined using equation (3). The designer can choose a particular altered tooth-sum (among the family) that gives the highest value. It is seen that altered tooth-sum gearing altered with positive values of Z_e yield contact ratios exceeding 2. In gears with HCR the length of contact will be longer than two times the base pitch, thus accommodating more pairs of teeth to mesh (a minimum of two pairs of teeth will always be in mesh). Besides smooth engagement, the prescribed GR ($\frac{Z_2}{Z_1}$) can also be achieved along with reduced tooth load resulting in lower stresses. The last column indicates the optimum value of X_1 computed using equation (3). For a GR of 1:1 with Z_e=+5 (Z_s'=105), X_1=−0.964 the highest CR achieved is 2.177 which is 24.11% higher and for a GR of 1:2 with Z_e=+4

(Z_s'=104), X_1=−0.60 it is 2.075 which is 19.4% higher compared to standard gearing (both are Z+ gearing shown in italics). The limitation of this equation is that it holds good only for number of teeth greater than 41 either on the pinion or on the gear, because this number of teeth decides whether the root circle is greater or base circle is greater that affects the effective length of involute or length of path of contact.

Table 2. Family of Z± gearing offering High Contact Ratio (HCR) for Z_s=100, GR 1:1 and 1:2 (considering module m=2mm and pressure angle α=20° deg).

GR and tooth-sum Z_s (Z_1xZ_2)	**CR for Std. gearing**	**No. of teeth altered Z_e**	**Altered tooth-sum gearing Z_s' (Z_1'xZ_2')**	**Profile shift co-efficient, X_1**					
				−1.250	**−1.000**	**−0.750**	**−0.500**	**−0.250**	**X_1 obtained by equation (3) for highest CR**
GR=1:1 Z_s=100 (50x50)	1.755	+3	103 (51x52)	-	-	2.011	2.007	-	**2.013 (X_1=−0.659)**
		+4	104 (52x52)	2.023	2.076	2.084	2.048	-	**2.086 (X_1=−0.829)**
		+5	105 (52x53)	-	2.176	2.152	-	-	***2.177 (X_1=−0.964)***
GR=1:2 Z_s=100 (33x67)	1.738	+3	103 (34x69)	-	-	2.012	2.001	-	**2.013 X_1=−0.710)**
		+4	104 (34x70)	-	-	-	2.059	-	***2.075 (X_1=−0.60)***
		+5	105 (35x70)	-	-	-	-	2.017	**2.070 (X_1=−0.35)**

Bold indicates value of highest CR obtained using X_1 (in parenthesis) computed from equation (3).

7.Conclusion

In view of the fact that contact ratio has been reported to have significant effect on tooth load sharing, noise and vibration levels, an alternative design approach is presented to design non-standard gearing with high CR by altering the tooth-sum. This produces quieter gears without expensive manufacturing costs or structural modifications to the gear housing, especially in spur gear applications. A gear pair with a tooth sum of 100 is altered with +1 tooth to +5 teeth resulting in altered tooth-sums 101 to 105 for GRs 1:1 and 1:2. With judicious selection of number of teeth altered and distribution of resulting total profile shift X_e such that X_1 being computed using equation (3) and X_2=X_e−X_1, it is possible to design gears having highest contact ratio to address the above issue. However this equation comes with a limitation that it holds good for gears having number of teeth greater than 41 either on the pinion or on the gear. Since highest CR can be achieved from the proposed equation without resorting to any structural changes, it may be concluded that altered tooth-sum design can be considered as a novel, promising and unique approach to gear design.

8.Acknowledgements

The authors wish to thank all the eminent authors in the references whose valuable effort and results has been a source of immense information in bringing out this research article.

IOP Conf. Series: Materials Science and Engineering 1189 (2021) 012033 doi:10.1088/1757-899X/1189/1/012033

9.References

[1] Niemann. G, "Machine Elements", Vol. II, Allied publishers, New Delhi, 1980.
[2] Merritt. H.E, "Gear Engineering", 3rd Edition, Pitman, London, 1962, pp 124-125.
[3] Gitin M. Maitra, "Hand Book of Gear Design", TMH, New Delhi.
[4] Dr. Joseph Gonsalvis and Sachidananda. H.K, "Altered Tooth-sum Gearing for High Contact Ratio Gearing", Proceedings of the ETIME - 2006, BMSCE, Bangalore.
[5] Dr. Gonzalo Gonzalez Rey, "Higher Contact Ratios for Quieter Gears", Gear solutions, January 2009, pp 22-27.
[6] Moldovean Gheorghe, Gavrila C Catalin and Huidan Livia, "Ways to Increase the Contact Ratio for Spur Gears", Annals of Oradea University, Fascicle of Management and Technological Engineering, Vol. XI (XXI), 2012, NR1, pp 2.82-2.88.

IOP Conf. Series: Materials Science and Engineering 1189 (2021) 012034 doi:10.1088/1757-899X/1189/1/012034

Acoustic emissions from a high-speed propeller

R Soundararajan[1] **and T J S Jothi**[1,2]

[1]Department of Mechanical Engineering, National Institute of Technology Calicut, Kerala, India.
[2]Author to whom any correspondence should be addressed.

Abstract. This paper investigates the far-field noise signature of propellers rotating at a speed range of 2500-6000 rpm. The experiment is conducted on a two-bladed propeller with a tip to tip diameter of 25.4 cm and a pitch of 17.8 cm. The propeller is connected to a brushless DC stepper motor of 900 rpm/V capacity that helps in achieving the above-mentioned propeller speed. The experiments are conducted inside the anechoic chamber to simulate the free-field environment. The acoustic spectra showed that the propeller noise at higher rpm consists of both tonal and broadband frequency components. The tonal noise and its harmonics were found to be lower than Blade Passing Frequency. The tonal peaks are more pronounced in higher rpm when compared to the lower rpm. Acoustic measurements also revealed the distinct variations in the noise levels with rpm. The Overall Sound Pressure Level tends to increase as the speed of the propeller increases. The rate of increase in noise levels is dominant at higher rpm compared to that at lower rpm. The propeller noise is 7 dB higher than the brushless motor noise at higher speeds. The obtained results are compared with previous research papers.

1. Introduction

Propellers are one of the crucial components of propulsion of unmanned aerial vehicles (UAV). Propeller is one of the cheapest parts of UAV, but it has a significant contribution to its efficiency and its noise. The aviation industry is mainly focusing on designing a more quitter propeller without affecting much of the propeller efficiency. Now, most of the UAVs use electric motors, which produce less noise than conventional engines. UAVs with reduced noise more suitable for the working environment. There are more UAVs and drones used in military applications. In such a case, when the propeller noise is less, it becomes difficult to trace the detect the UAV and the survivability increases. In civil applications, the propeller noise has to be low for the reduction of noise pollution and passenger comfortability. Maintaining an appropriate noise level is vital in airports. Propeller noise reduction is an essential aspect of propeller design. The propeller noise depends on various parameters like diameter, chord length, pitch, aerofoil type, speed, and the number of blades. Ffowcs-Williams and Hawkings [1] in their work have analytically predicted the noise produced by the arbitrarily moving surface using the Acoustic analogy proposed by Lighthill [2,3]. They have derived the density field for turbulence for moving objects due to the arbitrarily moving multipole sources. They have derived equations to represent the surface distribution of surface sources in arbitrary motion. They have also made further deductions on high speed moving surfaces. Farassat et al. [4 - 6], based on the Ffowcs-William-Hawking equation solutions, has derived the formulas for the noise of the rotating blade. Yang et al. [7] have numerically studied the effect of geometric parameters of the propeller, such as diameter, chord width, and pitch angle on the acoustic and aerodynamic performance. They have concluded that the diameter has a significant effect on the aerodynamic properties, especially in the propeller blade tip. When the diameter or the pitch is varied, the SPL reduces by almost 0.6 dB. However, the pitch angle does not affect SPL when it is above 71 degrees. Further, the chord length has its impact on noise when only it decreases. Pagano et al. [8] have carried out an optimization of a propeller in pusher configuration without much reduction in the propeller

[2]Author to whom any correspondence should be addressed. Email Id: tjsjothi@nitc.ac.in

efficiency. During optimization, they have considered tonal noise due to the blade tip thickness, and the broadband noise due to leading-edge and trailing edge. They have found out the tonal noise is due to the flow past the blade tip and the interaction of blade exhaust with the inner part of the blade. Marino [9] has carried out both numerical and experimental investigations on the acoustic emissions in propellers with varying advance ratios. On varying the microphone along axial positions, they found that the acoustic pressure was maximum near the tip region where the trailing vortex was higher. Chirico et al. [10] conducted a computational study using Computational Fluid Dynamics (CFD) solver to acoustically analyze various propeller designs. They used a validated CFD method for tonal noise estimations. Unequally spaced hub design appeared to be noisier than the baseline design. While operating at low speeds, offloaded tip design was producing less noise compared to the baseline design. This paper presents an experimental investigation of aeroacoustic properties of small diameter propeller. The noise study is mainly concerned about the acoustic characteristics of the propeller at different speeds from 2500 to 6000 rpm. The various acoustic spectra of the propeller with the motor are studied. An attempt is carried out to explore the various components of propeller noise at different speeds. The tonal and broadband characteristics of the propeller motor system are studied. OASPL of the propeller rotating at different rotational speeds is compared.

2. Experimental Setup

Experiments are carried out in an anechoic chamber of dimensions 2.6 m x 2.6 m x 3 m for a free-field environment. The chamber is soundproofed with polyurethane wedges. The cut-off frequency of the anechoic chamber is 300 Hz. The experiment is conducted on a two-bladed propeller with a tip to tip diameter of 25.4 cm and a pitch of 17.8 cm. The propeller is connected to a brushless DC stepper motor of 900 rpm/V capacity that helps in achieving the required speed. The speed of the motor is controlled by 40 amps Electronic Speed Controller (ESC). The motor is powered by a 3S LiPo battery of 11.1V capacity. An IR Sensor module is used to measure the instantaneous speed of the rotating propeller. The acoustic data are recorded using a quarter-inch free-field condenser microphone (PCB make, Model number 378C01). The sensitivity of the microphone is 2mV/Pa. The experimental setup and the schematic diagram are shown in Figure 1.

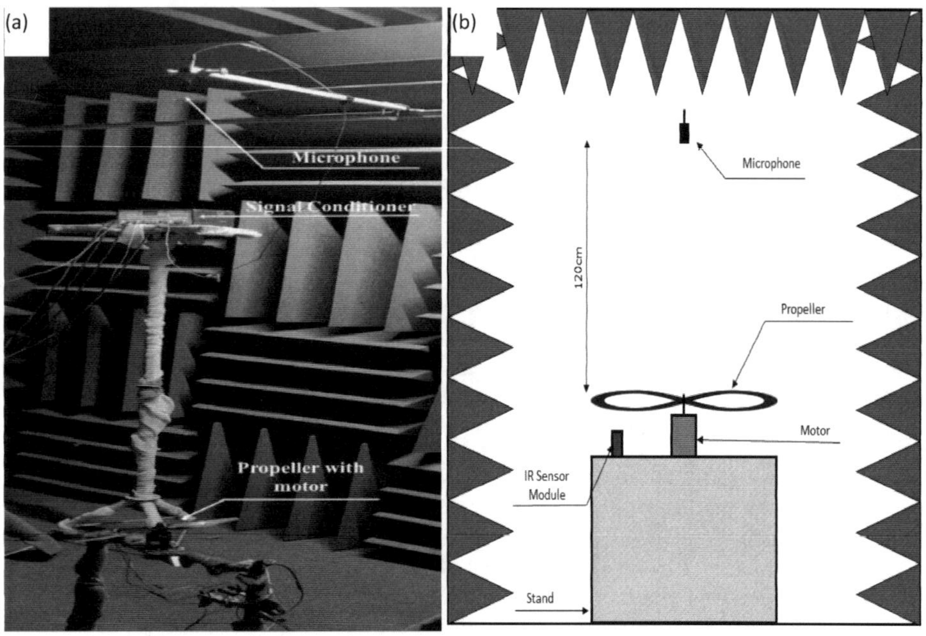

Figure 1: (a) Photograph and (b) Schematic diagram of the experimental setup

3. Experiment Methodology

The motor was mounted on a stand made out of mild steel links. The Infrared (IR) Sensor module was attached to the stand at a distance of 7 cm. The motor is calibrated with the ESC. The IR Sensor was calibrated with a

Digital Optical Tachometer. Both the speed of the motor and IR sensor is controlled by an Arduino Uno. The microphone is placed vertically above the motor at a distance of 120 cm from the tip of the propeller hub. The microphone data was captured at a rate of 150 kSa/s. The data acquisition of the microphone is made by using LabView Software. The Power Spectral Density and Overall Sound Pressure Level (OASPL) calculations were carried out in MATLAB. The uncertainty in the position is 0.1 mm, and the uncertainty in the speed measurements is 50 rpm.

4. Result and Discussions

Figure 2 shows the comparison of the acoustic spectra of the propeller-motor system and only the motor rotating at 6000 rpm. The propeller noise shows larger number of tones with a fundamental tonal frequency at the Blade Passing Frequency (BPF) 146.7 Hz, and the subsequent harmonic tones at higher frequencies. The main reason for tonal noise could be due to the Blade Vortex Interaction [11]. The vortex formed by the fast-moving blade is cut by the rotor blade. This acts as a source for tonal noise generation.

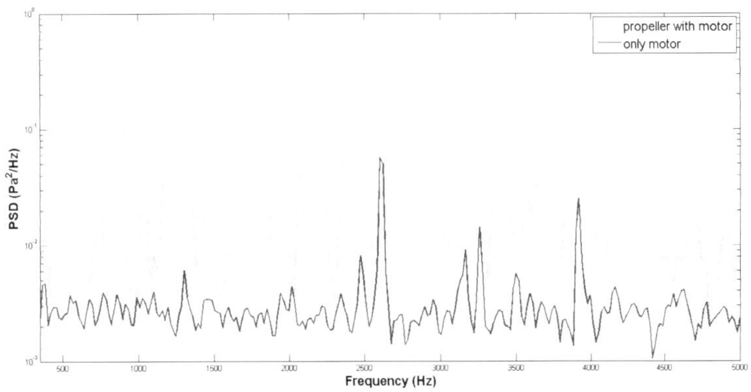

Figure 2: Acoustic spectra of the propeller with motor and only motor at 6000 rpm

As the frequency increases above 2.5 kHz, tones are not the harmonic. This is due to the contribution of the motor noise [12]. The frequency region in the range of 1.5 – 5.0 kHz shows a broadband noise that might be due to the laminar boundary layer vortex shedding. The Tollmein-Sichlichting waves are amplified by the bubble shear layer. This, when convected along the blade span, leads to the production of a dipole noise. Since there is a variation of velocity along the radius of the propeller, this tonal noise appears as a broadband bump at the higher frequency [13].

Table 1: OASPL in dB for propeller and only motor at four different speeds

Speed (rpm)	Propeller noise (dB)	Motor noise (dB)
2580	38.91	38.35
3500	39.88	38.61
4930	45.42	38.83
6000	46.89	39.46

The OASPL values of the propeller noise and the only motor were calculated, and is tabulated in Table 1. The OASPL was calculated for a frequency domain of 300 – 5000 Hz. The OASPL variation of the motor is minimal with speed. The OASPL value of the propeller with motor increases with rpm. The rate of increase in noise levels is dominant at higher rpm compared to that at lower rpm. The propeller noise is around 7 dB higher than the motor noise at higher speeds, whereas at lower speed propeller OASPL is about 1 dB higher than the motor noise. There is a significant increase in the OASPL as the speed increases. This might be due to the more contribution of tonal noise on the OASPL.

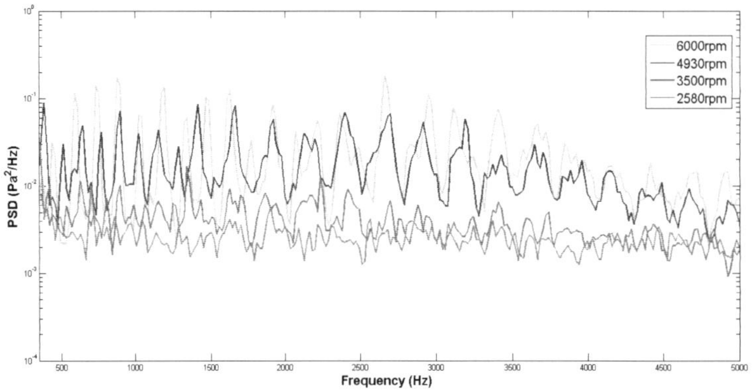

Figure 3: Acoustic spectra of the propeller with motor at different rpm.

Figure 3 shows the far-field acoustic spectra of the propeller with the brushless motor at 6000, 4930, 3500, and 2580 rpm by keeping microphone at a constant distance of 120 cm vertically above the propeller. The plot contains both tonal and broadband noise. The fundamental tonal noise and its harmonics are appearing in low frequencies, and the broadband noise is appearing in the high-frequency region. The tonal noise and its harmonics are near the BPF. The number of tonal peaks for 4930 rpm is less than that of 6000 rpm. This is because, as the speed of the propeller decreases, the interaction of the vortex by rotor blade decreases. Hence the number to tones subsequently decreases with a reduction in speed. The tonal noise and broadband noise produced by a propeller at low speeds are significantly less compared to that of high speeds, as the speed of the propeller decreases, the change of velocity along the span of the propeller decreases. This leads to the formation of a less significant dipole. This might be a reason for the appearance of non-distinguishable broadband noise at lower speeds. This also explains the smaller amplitude of broadband noise as the speed decreases.

5. Conclusion

Experiments were carried out in the paper to study the acoustics of propeller noise at different speeds. The speed of the propeller is varied between 2500 to 6000 rpm. The propeller noise consists of both tonal and broadband noise components. The tonal noise appears in the low-frequency region (300 - 1500 Hz). Broadband noise was occurring in the high-frequency region (> 1.5 kHz). The tonal and broadband noise increases with an increase in speed. The OASPL increases with an increase in speed. The tonal noise contributes more to the OASPL. The study also revealed that the propeller noise is 7 dB higher than the brushless motor noise at higher speeds.

6. References

[1] Ffowcs Williams J E and Hawkings D L 1969 Sound generated by turbulence and surfaces in arbitrary Motion *Philos. T. R. Soc.* A **264** pp 21-342.

[2] Lighthill M J 1952 On sound generated aerodynamically. I. General theory *Proc. of the Royal Soc. of London* A **211** pp 564- 587

[3] Lighthill M J 1954 On sound generated aerodynamically. II. Turbulence as a source of sound *Proc. of the Royal Soc. of London* A **222** pp 1-32.

[4] Farassat F and Succi G P 1980 A review of propeller discrete frequency noise prediction technology with emphasis on two current methods for time domain calculations *J. Sound Vib.* **71** pp 399–419.

[5] Farassat F 1981 Linear acoustic formulas for calculation of rotating blade noise *AIAA Journal* **19**:1122–30.

[6] Farassat F and Brentner K 1998 The acoustic analogy and the prediction of the noise of rotating blades *Theoret. Comput. Fluid Dynamics* **10** pp 155–170.

[7] Yang L, Huang J, Yi M, Zhang C, and Xiao Q 2017 A numerical study of the effects of design parameters on the acoustics noise of a high efficiency propeller *Acoust. Phys.* **63** pp 699-710.
[8] Pagano A, Barbarino M, Casalino D, and Federico L 2010 Tonal and broadband noise calculations for aeroacoustic optimization of a pusher propeller *J. Aircr.* **47** pp 835–48.
[9] Marino L 2010 Experimental analysis of UAV-propeller noise 16th AIAA/CEAS Aeroacoustics Conf..
[10] Chirico G, Barakos G and Bown N 2018 Numerical aeroacoustic analysis of propeller designs *The Aeronautical Journal* **122** pp 283-315.
[11] Brooks T F, J Ralph Jolly Jr and Marcolini M A 1988 *Helicopter main-rotor noise: Determination of source contributions using scaled model data* NASA Reference Technical Paper 2825.
[12] Sinibaldi G and Marino L 2013 Experimental analysis on the noise of propellers for small UAV *Appl. Acoust.* **74** pp 79-88.
[13] Angus L, Wong K C and Auld D 2008 Broadband noise reduction on a mini-UAV propeller. *14th AIAA/CEAS Aeroacoustics Conference*

Numerical Investigation of Single-Phase Heat Transfer in Converging and Diverging Microchannel

R Prajwal Athreya[1,a], **J Tejas**[1,b], **Mascarenhas Venson**[1,c] and **N S Srikanth**[2,d]

[1]UG scholars, Department of Mechanical Engineering, NIE Mysuru.
[2]Assistant Professor, Department of Mechanical Engineering, NIE Mysuru.
[a]prajwalathreyar@gmail.com, [b]tejasjayanna0@gmail.com, [c]vensonvm@gmail.com, [d]srikanthns43@gmail.com

Abstract. Miniaturization of electronics has been the new trend in recent times. This has resulted in an exponential increase of heat flux and hence demands novel cooling methods. Insufficient cooling results into degraded performance, reduced power and, in the long run, may result in device failure. Microchannel cooling technology, due to its several advantages/challenges has been the central point of many researchers in the field of electronics cooling. The current paper concentrates on the investigation of single-phase heat transfer characteristics in converging and diverging microchannel. The numerical investigation of converging and diverging microchannel has been performed using commercially available solver ANSYS FLUENT. The heat transfer coefficient, Nusselt number, and outlet temperature were monitored for the various combinations of converging and diverging microchannels. Also, the results have been compared with a straight microchannel. It was observed that for the Reynolds number ranging from 250-1000 converging microchannels showed a higher value of heat transfer coefficient than the straight channel. Also, it was found that the heat transfer coefficient was directly proportional to the slope of the convergence. The value of h was 10491 W/m^2K for the slope of $m=5\times10^{-3}$. The results for diverging microchannel have also been extensively tabulated in the paper. It was also noticed that the pumping power was reduced in diverging microchannels when compared to a straight channel.

1. Introduction

The fluid flow through microchannel is one of the most efficient processes in heat exchange technology. At present, it is one of the key technologies in the field of chip cooling. Microchannel provides a high surface to volume ratio thus has a high heat transfer rate. The improvements of single-phase flow in microchannels can be divided into two aspects: heat transfer and fluid flow. The problem of obtaining a compact, high performance forced fluid cooling of planar IC was investigated by Tuckerman and Pease [1] in 1981. It was a pioneering experimental study in the field of microchannel heat transfer.

A review of microchannel heat exchangers was given by Goodling [2] in 1993. An experimental study with water as a coolant in microchannel heat exchangers was conducted by Peng et al [3] in 1994. They studied the heat transfer characteristics for the forced convection flow of water through rectangular microchannels. Channel hydraulic diameter was ranged from 133μm to 367μm and aspect ratios from 0.333 to 1. Reynolds number for the flow ranged from 200 to 1500. They developed the correlations for the Nusselt number and friction factor for the flow of water in the channel. It was also found that aspect ratio had a considerable effect on the heat transfer rate and pressure drop both in turbulent and laminar flows. Slowly the research began to shift towards the improvement of flow characteristics through surface and geometrical enhancement of microchannels. Kandlikar [5] in his paper of 2006 compares the single phase flow characteristics of water in plain microchannel and enhanced silicon microchannel with offset fin geometry. They employed flow restrictors at the inlet to act as physical pressure drop element. They developed a parameter called "pumping power coefficient" to evaluate the effectiveness of enhanced microchannels. It was found that the maximum COP for the plain microchannels is 11, whereas for the enhanced microchannel it was 290. Hence it was concluded that enhanced microchannel was more effective at heat dissipation despite the increase in the pressure drop. With the availability of commercial solvers for CFD, in the early 21st century, slowly there were a few rising developments in the analytical studies of 2D and 3D fluid flow in microchannels. Varying geometry of the microchannels exposes varying amounts of effective heat transfer area. Thus geometry of the channel affects the pressure drop and heat transfer performance which was clearly shown by Shah and London [4]. Dharaiya et al [6] studied the entrance header effects in small microchannels. Numerical simulations for three dimensional model of microchannel using CFD software tool FLUENT were carried out. Chandra et al [7] studied single phase flow through a series of uniform microchannels connected via converging- diverging sections with and without throat. It was inferred that converging-diverging part without throat shows better performance over with throat configuration.

In the present work, converging/diverging microchannels have been considered for the study of its heat transfer behaviour. Constant heat flux boundary condition has been applied to the bottom wall. Extensive numerical analysis has been performed by varying the angle of convergence and divergence. The heat transfer coefficient for different cases has been tabulated. The results have been compared with straight rectangular microchannel.

2. Nomenclature

a_i= height of the converging microchannel at inlet
b_i= width of the converging microchannel at inlet
a_o= height of the converging microchannel at outlet
b_o= height of the converging microchannel at outlet
L= length of the microchannel
m = slope of convergence/ divergence= [(ai-ao)/L]
q" = total heat flux

D_h = hydraulic diameter
Nu = Nusselt number
Re = Reynolds number
α = aspect ratio = [a/b]
h = heat transfer coefficient
ρ= density of the fluid
μ = dynamic viscosity

3. Governing Equations

3.1. Continuity Equation

Assuming the flow to be steady and incompressible continuity equation becomes

$$\frac{\partial u}{\partial x}+\frac{\partial v}{\partial y}+\frac{\partial w}{\partial z}=0 \qquad (1)$$

3.2. Momentum Equation

Momentum equation for x-direction is given below (2). Similarly, it can be written for y and z directions.

$$u\frac{\partial u}{\partial x}+v\frac{\partial u}{\partial y}+w\frac{\partial u}{\partial z}=(-\frac{1}{\rho})\frac{\partial p}{\partial x}+\nu\nabla^2(u) \qquad (2)$$

IOP Conf. Series: Materials Science and Engineering 1189 (2021) 012035 doi:10.1088/1757-899X/1189/1/012035

3.3. Energy Equation

The contributions from radiation and other volumetric heat sources are accounted by the source term S_h.

$$\frac{\partial}{\partial t}(\rho E) + \nabla.(\vec{v}(\rho E + p)) = \nabla.(k_{eff}\,\Delta T) + S_h \tag{3}$$

4. Computational Domain and Numerical Procedure

The heat transfer and fluid flow in microchannels with varying cross sections are studied using commercially available CFD solver, ANSYS FLUENT. Water was considered as the working fluid. It enters the microchannel with a fully developed velocity profile. Figure 1 shows the geometry of the convergent/divergent microchannel used in the present study.

Temperature of 300 K was maintained at the inlet. Initially, numerical simulation was performed for a straight microchannel with constant heat flux boundary condition at the bottom wall and the obtained results were compared with the scheme proposed by Shah and London [4].

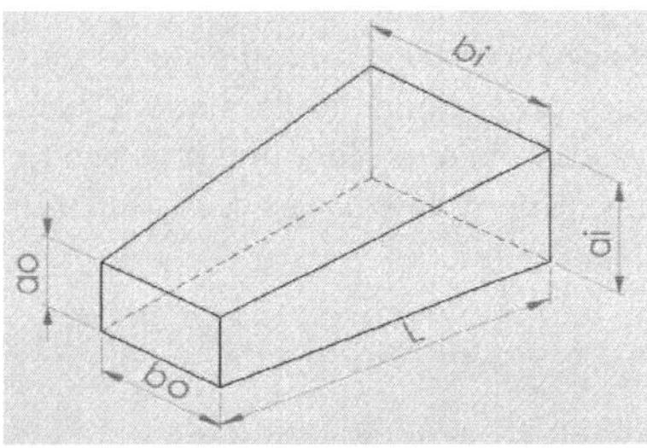

Figure 1. Cross-sectional area of converging/diverging rectangular microchannel.

Table 1. Comparison of Nu, from current numerical work with Shah and London [4].

α (a/b)	0.5
Re	100
Nu[4]	3.744
Nu obtained	3.8008

Then the divergence/convergence effects on the microchannel of various cross sections were investigated. This comparison helps us to appreciate the effect of convergence/divergence angle on the heat transfer characteristics in microchannels. Very fine mesh is created so that the heat transfer and fluid flow effects are more accurate. Hexa/Prism mesh scheme was used as a result of simple geometry. Each of the geometries possesses approximately 4×10^6 grid elements. To ensure the best mesh spacing, grid independence test was carried out for the present computational model. At the inlet, Reynolds number was varied from 100, 250, 500, 750, 1000 to maintain laminar flow through microchannels and the hydraulic diameter is constant for all cases. Pressure outlet and a heat flux of $50kW/m^2$ were specified at the bottom wall. Side and top walls were given symmetry. To achieve steady state analysis, pressure-based solver was used for simulation. SIMPLE method was used to introduce pressure into the continuity equation. In order to predict the heat transfer effects, the energy equation was activated during the analysis. The simulations were performed for a convergence criterion of 10^{-6}. The microchannels used for the current study are shown in Table 1.

Table 2. The dimensions of micro-channels used for the current study.

Channel Type	Inlet Dimension (b × a) (μm × μm)	Outlet Dimension (b × a) (μm × μm)	m $(\times10^{-3})$
Straight	500×400	500×400	0
Converging/ Diverging (1)	550×450	450×350	1
Converging/ Diverging (2)	600×500	400×300	2

Converging/ Diverging (3)	650×550	350×250	3
Converging/ Diverging (4)	700×600	300×200	4
Converging/ Diverging (5)	750×650	250×150	5

5. Results and Discussions

This section explains the various results obtained by the numerical analysis which were performed across straight, converging & diverging microchannels. Also, it showcases various contours and graphs obtained for the different types of microchannel.

5.1. Validation of numerical model

Initially, the numerical results were validated for the rectangular straight channel with Shah and London [4]. Table 1 shows that the fully developed Nusselt number obtained from current numerical analysis is in good agreement with Nu obtained by Shah and London [4]. V.V. Dharaiya and S.G. Kandlikar [6] showed the variation of fully developed Nusselt number with different aspect ratio (α) of the rectangular microchannel. Figure 2 shows that the numerical model used to predict Nu for one wall constant heat flux in current work is in good agreement to values obtained by V. V. Dharaiya and S. G. Kandlikar [6].

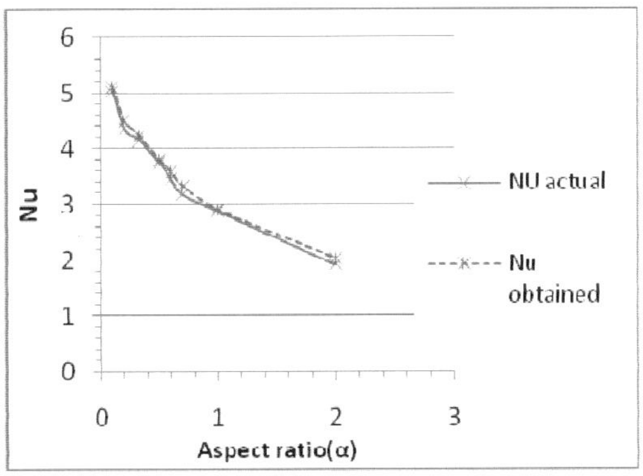

Figure 2. Comparison of fully developed Nusselt number of published data [6] with current numerical results.

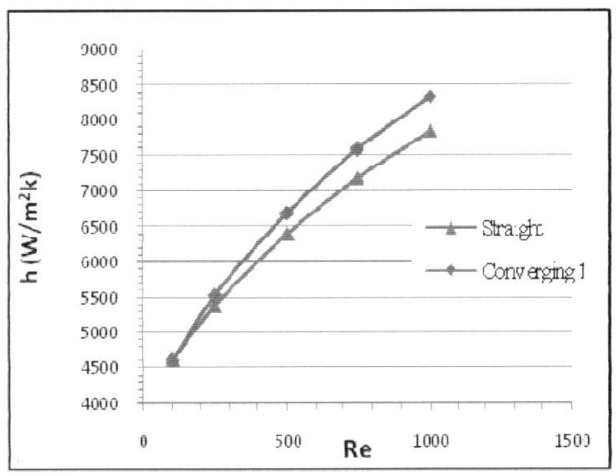

Figure 3. Comparison of h v/s Re for straight and converging microchannel.

5.2. Effect of convergence on heat transfer coefficient (h) for a rectangular microchannel

The effect of slope of convergence on heat transfer coefficient has been investigated thoroughly for rectangular shaped microchannel. Hydraulic diameter of 450μm has been considered and a constant heat flux of 50kW/m2 was maintained at the bottom wall. The velocity at the inlet was varied from 0.2 to 2.2 m/s and the heat transfer coefficient was obtained for every case. The results of heat transfer coefficient for the straight channel were compared with converging microchannel of slope 1×10-3. It can be clearly observed from figure 3 that h is considerably highe\r for the converging channel compared to the straight channel. Further numerical analysis were performed to confirm the increase of heat transfer coefficient in converging microchannels. An extensive study has been carried out for the converging microchannel by varying the slope and observing its effect over heat transfer coefficient.

5.3. Effect of slope of convergence (m) on heat transfer coefficient (h)

Various converging microchannels (Table 2) were considered in the current study and for every case, heat transfer coefficient was tabulated for different Re values. Figure 4 infers that h increases as Re increases for all converging microchannels. Also, the common trend that can be observed from figure 4 that converging channels show a higher value of heat transfer coefficient over the straight microchannel. It can be observed in figure 5 that the heat transfer coefficient is increasing linearly with increase in slope of the converging microchannel. Also, the best fit curve has been shown in figure 5. Figure 7 depicts the variation of h with slope of convergence, which shows that h increases with increase in Reynolds number. It was inferred that as the cross-section of the channel decreases along the length of the channel, the velocity of the liquid flow increases in order to maintain a constant flow rate. This is in accordance with the principle of conservation of mass. Since h is directly dependent upon the velocity gradient (Figure 9) of the fluid flow, there is an overall increase in the heat transfer coefficient and pressure drop for the converging microchannels. As the amount of convergence of the channel increases, so does the velocity of the flow to a greater amount, consequently causing the increase in h and pressure drop..

5.4. Comparison of converging and diverging microchannels

Figure 8 shows that there is a decrease in h for the diverging microchannel compared to straight and converging channels. This is because as the cross-sectional area of the channel increases, there is a decrease in the velocity along the channel length. This results in the decrease of pressure drop and heat transfer coefficient. Comparison of pressure drop for converging and diverging microchannels is shown in figure 6.

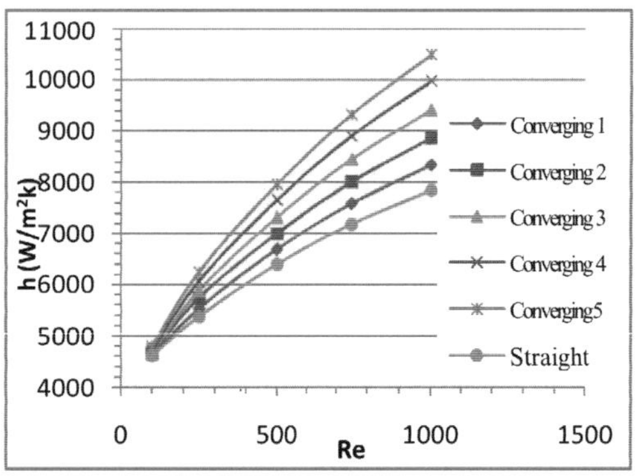

Figure 4. Comparison of h v/s Re for microchannels with different convergence angles.

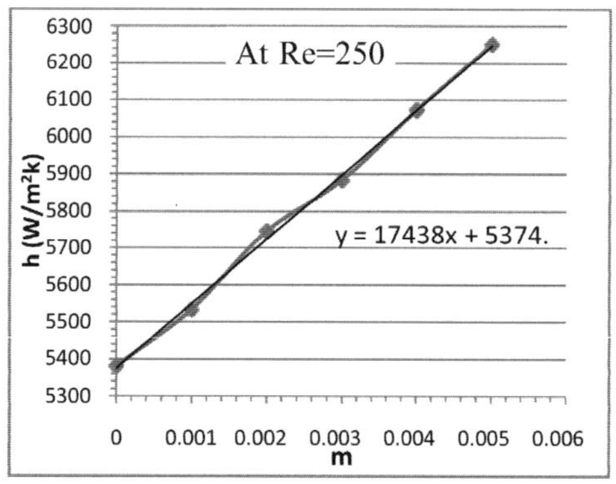

Figure 5. Graph of h v/s m at a Reynolds number of 250.

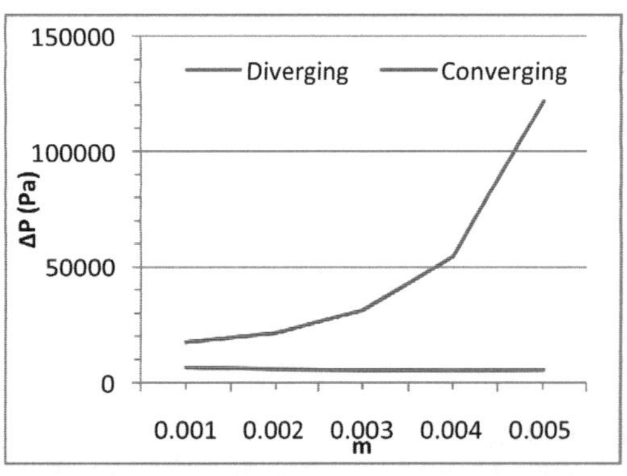

Figure 6. Comparison of pressure drop for converging and diverging microchannel.

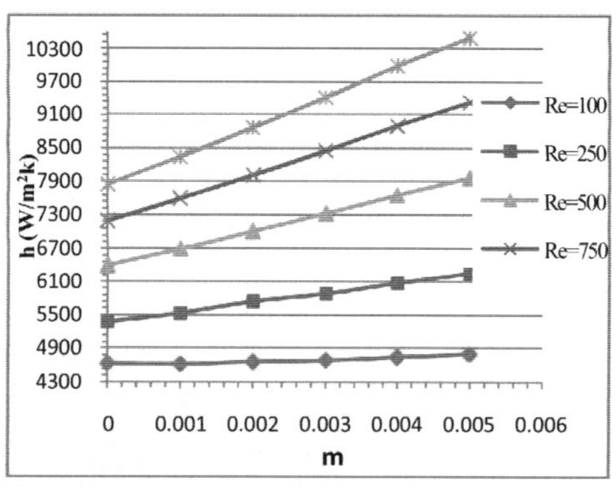

Figure 7. Graph of h v/s m for varying Reynolds number.

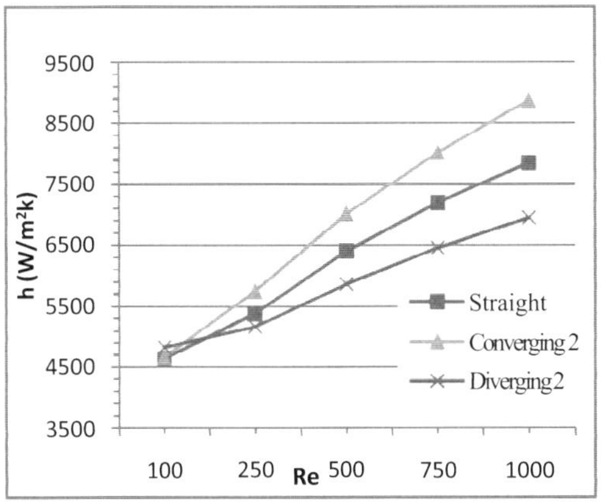

Figure 8. Graph of h v/s Re for straight, converging and diverging microchannels.

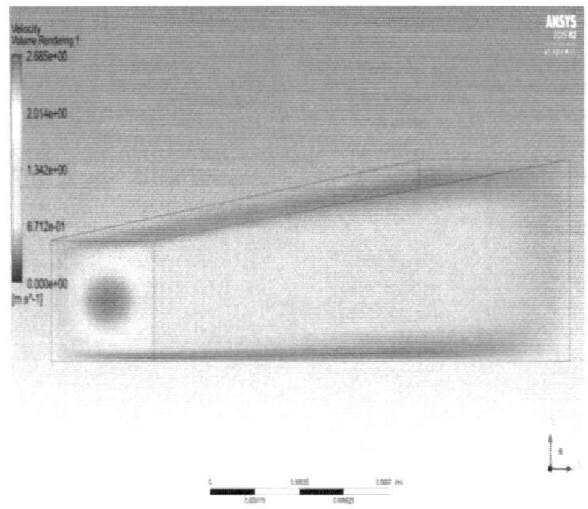

Figure 9. Velocity gradient of the fluid flow in converging microchannel.

6. Conclusions

Following conclusions were drawn from the study:

- Converging microchannel has a better heat transfer coefficient compared to a straight channel both at low and high flow rates.
- At a given flow rate, the heat transfer coefficient can be increased by increasing the angle of convergence of the channel.
- Increasing the angle of convergence resulted in higher pressure drop along the channel, which results in increase in pumping power.
- Diverging microchannel had a lower heat transfer coefficient. Also, there was not much change in the heat transfer coefficient on increasing the divergence angle.
- Diverging microchannel had a lower pressure drop and hence lower pumping power was required.
- At very low flow rates (Re < 250) there is not much difference in the performance of straight and converging.

7. Future Work

Optimization of converging/diverging microchannel for a range of characteristic parameters and validating it with experimental data. The investigation of heat and flow characteristics in straight/converging/diverging microchannel in two phase flow are considered for the future work.

8. References

[1] Tuckerman D. B and Pease R. F. W., 1981, High-performance heat sinking for VLSI, *IEEE Electron Dev. Lett*, **2**, 126–9.

[2] Goodling J. S, July 12–13, 1993, Microchannel heat exchangers: a review, *Proc. High Heat Flux Engineering II, San Diego, CA*, 66–82.

[3] Peng, X. F, Peterson G. P and Wang B. X, 1994, Frictional flow characteristics of water flowing through rectangular microchannels, *Experimental Heat Transfer*, **vol. 7**, 249–64.

[4] Shah R. K and London A. L. 1978, Laminar forced convection in ducts, *Adv. Heat Transfer, Academic Press, New York, NY.*

[5] M.E. Steinke, S.G Kandlikar, Single-phase liquid heat transfer in plain and enhanced microchannels, *ASME 4th Int. Conf. Nanochannels, Microchannels, Minichannels, Parts A.*

[6] Dharaiya V. V and Kandlikar S. G, 2012, Numerical investigation of heat transfer in rectangular microchannels under H2 boundary condition during developing and fully developed laminar flow. *Journal of Heat Transfer*, 134(**2**), 020911.

[7] Chandra A. K, Kishor K, Mishra P. K and Alam M. S, 2015, Numerical simulation of heat transfer enhancement in periodic converging-diverging microchannel, *Procedia Engineering*, **127**, 95–101.

IOP Conf. Series: Materials Science and Engineering 1189 (2021) 012036 doi:10.1088/1757-899X/1189/1/012036

Experimental Investigation of Forced Convection Heat Transfer from Vertical Grooved Plates

[1]Abdul Razak Kaladgi, [2]Vishwanath K C, [3]Sreenivasalu Reddy [4*]Chandrashekar A
[1]Department of Mechanical Engineering, P.A. College of Engineering, Mangalore – 574153, Karnataka
[2, 3] Department of Mechanical Engineering, RajaRajeswari College of Engineering, Bangalore-560074, Karnataka
[4]Department of Mechanical Engineering, Bangalore Institute of Technology, Bangalore – 560004, Karnataka
*Correspondence: acsmech@gmail.com

Abstract: Increasing the heat transfer rate has become prime task in electronic industry because of miniaturization of equipments. Recently many investigators carried out numbers of experiments for achieving maximum heat transfer rate, the present investigation mainly focus on increasing the thermal dissipation from the surface by providing grooves over the surface of aluminum plates with various ranges of length to width ratio and length to depth ratio. The experiment was carried out for varying Reynolds number and varying the amount of heat flux given to the plate. It has been observed that variation in L/D ratio and flow rate affects the heat transfer rate and friction factor.

Keywords: Nano fluids; Heat flux; Groove flat plate, air, Reynolds number, heat transfer coefficient.

1. Introduction

Plates have received considerable attention because of the fact that they have been used widely in industrial applications. They have found extensive use in heat exchangers. When improvement in the process of heating or cooling is required, then better design of plate compactness and spatial geometry is very essential. The use of heat transfer enhancement has become widespread during the last 50 years. The goal of heat transfer enhancement is to reduce the size and cost of heat exchanger equipment, or increase the heat duty for a give size heat exchanger. This goal can be achieve in two ways: active and passive enhancement. Of the two, active enhancement is less common because it requires addition of external power to cause a desired flow modification. On the other hand, passive enhancement consists of alteration to the heat transfer surface or incorporation of a device whose presence results in a flow field modification. One of the popular enhancements is the introduction of grooves on the surface of the heat exchanger. The apparent advantages of grooved plates are that they increase the heat transfer rate by providing additional surface area.

Inspite of the above facts, only a very few studies have been reported in the literature on the flat plate with grooves.

Ashish Dixita & Anil Kumar Patila [1] carried out an experimental study on different types of grooves which were made on the extended surfaces. They found that the heat transfer was more for fins having inclined grooves. Chang, S. W., Su, L. M., Yang, T. L., and Chiou, S. F [2] made an experimental study on fin channels having 900 staggered ribs and developed correlation to study the influence of Reynolds number and L/B on heat transfer, also suggested the optimum length to gap ratio that results in increased heat flux. The increment in heat transfer was up to 140–200 % using 90° staggered ribs. Kadir Bilen, Murat Cetin, Hasan Gul, Tuba Balta [3] conducted an experiment on tubes having grooves of different geometry. The ratio L/D was fixed in their study. They mainly developed a correlation for heat transfer and friction factor. The results of their study revealed that heat transfer enhancement does occur for tubes having grooves as compared with tubes without grooves. Faheem Akthar, Abdul Razak R Kaladgi and

Mohammed Samee A Dafedar [4] conducted an experimental investigation of the natural convection heat transfer over circular dimpled surfaces .The arrangement was inline. The various heat transfer parameters considered were Nusselt number, heat transfer coefficient and heat transfer rate. They concluded that heat transfer enhancement occurs for the dimpled surfaces as compared to flat plates. In an another study, Amjad Khan, Mohammed Zakir Bellary, Mohammad Ziaullah, Abdul Razak Kaladgi [5] carried out an experimental study on plates having dimples of circular shapes. The analysis was carried out for different arrangements like centrally increasing the diameter of dimples in the direction of flow; centrally decreasing the diameter of dimples in the direction of flow etc. They concluded that heat transfer enhancement does occur for plates having dimples but at the cost of pressure drop. N. K. Ghaddar, K. Z. Korczak, B. B. Mikic, A.Y. Patera [6], numerically studied the heat transfer characteristics in the channels having grooves on one plate. They solved the Navier-stokes and energy equations numerically and showed complex flow patterns of flow recirculation flow separation etc. Tang xinyi and zhu dongsheng [7] conducted experimental and computational study (using software) on rectangular channels having ribs and grooves. They used the SST turbulence model to solve the equations and carry out the simulations. They found that the Nu ratio and Cf were more for the channels having ribs and grooves as compared to the channels having only ribs. Apurba Layek, J. S. Saini, S. C. Solanki [8], carried an experiment to study the effect of heat transfer on a duct having ribs and grooves on one surface. They observed an increment in the Nusselt number by about 3.24 times as compared to smooth surface. Ali Najah Al-Shamani, K. Sopian, H. A. Mohammed, Sohif Mat, Mohd Hafidz Ruslan, Azher M. Abed [9] computationally studied heat transfer characteristics of a channel having ribs and grooves on one wall. They used the STD K-epsilon turbulence model to solve the governing equations. They carried the investigation on various ribs and groove combination. They also used different nano fluids with different concentrations. They found that trapezoidal rib groove of good height gives the maximum heat transfer rate. Smith Eiam-saard and Pongjet Promvonge [10] conducted an experiment to study heat transfer characteristics in a duct using 3 types of groove and rib combination. They found an increment in the heat transfer as compared to smooth duct.

2. Experimental Setup

A forced convection apparatus was adapted for this study. Figure 1 show the experimental set up used for the present study. It consists of a rectangular duct to keep the grooved plates, a blower to supply air. The air from blower passes through a flow passage consisting of pipes and finally through the rectangular duct, orifice meter was used to measure flow rate through the passage, a band heater placed inside the duct heats the air and is controlled by dimmer start. Temperature of the air at inlet and outlet were measured using thermocouples.

Figure.1: Experimental set up

The experimental set-up consists of following instruments: -

- Dimmer stat (0-250 Volt AC Supply)
- Voltmeter (0 - 400 Volt)
- Ammeter (0 - 10 Ampere)
- Digital Temperature Indicator of 8 channels (Range 0oC – 400 oC)
- Calibrated k-type Thermocouples
- Four Test Plates
- Orifice plate, manometer, Blower (13000 rpm)

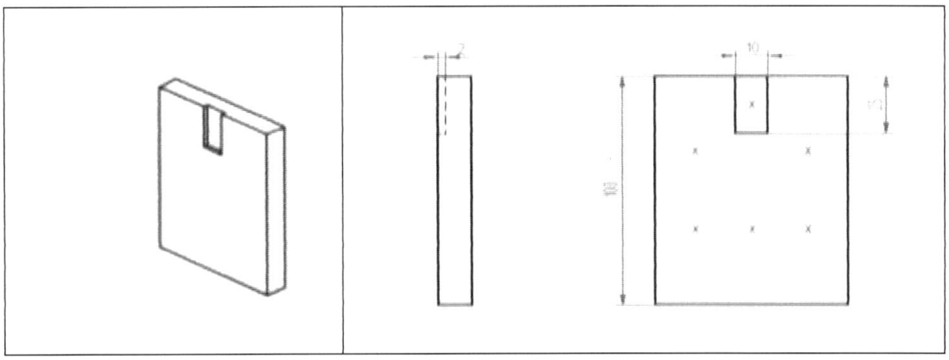

Fig. 2a: Schematic diagram of plate 1 with location of thermocouples (Shown by cross)

Fig. 2b. Schematic diagram of all the plates

The test plates were made of aluminium and rectangular grooves of different lengths of 25 mm, 50 mm, 75 mm, and 100 mm were made onto each of them. The dimension of the plate is 100 mm in length, 60 mm in breadth and 10mm in thickness. The groove breadth is 10 mm. Figs. 2a and 2b shows the schematic diagram of the plates used for the present study.
The cross symbol (X) in figure 2a shows the measurement points of the temperature. The measurement points on the surface of the plate other than the groove were positioned in an equal distance from both top and bottom of the plate. The temperature measurement points in the groove were placed in an equidistant manner according to the groove length. Thermocouples (k-type) were used to measure the temperature. The test plates are fixed onto a wooden box with the groove side open to atmosphere. The wooden box itself acts as an insulator, is filled with glass wool (insulator), in which the strip heater is held. The rectangular plate is kept on the wooden box such that the back side is on the heater and front side open.

3. Results and Discussion

The aim of the experiment is to measure the temperatures along the surface of the plate for the desired constant heat flux (The heater is adjusted for the desired power input with the help of dimmer stat). The experiment is allowed to run till the steady conditions are reached, and the temperature reading was noted.

The parameters varied during the experimentation were:

(i) Heat Input: The experiment were carried out for different heat inputs.

(ii) Aspect ratio: The length of the groove to the width of the groove is altered during the experiment.

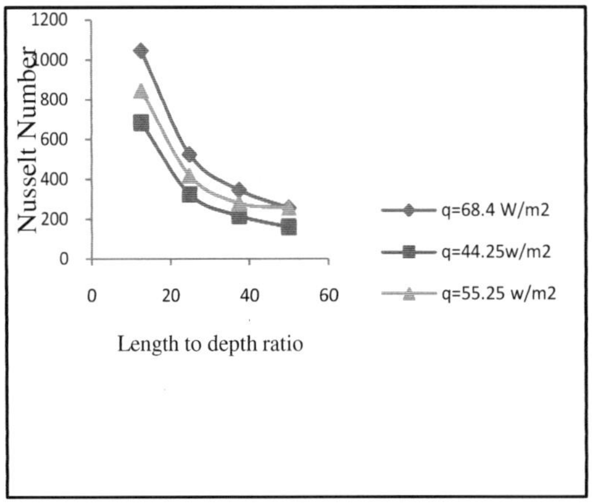

Figure 3.:Nusselt's number Vs of L/D ratio

Figure 3 shows the effect of length to depth ratio on Nusselt number of the plate. The results were plotted for different heat flux conditions. It was observed that Nusselt number increases as the size of groove decreases due to direct flow impingement on the downstream boundary and strengthened flow mixing by vortices at the downstream [11,12]. The formation of vortex pairs periodically shedding off from the grooves, a large up wash regions with some fluids coming out from the central regions of the grooves are the main causes of enhancement of Nusselt number & is more pronounced near the downstream rims of the dimples [13].

Figure 4 shows the variation of Nusselt number for different discharges. It can be seen that Nusselt number increases, as the value of discharge increases as expected & hence increases the heat transfer coefficient. This may be attributed to the fact that due to increased flow rates, the velocity increases and hence increases the heat transfer rate of the plate. It can also be observed that the Nusselt number is highest for L/D ratio of 12.5 and lowest for L/D ratio of 50. So length of the groove affects significantly the heat transfer rate.

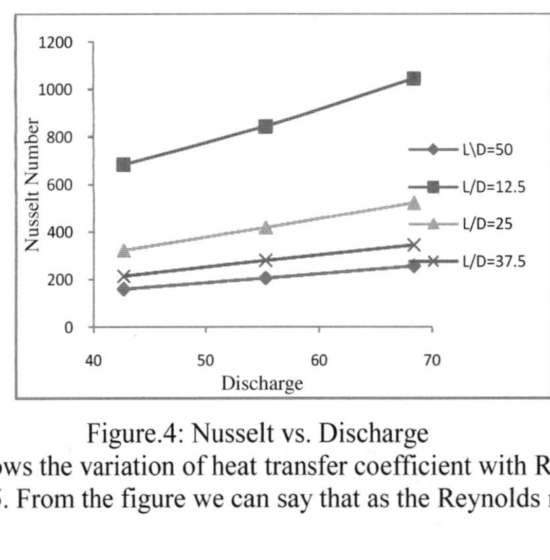

Figure.4: Nusselt vs. Discharge

The Figure 5 shows the variation of heat transfer coefficient with Reynolds number for an L/D ratio of 12.5. From the figure we can say that as the Reynolds number increases, the heat transfer coefficient increases as expected, because at high Reynolds number, turbulence mixing takes place and because of turbulence the heat transfer coefficient increases.

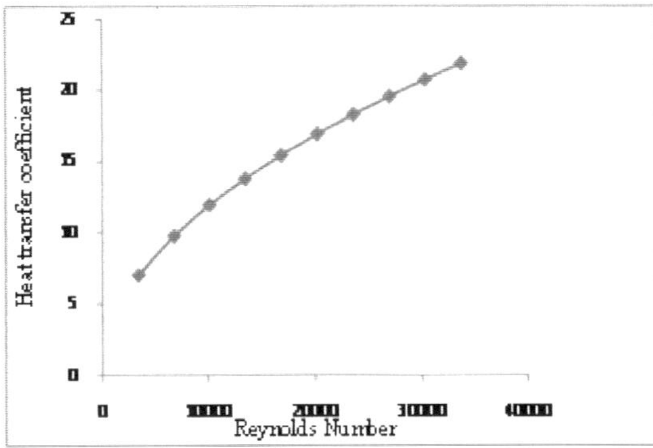

Fig 5. Heat transfer coefficient vs Reynolds number

Fig. 6 shows the variation of friction factor with Reynolds number. It can be seen that as the Reynolds number increases the value of friction factor decreases as expected.This may be attributed to the fact that increases in Reynolds number increases turbulence and decreases friction in the flow.

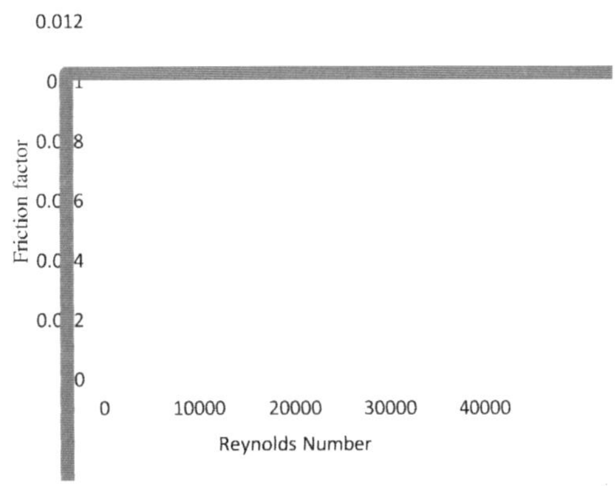

Figure 6. Friction factor vs. Reynolds number

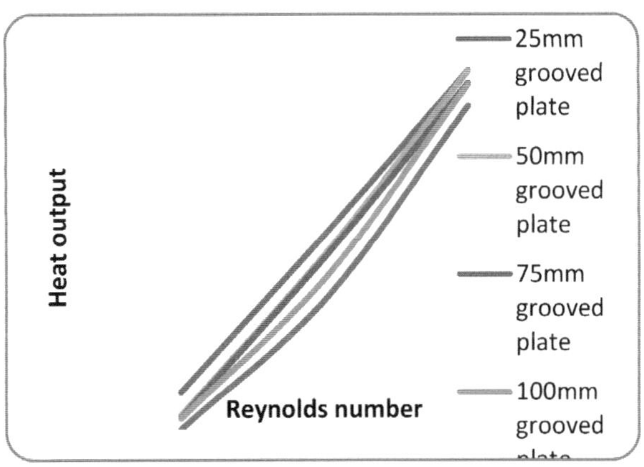

Figure 7. Friction factor vs. Reynolds number

Figure 7 shows variation of heat transfer rate with respect to Reynolds number. It can be seen that heat transfer rate increases as Reynolds number increases as expected. Also it can also be observed that the heat transfer rate is higher for a plate having 25 mm length groove and it decreases as groove length increases. Finally, it can be concluded that grooved plate's helps in better enhancing the heat transfer rate compared to flat plates without grooves.

4. Conclusions

The present study deals with the heat transfer characteristics of plates with rectangular grooves placed vertically within horizontally confining walls. The steady state laminar flow forced convection heat transfer analysis of plate with rectangular groove having different length to depth ratio was undertaken. The discharge and heat flux was varied in this experiment and the following conclusions were drawn

1. Length to width ratio has major impact on heat transfer characteristics.
2. With increasing length to depth ratio of grooves the Nusselt number decreases.
3. With increase in Reynolds number, the coefficient of friction decreases but the rate in decrease is very low as Reynolds number increases.
4. The Nusselt number and heat transfer coefficient increases with increase in the flow rate.
5. Among other plates, the plate of 25mm groove length has highest heat transfer rate.

REFERENCES

[1] Ashish Dixita & Anil Kumar Patila' Heat Transfer Characteristics of Grooved Fin under Forced Convection' Heat Transfer Engineering, vol 36:16, pp.1409-1416, 2015

[2] Chang, S. W., Su, L. M., Yang, T. L., and Chiou, S. F., Enhanced Heat Transfer of Forced Convective Fin Flow With Transverse Ribs, International Journal of Thermal Science, vol. 43, pp. 185–200, 2004.

[3] Kadir Bilen , Murat Cetin , Hasan Gul , Tuba Balta ' The investigation of groove geometry effect on heat transfer for internally grooved tubes, Applied Thermal Engineering, vol. 29 pp.753-761, 2009.

[4] Faheem Akthar, Abdul Razak R Kaladgi and Mohammed Samee A Dafedar, Heat transfer enhancement using dimple surfaces under natural convection—an experimental study, Int. J. Mech. Eng. & Rob. Res. Vol. 4, No. 1, 2015.

[5] Amjad Khan, Mohammed Zakir Bellary, Mohammad Ziaullah, Abdul Razak Kaladgi, An Experimental Approach to Determine the Effect of Different Orientation of Dimples on Flat Plates, American Journal of Electrical Power and Energy Systems, vol.4, pp. 29-33, 2015.

[6]] N. K. Ghaddar, K.Z. Korczak, B.B. Mikic, A.Y. Patera, Numerical investigation of incompressible flow in grooved channels. Part 1. Stability and self-sustained oscillations, J. Fluid Mech. Vol-163 pp.99–127, 1986.

[7] TANG Xinyi and ZHU Dongsheng 'Experimental and Numerical Study on Heat Transfer Enhancement of a Rectangular Channel with Discontinuous Crossed Ribs and Grooves' Chinese Journal of Chemical Engineering, Vol 20, pp.220-230 ,2012.

Effect of copper oxide nano fluids as coolant on thermal performance of spiral heat exchanger

[1]Abdul Razak Kaladgi, [2]Vishwanath K. C., [3]Madhu P., [4*]Chandrashekar A., and [5]Chaluvaraju B. V

[1]Department of Mechanical Engineering, P.A. College of Engineering, Mangalore – 574153, Karnataka
[2]Department of Mechanical Engineering, Raja Rajeswari College of Engineering, Bengaluru – 560074, Karnataka
[3]Department of Mechanical Engineering, B. N. M. Institute of Technology, Bengaluru – 560070, Karnataka
[4]Department of Mechanical Engineering, Bangalore Institute of Technology, Bengaluru – 560004, Karnataka
[5]Department of Physics, Bangalore Institute of Technology, Bengaluru – 560004, Karnataka
*Correspondence: acsmech@gmail.com

Abstract: Over the past couple of years, the use of nano fluids to enhance the heat transfer has been reported by several researchers. The experimental investigation was done on the forced convection heat transfer of spiral heat exchangers using copper oxide (CuO) nano fluids (of fixed volumetric concentration) as a coolant. The various heat transfer parameters considered for the study which are Nusselt number, heat transfer coefficient, heat transfer rate and Reynolds Number. Studies revealed that, the heat transfer enhancement does occur in the heat exchanger (HE) using nanofluids but at the cost of pumping power.

Keywords: Nano fluids; Heat flux; Turbulent flow; Spiral heat exchanger; Reynolds number

1. Introduction

Spiral heat exchanger (SHE) is a compact-type heat exchanger. Its flow path can be set to counter current analogous to counter flow exchanger to obtain maximum rate of heat transfer (HT). A high coefficient of convection heat transfer (CHT) is obtained in this heat exchanger (HE) due to the presence of spiral channels. It also incorporates significant less amount of space due to spiralling which helps in installation and shipping. Another interesting characteristic of SHE's is that, the presence of low fouling which occurs due to the presence of particular geometry of the SHE which causing a continual change in flow path direction and producing higher turbulence and smaller stagnant zones. Many such factors make this HE a superb HE [1]. The concept of SHE may be said to be introduced in the late nineteenth century, and in 1930s, this device was modified and reinvented in Sweden [2]. After that, more research was carried out on SHE's. For example, Bahiraei and Ahmadi [3] numerically investigated the thermal-hydraulic efficacy of a SHE filled with H2O-Al2O3 nanofluid under turbulent flow conditions. They noticed that, as an increase in the volume fraction of nanofluids and Reynolds number (Re), the average heat flux increases while it decreases with channel spacing. In addition, the coefficient of CHT and overall heat transfer coefficient (U) increases with volume fraction of nanofluids and Re. By increasing either volume fraction of nanofluids or Re and decreasing the spacing, the pumping power increases. In a similar study, Bahiraei et al. [4] used graphene nanofluid in the SHE and numerically evaluated the thermal-hydraulic efficacy of a SHE. They found that the effectiveness, Number of Transfer Units (NTU), performance index and pressure loss increases with increase in Re and nanoparticles concentration (φ). Rajavel [5] analysed mathematically and quantitatively, the thermal performance features of a SHE. The effect of geometrical and process parameters on the HT characteristics of the HE was investigated. In addition, they validated the computer model at various flow rates. A correlation for estimating the Nusselt number (Nu) was also extracted from the results. Rostami et al. [6] numerically investigated the influence of channel spacing on the thermal performance of SHE filled with hybrid nano fluids. They found that the outlet temperature and thermal efficacy can be improved either by reducing the channel spacing or increasing the nanoparticle φ, at constant flow velocity. Also, the pressure loss increases with decrease in spacing. Khorshidi and Heidari [7] quantitatively investigated the effectiveness of a SHE by building a framework for this HE. The analysis indicates that Nu increases by increasing the mass flow rate. They were able to get an increment of nearly 100 % in the mean Nu which showed a remarkable improvement. Nguyen and San [8] proposed a new method of developing a SHE using the 3d scanning method. Using this method, they were able to improve the effectiveness of the HE by selecting a proper material. This material reduced the wall thermal resistance and improved the performance of the SHE. Sterger et al. [9] showed that double spiral is better than other types of heat exchangers in terms of reducing the heat loss to environment. Wu [10] presented the calculation process and derived formulas to construct the shape of a SHE. Picon-Nunez et al. [11-12] done a numerical and analytical study and provided a method to estimate the size of spiral heat exchangers in single-phase flow conditions. Targett et al. [13] showed the distribution of temperature in the fluid domain of a double-spiral heat exchanger. Naphon and Wongwises [14] calculated average in-tube heat transfer coefficients (HTC) in a SHE under dehumidifying conditions. They also considered the effect of inlet conditions of both the fluids. Ho et al. [15] and Wijeysundera et al. [16] analysed heat transfer performance of a compact spiral-coil heat exchanger. They found an increment in heat transfer rate of the heat exchanger using nanofluids. Wilhelmsson [17] showed various applications of spiral heat exchangers. Segundo et al. [18] conducted optimization studies on SHE by means of wind-driven optimization and a new version of this approach by adding a statistical distribution to adjust the evolution parameters to themselves. The parameters adopted for optimisation were the HE's length, width and thickness as well as the channel space between the two fluids. The main objective was to minimize the price and maximize the overall heat transfer coefficient (U). Using this technique, they were able to achieve a cost reduction of nearly 23 percent and a rise of almost 16.4 percent in U. Turgut and Coban [19] optimized SHE's and heat pipes for thermal design using a optimization algorithm called global best algorithm. Both single variable and multivariable objective optimisation were carried out. The technique showed good results as compared to other studies in terms of consistency and performance. They found that the operating expenses were extremely dependent on hot and cold side pressure drops. Bidabadi et al. [20] carried single objective and multi-objective optimization (using genetic algorithm) to investigate the optimum performance of SHE. Various geometrical parameters (with logical constraints) were used for optimization. With single objective optimization, they were able to achieve a reduction in pressure loss (by about 50 %) and increment in overall heat transfer coefficient (by about 13 %). Using multi objective optimization which was carried out to obtain a trade-off between these two parameters, they were able to achieve an improvement in heat transfer coefficient (by about 60 %) with a reduction in pressure loss by about 20 %. From the literature it is clear that many researchers have conducted analysis study on SHE due to its high performance. But most of the studies were done using water as a coolant. Nanofluids with high thermal conductivity can be used further to improve its performance. Only few studies have used nanofluids for investigating the performance of SHE. To the author's best knowledge, the study was not carried out using CuO nano fluids of various concentrations. So in this paper an experimental study was carried out to analyze the thermal performance parameters of the spiral heat exchanger. The study was carried for steady forced flow conditions using copper nanofluids of various concentrations (0.2 % to

IOP Conf. Series: Materials Science and Engineering 1189 (2021) 012037 doi:10.1088/1757-899X/1189/1/012037

0.6 %) as the coolant. The flow rate on one side of the heat exchanger (hot water) was kept constant during the study.

2. Methodology

Fig. 1a Experimental setup

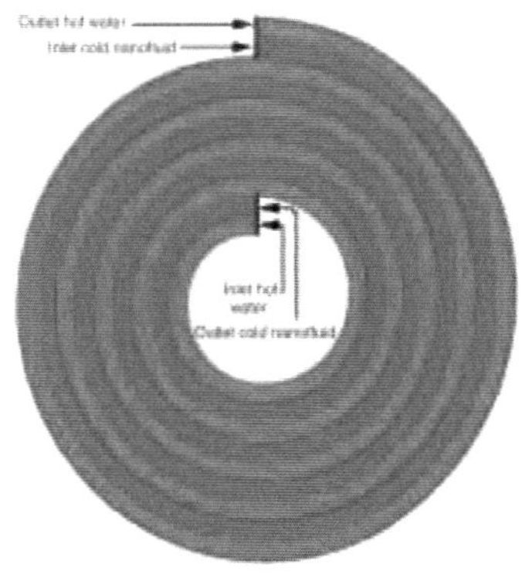

Fig. 1b. Schematic diagram of 2D geometry of *SHE* [6]

The schematic diagram of experimental set-up is shown in Figure 1.a and the 2D geometry of *SHE* is shown in Figure 1.b. The set-up is a well instrumented single-phase heat exchanging system which has a hot water stream flowing inside the tube-side is cooled with a cold water stream flowing in the shell-side. The main parts of the cycle are SHE, centrifugal pump, storage tank, and heater. The heat exchangers include a copper coiled tube and an insulated shell. The water in storage tank is heated using an electric heater, reaching to a prescribed temperature, pump is started to circulate the hot water in the cycle. A ball valve is used to control the flow rate of coolant water and hot water respectively. To measure the flow rate of the cold stream/hot stream, a rotameter is installed upstream of the heat exchanger. The inlet and outlet temperatures of hot and cold water were recorded manually using four glass alcohol thermometers inserted in the small holes made in the inlet and outlet tubes of each heat exchanger and sealed to prevent any leakage. Also, all the pipes and connections between the temperature measuring stations and heat exchanger were duly insulated.

3. Results and Discussion

Figure 2 depicts the effect of Re on Nu. It should be known that the flow rate on one side of the heat exchanger (hot water side) is kept constant, so that the increase in Re is applicable to the other side (nano fluid side). It can be observed that, as expected and very much clear from the figure that as Re increases, Nu also increases. Because at higher Re, the boundary layer gets thinner and the rate of convection of fluids amplifies. The convective coefficient therefore rises. Hence, the heat transfer rate increases. But, it can be seen that the rate of increase of Nu with Re decreases as Re increases. So, it can be addressed that there is a limit beyond which the Nu cannot be increased just by incrementing Re. It can also be seen that the value of Nu is more when CuO nano fluids are used instead of water and it increases with increase in volume concentration of nanofluids. Hence the hot water, which has a constant mass flow rate, experiences a higher temperature decline in the counter flow configuration and demonstrates a higher heat transfer rate in the heat exchanger.

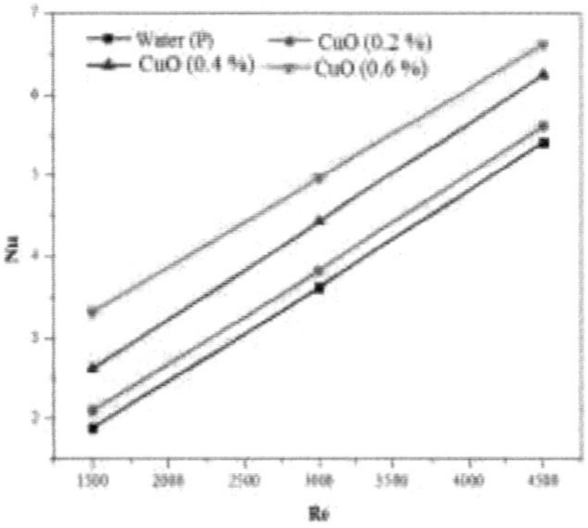

Figure 2 Effect of *Re on Nu*

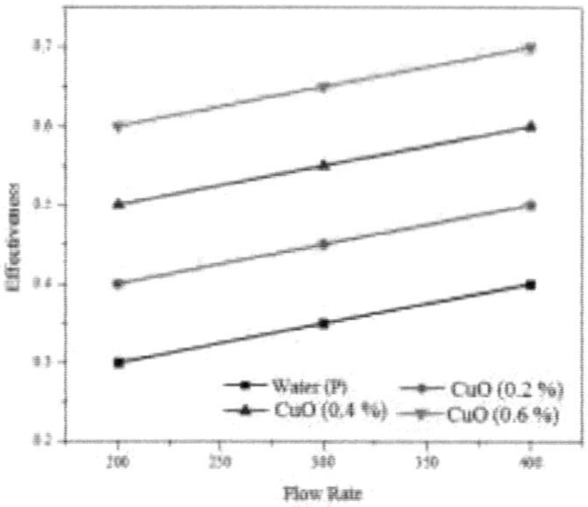

Figure 3 Effect of flow rate *on* Effectiveness, Ɛ

The effectiveness of HT's is a significant criterion in the analysis that compares the actual transfer rate of heat with probable highest heat transfer rate. In the present SHE being studied, lowest possible heat capacity rate for all situations are linked to hot water and therefore the effectiveness is given by as follows [4]:

$$\epsilon = \frac{C_c(T_{co} - T_{ci})}{C_h(T_{hi} - T_{ci})} \ldots\ldots\ldots\ldots\ldots\ldots\ldots (1)$$

Figure 3 shows the effect of flow rate on effectiveness (Ɛ) of the HT for both the cases (cold water and nanofluids). It can be noted that as flow rate increases, the Ɛ amplifies. Flow rate increases due to Leftovers invariant i.e. Ch(Thi-Tci), while the numerator rises because of the increase in heat transfer rate at higher flow rate. Hence Ɛ increases with flow rate. Also, it can be seen that, as predicted and very much evident from Figure 3 that Ɛ was found to be more for the HT using nano fluids as compared to water for the same flow rate and also, it increases with volume concentration of nanofluids. Because higher volume concentration nanofluids have higher thermal conductivity which increases the heat transfer rate. Hence the numerator part of equation 1 increases and increases the Ɛ.

NTU is among the main heat transfer features that determine the size of the heat transfer area and heat exchanger efficiency. Figure 4 shows the effect of Re on NTU. It can be seen that, as Re increases, NTU decreases indicating the smaller size requirement for the fluids at high Re. Also NTU was more for the SHE using nano fluids as compared to water, indicating more space requirements for these heat exchangers. The NTU value increased with concentration of the nanofluids. Hence nano fluids of higher concentration require a bigger size of HE to transfer the same amount of heat transfer.

IOP Conf. Series: Materials Science and Engineering 1189 (2021) 012037 doi:10.1088/1757-899X/1189/1/012037

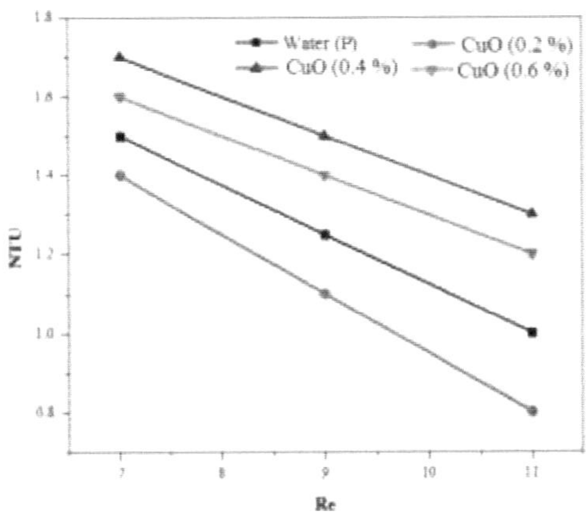

Figure 4 Effect of *Re on NTU*

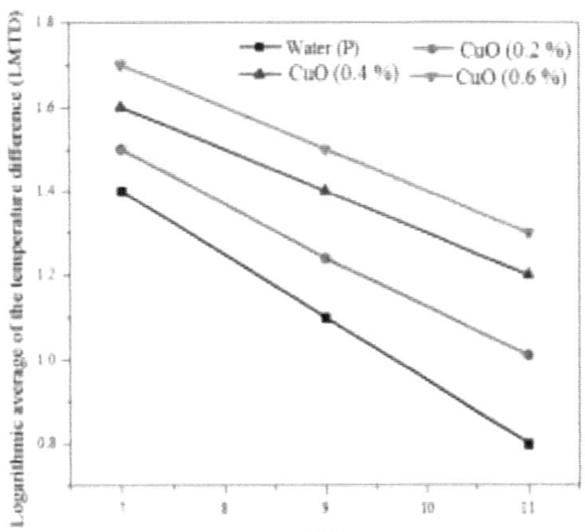

Figure 5 *LMTD vs NTU*

Figure 5 shows the graph of LMTD vs NTU. It can be seen that, as the NTU value increases, the temperature difference between the cold flow and the hot flow decreases. It can also be seen that NTU value is low for nano fluids as compared to water for a given Re indicating a smaller size heat exchanger requirement when nano fluids are used. Figure 6 shows the consequence of Re on Heat transfer rate of the SHE. It can be seen that, as Re increases, heat transfer rate also increases as expected. It can also be noted that this increase in the heat transfer rate decreases as Re further increases. So it can be easily concluded that there is upper limiting value of Re beyond which the increase in heat transfer rate is negligible. Also, the heat transfer rate was higher for nanofluids as compared to water, as expected. This is because of the high thermal conductivity of the nano fluids as compared to water. It can also be seen that heat transfer rate was more for nano fluids of high concentration due to high thermal conductivity.

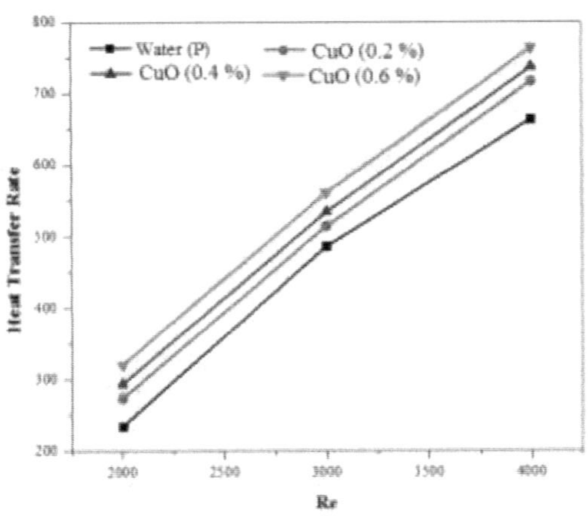

Figure 6 Heat transfer rate vs Re

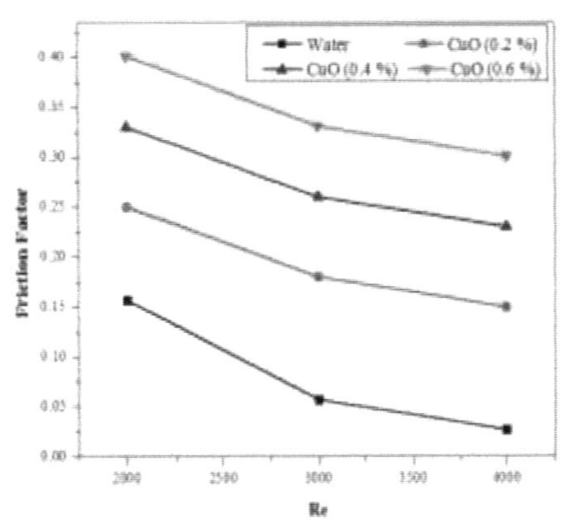

Figure **7** Effect of *Re on Friction factor*

Apart from the thermal features, hydraulic features like the friction factor or the pumping power must also be taken into account for the HE's, since they indicate the power usage rate for the heat exchanger process. When nano particles were added, exaggerate the friction factor (pressure drop) as well as raise the heat transfer rate. The drop in pressure upsurge is an unwanted problem for HE's as they increases the energy usage. Figure 7 depicts the effect of Re on friction factor for both the cases (nano fluids and cold water) maintaining hot water flow rate at one side as constant. It is evident and abundantly clear from Figure 7 that the friction factor amplifies by rising Re. This is because with rise in Re, the gradients of the velocity rise and hence the friction factor and pressure drop rises. In fact, greater Re means higher speed (turbulence) and thus greater flow rate. Hence, pumping capacity is more at higher Re. Also, it can be seen that the friction is more for nanofluids as compared to water for the same Re due to high viscosity of the nano fluids.

4. Conclusions

An experimental investigation was carried out and evaluated the thermal performance parameters of SHE using both cold water and nano fluids (of different concentrations). The flow rate of hot water, at the other side, was kept constant. The experiment was carried under steady forced flow conditions. It was observed that, the following conclusions were drawn from the analysis as follows:

1. Nu and HTC increases with Re. The increase was higher for nano fluids especially for higher volume concentration (0.6 %) as compared to water,
2. The effectiveness of the heat exchanger increases as Re increases and it was higher for nano fluids of higher volume concentration,
3. As Reynolds number increases, NTU decreases which indicating a small size heat exchanger requirement at high Reynolds number. Also, the NTU was higher for nano fluids of higher concentration indicating a large size HE requirement if nano fluids are used as a coolant as compared to water,
4. As NTU value increases, the temperature difference between the cold flow and the hot flow decreases and
5. As Re increases, friction factor (pressure drop and pumping power) increases. The friction factor was more for nano fluids as compared to water due to the high viscosity of nano fluids which indicating that the heat transfer enhancement does occur in nano fluids but at the cost of pumping power.

REFERENCES

[1] D. K. Nguyen and J. Y. San, Heat Transfer and Exergy Analysis of a Spiral Heat Exchanger, Heat Transf. Eng., vol. 37, no. 12, pp. 1013–1026, 2016, doi: 10.1080/01457632.2015.1104159

[2] G. F. Hewitt, G. L. Shires, and Y. V. Polezhaev, International encyclopedia of heat & mass transfer, 1997.

[3] M. Bahiraei and A. A. Ahmadi, Thermohydraulic performance analysis of a spiral heat exchanger operated with water–alumina nanofluid: Effects of geometry and adding nanoparticles, Energy Convers. Manag., vol. 170, no. January, pp. 62–72, 2018, doi: 10.1016/j.enconman.2018.05.061

[4] M. Bahiraei, H. Kiani Salmi, and M. R. Safaei, Effect of employing a new biological nanofluid containing functionalized graphene nanoplatelets on thermal and hydraulic characteristics of a spiral heat exchanger, Energy Convers. Manag., vol. 180, no. October 2018, pp. 72–82, 2019, doi: 10.1016/j.enconman.2018.10.098.

[5] R. Rangasamy, Experimental and numerical studies of a spiral plate heat exchanger, Therm. Sci., vol. 18, no. 4, pp. 1355–1360, 2014.

[6] S. Rostami, A. Aghaei, A. Hassani, J. Hossein, M. Hezaveh, and M. Sharifpur, Thermal – hydraulic efficiency management of spiral heat exchanger filled with Cu – ZnO / water hybrid nanofluid, J. Therm. Anal. Calorim., no. mm, 2020, doi: 10.1007/s10973-020-09721-9.

[7] J. Khorshidi, S. Heidari, and others, Design and construction of a spiral heat exchanger, Adv. Chem. Eng. Sci., vol. 6, no. 2, p. 201, 2016.

[8] D.-K. Nguyen and J.-Y. San, Decrement in heat transfer effectiveness due to solid heat conduction for a counter-current spiral heat exchanger, Appl. Therm. Eng., vol. 103, pp. 821–831, 2016.

[9] M. R. Strenger, S. W. Churchill, and W. B. Retallick, Operational characteristics of a double-spiral heat exchanger for the catalytic incineration of contaminated air, Ind. Eng. Chem. Res., vol. 29, no. 9, pp. 1977–1984, 1990.

[10] W. Dongwu, Geometric calculations for the spiral heat exchanger, Chem. Eng. Technol. Ind. Chem. Equipment-Process Eng., vol. 26, no. 5, pp. 592–598, 2003.

[11] M. Picon-Nunez, L. Canizalez-Davalos, and J. M. Medina-Flores, Alternative sizing methodology for compact heat exchangers of the spiral type, Heat Transf. Eng., vol. 30, no. 9, pp. 744–750, 2009.

[12] M. Picon-Nunez, L. Canizalez-Davalos, G. Martinez-Rodriguez, and G. T. Polley, Shortcut design approach for spiral heat exchangers, Food Bioprod. Process., vol. 85, no. 4, pp. 322–327, 2007.

[13] M. J. Targett, W. B. Retallick, and S. W. Churchill, Solutions in closed form for a double-spiral heat exchanger, Ind. Eng. Chem. Res., vol. 31, no. 3, pp. 658–669, 1992.

[14] P. Naphon and S. Wongwises, An experimental study on the in-tube convective heat transfer coefficients in a spiral coil heat exchanger, Int. Commun. Heat Mass Transf., vol. 29, no. 6, pp. 797–809, 2002.

[15] J. C. Ho, N. E. Wijeysundera, S. Rajasekar, and T. T. Chandratilleke, Performance of a compact, spiral coil heat exchanger, Heat Recover. Syst. CHP, vol. 15, no. 5, pp. 457–468, 1995.

[16] N. E. Wijeysundera, J. C. Ho, and S. Rajasekar, The effectiveness of a spiral coil heat exchanger, Int. Commun. heat mass Transf., vol. 23, no. 5, pp. 623–631, 1996.

[17] B. Wilhelmsson, Consider spiral heat exchangers for fouling applications: The inherent'self-cleaning'action reduces downtime, Hydrocarb. Process. (International ed.), vol. 84, no. 7, pp. 81–83, 2005.

[18] E. H. De Vasconcelos Segundo, V. C. Mariani, and L. Dos Santos Coelho, Design of spiral heat exchanger from economic and thermal point of view using a tuned wind-driven optimizer, J. Brazilian Soc. Mech. Sci. Eng., vol. 40, no. 4, p. 212, 2018.

[19] O. E. Turgut and M. T. Coban, Thermal design of spiral heat exchangers and heat pipes through global best algorithm, Heat Mass Transf., vol. 53, no. 3, pp. 899–916, 2017.

[20] M. Bidabadi, A. K. Sadaghiani, and A. A. Vahdat, Spiral heat exchanger optimization using genetic algorithm, 2013.

IOP Conf. Series: Materials Science and Engineering 1189 (2021) 012038 doi:10.1088/1757-899X/1189/1/012038

Modelling and simulation of virtual NO_x sensor for diesel engine using thermodynamic model

Shoaib Ahmed[1], Halesh M R[2], Narasimha kollaparti[3], Aman P V[4], Satish V T[5]

[1] PG Research Student, Department of Electronics and Communication, SJCE Mysore.
[2] Assistant Professor, Department of Electronics and Communication, SJCE Mysore.
[3] Subject Matter Expert, KPIT Technologies, Bangalore.
[4] Subject Matter Expert, KPIT Technologies, Bangalore.
[5] Assistant Professor, Department of Mechanical Engineering, MCE Hassan.

Abstract: This paper presents a thermodynamic model for virtual NO_x sensor which is capable of predicting nitrogen oxides (NO_x) emissions in diesel engines. In-cylinder volume and pressure are the inputs from estimator. The proposed method is based on a two-zone thermodynamic model which divides the combustion chamber into burned and unburned gas zone. The contribution of proposed method arises from: In-cylinder volume and pressure are the inputs from estimator using slider crank mechanism. And the two-zone thermodynamic modeling framework is applied in a way that the thermodynamic equations can be solved in a closed form, which provides the basis for achieving high level of predictiveness using MATLAB which uses ADVISOR as prerequisite for the project. The model is validated using real data values. The model is suitable for two different direct injection diesel engines, i.e. a light and a heavy-duty engine. Also, it maintains a low pipe emission which minimizes fuel consumption.

Keywords: Thermodynamics, NO_x (Nitrogen Oxide), Burnt zone, Unburnt zone, Estimator, MATLAB and ADVISOR (Advanced Vehicle Simulator).

1. Introduction

Diesel engine powered automobiles are increasing in recent years, due to high thermal efficiency of diesel engines. These engines have become an environmental issue since it represents a major source of NO_x emissions [1]. The ADVISOR is used as prerequisite for understanding the exhaust gas temperature, exhaust catalyst efficiency and emissions for a selected vehicle. ADVISOR package developed by national renewable energy laboratory is being utilized for rapid analysis of performance and fuel economy of mentioned vehicles such as conventional, electric and hybrid vehicles. ADVISOR package runs in background of MATLAB environment, which provides a backbone for simulation and analysis for the defined drivetrain components. Based on survey carried, the well-known deficiencies always found with emissions [2]. To overcome these increasingly stringent emissions in diesel engine adoption of control strategies based on models is required. A real time-based NO_x model can be able to estimate engine out NO_x in diesel engines. The accurate measurement of in- cylinder NO_x being an input parameter for active control methods (like control closed loop combustion) and also for the control of exhaust treatment systems. Real time models for ignition delay and emission formation combined with closed-loop control and optimization of injection timing might be a powerful tool to control these future engines and to bring out the full potential of the emerging technologies. Another application of fast emission model is for engines fitted with a NO_x after treatment system that adds a reduction agent to the exhaust stream. In that case it is essential to know engine out NO_x level in order to control the reduction agent flow into the catalytic converter [3]. Some of the modern diesel engines are equipped with pressure sensors (Transducer), which gives input to combustion related data [4, 5]. The importance of measuring in-cylinder pressure used for obtaining information related to aging of engine components as well as adaption of the injection strategy with respect to the variation in the fuel quality [6]. Thus, focusing on this importance of pressure estimator, it can be useful for emerging a pressure estimator model. The research uses input for the combustion modelling and input for calculation of thermodynamic parameters inside the cylinder which enables virtual NO_x sensor with higher level of predictiveness. Over the past decade, diesel engine technology has increasing rapidly and evolved to the point where control solutions for obtaining high fuel efficiency and low NO_x emissions are possible. The strategies are being implemented to manage fueling system and air handling system. Also, this will be relying on sensor replacement and limitations of costs. A space of special interest is considered for this type of analysis inside the cylinder volume. The laws of thermodynamics are being an application for solving this kind of equations [7]. According to piston-cylinder geometry it is seen that straight-line motion can be converted into rotary motion. Hence the slider crank model has led to derive in-cylinder volume. The available piezoelectric transducer will lose its sensibility due to aging factor. Hence this allows for development in grey area of an automobile [8]. The model accuracy largely depends on tuning parameter. Hence, advanced virtual NO_x sensor models have general applicability for NO_x estimation [3]. The Estimation of NO_x at upstream and downstream opens a capability for getting SCR efficiency for further urea dosing strategy. As per the research, the main thermodynamic parameters require for NO_x prediction are rate of heat release (ROHR) and consequently the temperatures and concentrations in the combustion chamber. The properties of thermodynamic model can be assessed by physical depth in models which also reflects computational complexity of models [9]. On the system level, 0-D combustion models, [10,11], are often used as models that are effectively combine the requirements on multi-zone spatial resolution and much shorter computational times compared to the 3D CFD models.

2. Specifications of the work

The specifications used for modelling is given below. The simulation is carried on light duty vehicle and heavy- duty vehicle. Table 1 provides specification of light-duty vehicle. Similarly, Table 2 provides specification of heavy-duty vehicle.

Table 1: Light duty vehicle specification

Cylinders	4 inline
Displacement	1560Cms
Bore x Stroke	75mm x 88.3mm
Compression	18:1
Fuel injection system	CRDI
Connecting rod length	136.8mm
Maximum power	66.2kW @ 4000 rpm
Maximum Torque	215Nm @ 1750 rpm

Table 2: Heavy duty vehicle specification

Cylinders	6 inline
Displacement	6870cm^3
Bore x Stroke	108mm x 125mm
Compression	18:1
Fuel injection system	DI
Connecting rod length	182.5mm
Maximum power	162kW @ 2400 rpm
Maximum Torque	825Nm @ 1400-1700 rpm

3. Workflow of modelling and methodology

The modelling framework starts with estimation of in-cylinder volume which is obtained using slider crank mechanism. The instantaneous volume can be obtained by applying derivative. The estimation of in-cylinder volume is further used for instantaneous pressure estimation. The instantaneous values are inputs from estimator. Till the end of injection, the in-cylinder gas mixture is modelled as single zone. The measurement of temperature for considered ideal and non-perfect gas is followed. The main concentration is given for deducting temperature out of mass of fuel, heat addition terms using thermodynamic modeling for thermal NO_x calculation. After the combustion begins and burns the charge the combustion area is divided into burned and unburned zone. Thus, applying control volume concept on the combustion zone. Figure 1 represents the workflow of the model.

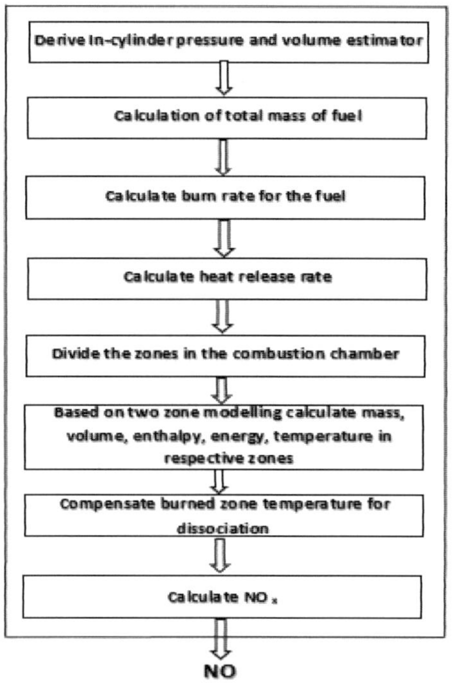

Figure. 1: Workflow for calculating NO_x

The proposed methodology is shown in figure 2, which indicates the Overall view of models used for modelling the NO_x estimator. The following are the models which are modelled as per the engine specification: Volume Estimator model, Pressure estimator model, Fuel rate model, Model for conserved mass in two-zones, Heat release model, Heat transfer model, Temperature model for thermal NO_x calculation, and NO_x estimator model.

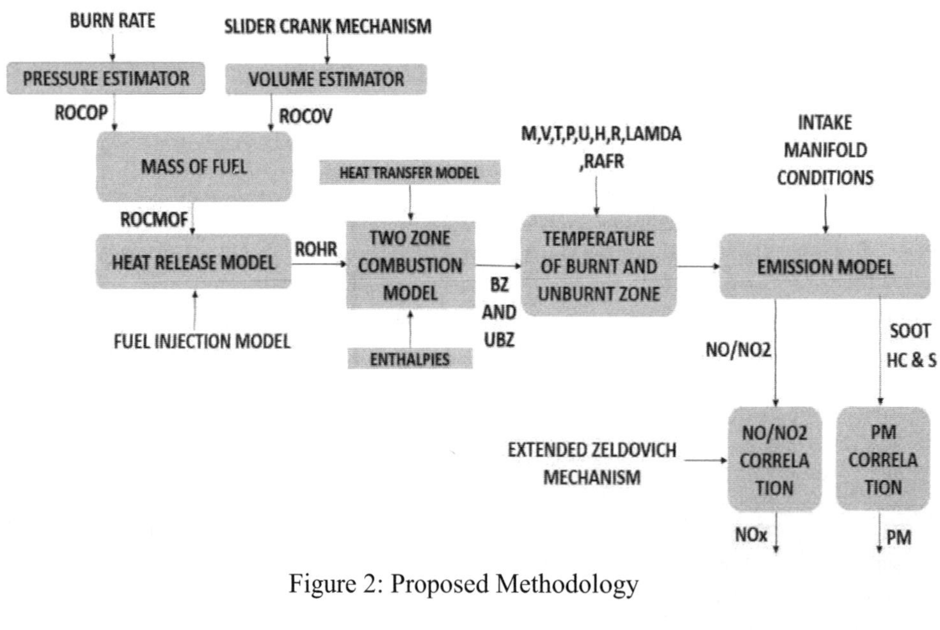

Figure 2: Proposed Methodology

4. Simulation of NO_x sensor

4.1 Volume model

It is assumed that volume can be derived using slider crank model from the compression ratio, bore, connecting rod and stroke. The energy equations are applied to the volume and crank-angle domain and the volume estimator obtained as given in equation 1:

$$\frac{dV}{d\theta} = A_p\left[(-r\sin\theta)\left(1 + \frac{\cos\theta}{\sqrt{\left(\frac{l}{r}\right)^2 - \sin^2\theta}}\right)\right] \quad (1)$$

4.2 Pressure estimator model

The energy equations are applied to the pressure and crank-angle domain and the pressure estimator derived as follows:

$$\frac{dP}{d\theta} = -\frac{\gamma}{V_{cyl}}\frac{dV}{d\theta}P_{cyl} + \frac{\gamma - 1}{V_{cyl}}\left[(1-\alpha)m_{fuel}Q_{LHV}\frac{dx_b}{d\theta}\right] \quad (2)$$

4.3 Model of conserved mass in two zone

The Conservation of mass is applied to the control volume and conservation of mass of burnt is equal to the mass transfer from unburnt to burnt zone, accord into this assumption it is given as:

$$\frac{dm_{u\to b}}{d\theta} = \frac{dm_{fuel}}{d\theta}\frac{(\lambda_{comb}L_{st} + 1)(1 + \lambda_u L_{st})}{L_{st}(\lambda_u - 1)} \quad (3)$$

Thus, the conservation of mass of burnt zone is obtained as:

$$\frac{dm_b}{d\theta} = \frac{dm_{u\to b}}{d\theta} \quad (4)$$

Assuming the unburned zone usually consists of trapped air, unburnt fuel and its equation is given as:

$$\frac{dm_u}{d\theta} = \frac{dm_{u,inj}}{d\theta} - \frac{dm_{u\to b}}{d\theta} \quad (5)$$

4.4 Heat release model

It is assumed that the rate of heat release occurs in burned zone only and can be obtained by heat addition parameter as follows:

$$\frac{dQ_{b,b}}{d\theta} = \frac{dm_{fuel}}{d\theta}Q_{LHV} \quad (6)$$

4.5 Temperature model for thermal NO_x calculation

The temperature responsible for thermal NO_x is calculated using following equations and separate equations are written for burnt and unburnt zones as given below:

$$\frac{dT_b}{d\theta} = \frac{\frac{dQ_{b,B}}{d\theta} + \frac{dQ_{ht,b}}{d\theta} + h_u\frac{dm_b}{d\theta} - h_b\frac{dm_b}{d\theta} + \frac{m_b R_b T_b}{P}\frac{dP}{d\theta}}{m_b R_b} \quad (7)$$

$$\frac{dT_u}{d\theta} = \frac{\frac{dQ_{ht,u}}{d\theta} + \frac{dQ_{evp,u}}{d\theta} + h_{inj}\frac{dm_{u,inj}}{d\theta} + \frac{m_u R_u T_u}{P}\frac{dP}{d\theta}}{m_u R_u} \quad (8)$$

IOP Conf. Series: Materials Science and Engineering 1189 (2021) 012038 doi:10.1088/1757-899X/1189/1/012038

4.6 NO_x estimator model

The equilibrium species calculation is being performed using Gibbs free energy equation. To reduce the complexity for computation. The pre-calculated values are being used as stored look-up table. The calculation of equilibrium species can be performed for many sets of reactions using extended zeldovich mechanism. Here the case of one reaction is considered for modelling basic NO_x estimator.

$$N_2 + O \rightarrow NO + N$$

The extended zeldovich mechanism [12] is obtained for the above reaction and initial NO formation rate is given as:

$$\frac{d[NO]}{d\theta} = \frac{6 * 10^{16}}{{T_b}^{1/2}} e^{\frac{-69{,}090}{T_b}} [O]_e^{0.5}[N_2]_e \quad (9)$$

5. Results

The simulation of NO_x estimator is done for light-duty vehicle and heavy-duty vehicle. The mass of air, mass of fuel, engine specifications are different to simulate on two different engines. Considering these factors, the simulation is carried in MATLAB platform.

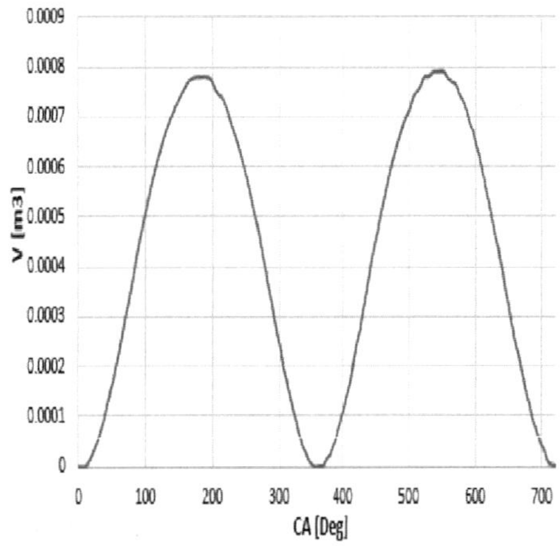

(a)

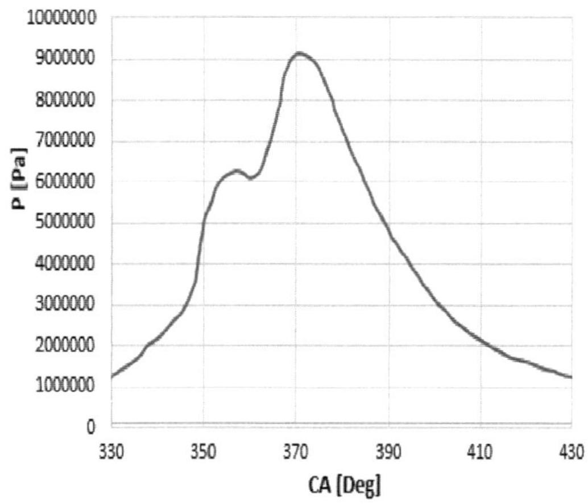

(b)

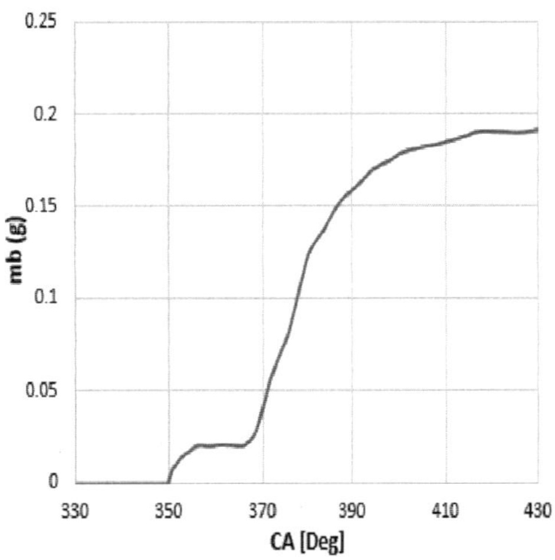

(c)

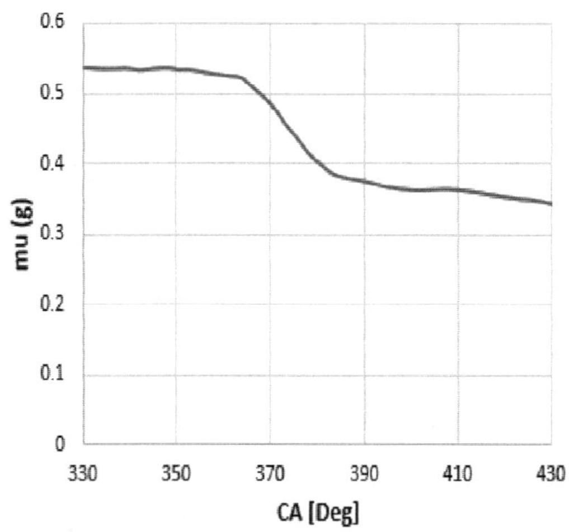

(d)

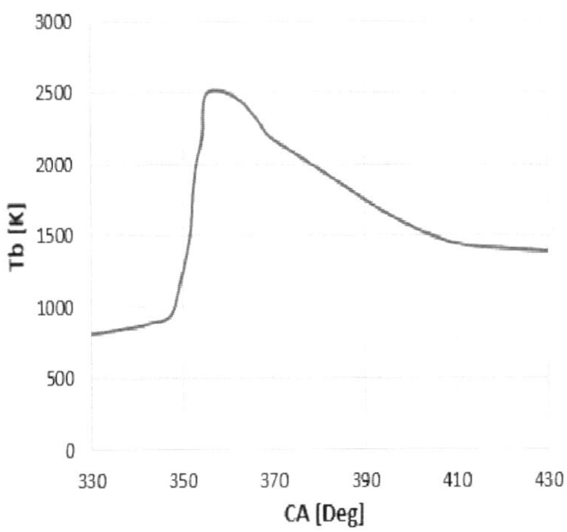

(e)

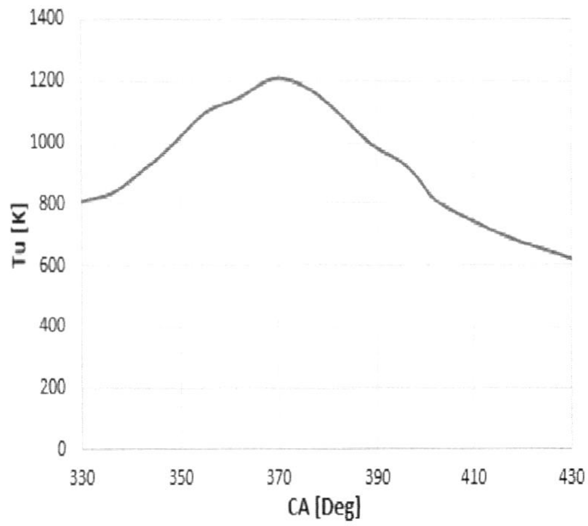

(f)

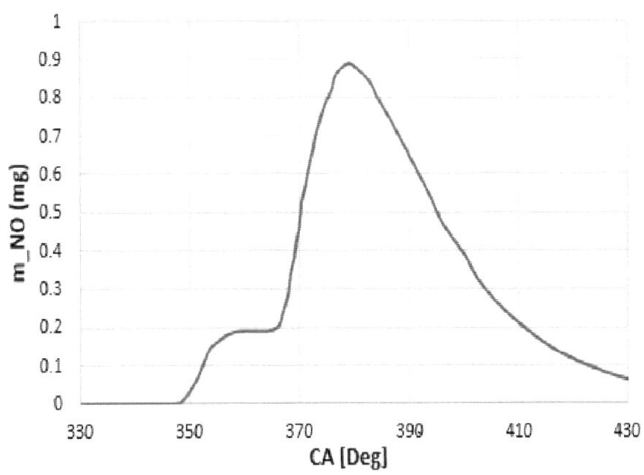

(g)

Figure 3: Simulation result of Light-duty engine operating point at N=3000rpm

(a) Volume estimator, (b) Pressure estimator, (c) Mass of burnt, (d) Mass of unburnt, (e) Burnt zone temperature, (f) Unburnt zone temperature, (g) NO_x estimator.

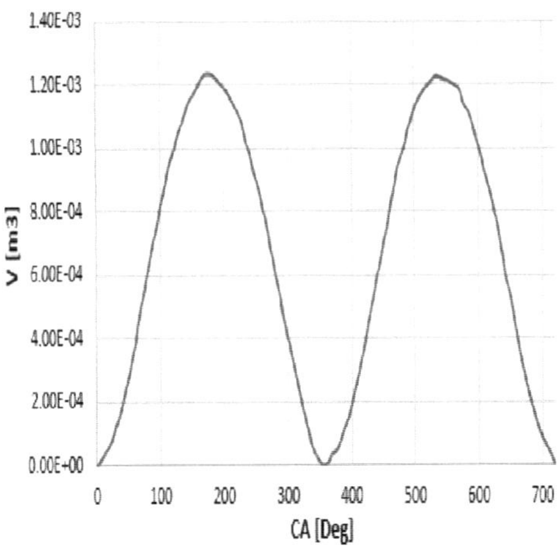

(a)

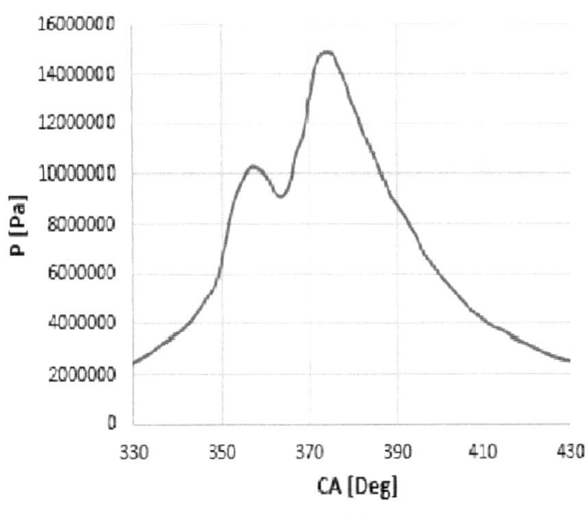

(b)

IOP Conf. Series: Materials Science and Engineering 1189 (2021) 012038 doi:10.1088/1757-899X/1189/1/012038

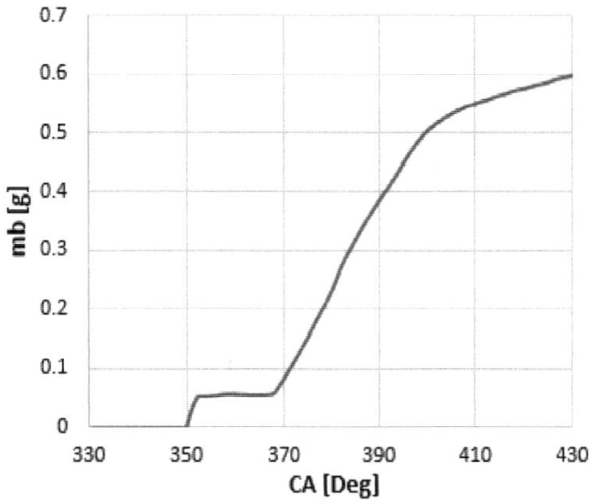

(c)

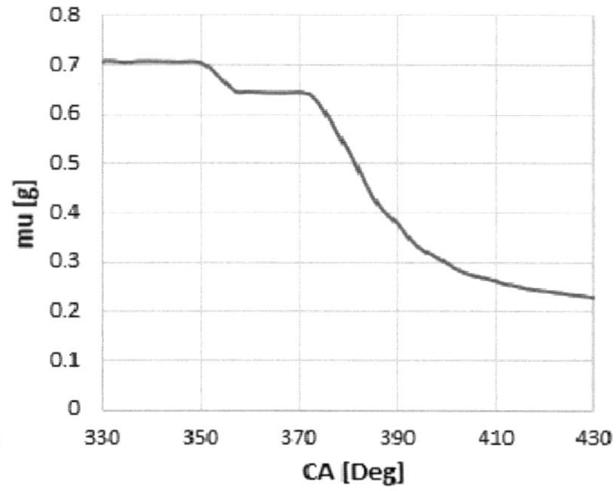

(d)

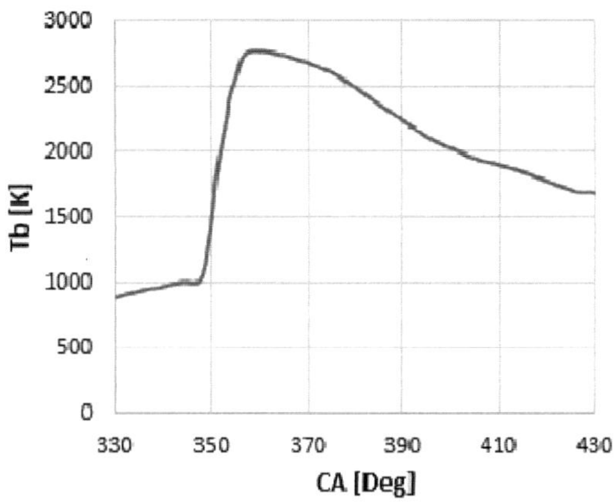

(e)

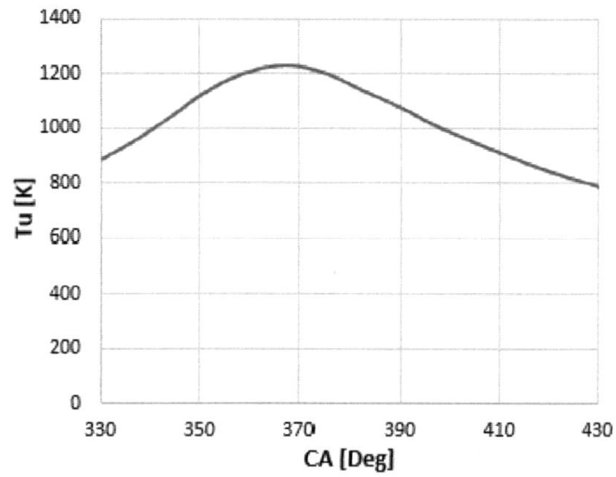

(f)

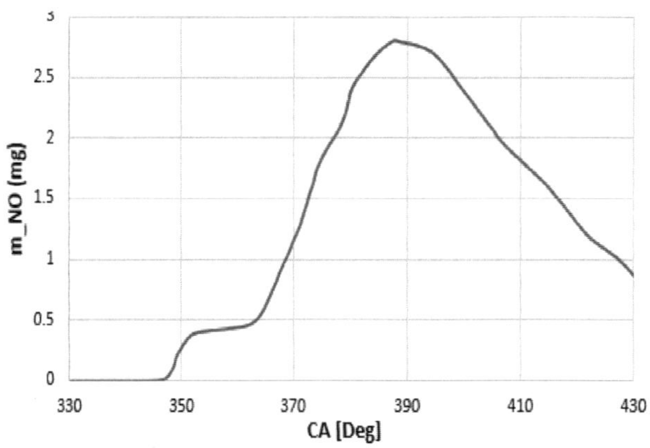

(g)

Figure 4: Simulation result of Heavy-duty engine operating point at N=3000rpm

(a) Volume estimator, (b) Pressure estimator, (c) Mass of burnt, (d) Mass of unburnt, (e) Burnt zone temperature, (f) Unburnt zone temperature, (g) NO_x estimator.

LD ENGINE: The light-duty engine is equipped with common rail direct injection system. The build in pressure for the light-duty engine is less comparative to heavy-duty engine. Sensible check is carried and analyzed the plots of mass of burnt zone which is getting increased as combustion begins with respect to provided combustion angle. Also, the unburnt mass getting decrease at the same angle since unburnt zone is supplying amount of charge. The combustion temperature starts increasing rapidly at same instant, it also considers heat transfer. Thus, the NO formation starts at same angle for the cycle.

HD ENGINE: The heavy-duty engine is equipped with direct injection fuel injection system for validation purpose which influences the air and fuel preparation, also the HD-engine operates with leaner mixture compare to LD-engine. Thus, it impacts on start of combustion temperature and concentration. The build in pressure for the heavy-duty engine is very high. Sensible check is carried and analyzed the plots of mass of burnt zone which is getting increased as combustion begins with respect to provided combustion angle. Also, the unburnt mass getting decrease at the same angle since unburnt zone is supplying amount of charge. The combustion temperature starts increasing rapidly at same instant, also it considers heat transfer. Thus, the NO formation starts at same angle for the cycle.

6. Conclusion

The analysis is carried using ADVISOR package. A NO_x sensor estimator is modelled for diesel engine which is able to measure engine out NO_x using thermodynamic modelling framework which is usually measured to maintain low tail pipe emissions. In-cylinder pressure and volume are obtained successfully using slider crank mechanism, and input from pressure and volume estimator is used for calculating fuel rate. Finally, the temperature is obtained which estimates engine out thermal NO_x. A two-zone thermodynamic modelling is carried in such a way that, it gives the mass of burnt and its temperature which signifies the amount of thermal NO_x produced in diesel engine using MATLAB. The application of model on two different injection engines is validated, which allows to obtain NO_x in heavy-duty engine which produces slighter higher amount of NO_x emission compare to light-duty engine as mentioned in equation (9). The reason for NO_x production is due to higher inlet temperature for this purpose equation (7) is modelled. The pressure estimator is modelled to correct the inlet pressure. Thus, by increasing inlet pressure the lower NO_x can be obtained. It also seen that increase in compression ratio increases pressure which effects the combustion temperature to a higher value which produces higher NO_x. The future work needs to be carried out to maintain low emissions while minimizing fuel consumption. Thus, it limits harmful gases from internal combustion engine.

7. References

[1] Querel, Carole, Olivier Grondin, and Christophe Letellier. "Semi-physical mean-value NO_x model for diesel engine control." Control Engineering Practice 40 (2015): 27-44.

[2] Brooker, T. Hendricks, V. Johnson, K. Kelly, B. Kramer, M. O'Keefe, S. Sprik, K. Wipke "A systems analysis tool for advanced vehicle modeling" T. Markel*, National Renewable Energy Laboratory, Golden, CO 80401, USA Journal of Power Sources 110 (2002) 255–266.

[3] Baskovic, Urban Zvar, and Tomaz Katrasnik. "Real-time capable virtual NO_x sensor for diesel engines based on a two-Zone thermodynamic model." Oil & Gas Sciences and Technology–Revue d'IFP Energies nouvelles 73 (2018): 11.

[4] Handler J., Flalko R ., Dorenkamp R., Stehr H., Hilzendeger J., Kranzusch S. "Volkswagen's New 2.0 l TDI Engine Fulfils the Most Stringent Emission Standards". MTZ 2008; 69.

[5] Payri F., Luján, Guardiola C., Pla B. (2015) A challengingfuture for the IC engine: New technologies and the controlrole, Oil Gas Sci. Technol. Rev. IFP Energies Nouvelles 70,15–30. doi:10.2516/ogst/2014002

[6] Willems, Frank, et al. Cylinder pressure-based control in heavy-duty EGR diesel engines using a virtual heat release and emission sensor. No. 2010-01-0564. SAE Technical Paper, 2010.

[7] Al-Durra, Ahmed, Marcello Canova, and Steve Yurkovich. "Application of extended Kalman filter to on-line diesel engine cylinder pressure estimation." ASME 2009 Dynamic Systems and Control Conference. American Society of Mechanical Engineers Digital Collection, 2009.

[8] Engineering colo state education volume model : https://www.engr.colostate.edu/~allan/thermo/page2/page2.html

[9] Park, Wonah, et al. "Prediction of real-time NO based on the in-cylinder pressure in Diesel engines." Proceedings of the Combustion Institute 34.2 (2013): 3075-3082.

[10] Hiroyasu, Hiroyuki, Toshikazu Kadota, and Masataka Arai. "Development and use of a spray combustion modeling to predict diesel engine efficiency and pollutant emissions: Part 1 combustion modeling." Bulletin of JSME 26.214 (1983): 569-575.

[11] Poetsch, Christoph, Herwig Ofner, and Eberhard Schutting. Assessment of a multi zone combustion model for analysis and prediction of CI engine combustion and emissions. No. 2011-01-1439. SAE Technical Paper, 2011.

[12] Heywood J.B. (1988) Internal Combustion Engine Fundamentals,McGraw-Hill, New York, USA.

8.Abbreviations

A_p = Area of piston

L_c = Connecting rod length

R = Crank radius

θ= Crank angle

γ = Ratio of specific heats

V_{cyl} = Cylinder volume

P_{cyl} = Cylinder pressure

α = Calibration parameter

m_{fuel} = Mass of fuel

Q_{LHV} = Lower heating value

x_b = Burn rate

m_b = Mass of burnt zone

m_u = Mass of unburnt zone

$m_{u,inj}$ = Mass of unburnt fuel due to injection

$m_{u\rightarrow b}$ = Mass transfer from unburnt to burnt zone

L_{st} = Stiochiometric ratio

λ = Relative air fuel ratio

λ_u = Relative air – fuel of unburnt

λ_{comb} = Relative air – fuel of combustion

Acknowledgments
I heartily express my profound gratitude to my project guide, Prof. Halesh M R, Assistant Professor, Department of Electronics and Communication, JSS Science and Technology University, Mysuru, for her overwhelming support, sustained encouragement, prodigious suggestions, timely valuable guidance and co-operation throughout the period of project. I would like to thank project guidance, Prof. Satish V T, Assistant Professor, Department of Mechanical Engineering, Malnad college of engineering, Hassan. I would like to thank managers of the KPIT, Technologies, Narasimha kollaparti, SME, Department of E-powertrain and Aman P V, SME, Department of E-powertrain for technical support and guidance. Once again, I thank overall for all the authors of this research for their continuous encouragement, support and suggestions for guiding in every aspect during the period of this research work.

IOP Conf. Series: Materials Science and Engineering 1189 (2021) 012039 doi:10.1088/1757-899X/1189/1/012039

Effect of flap deflection angle on flow characteristics of aerofoil

G Abhishek Vinod[1] and T J S Jothi[2,†]

[1]Department of Physics, National Institute of Technology Calicut, Kerala, India, 673601.
[2]Department of Mechanical Engineering, National Institute of Technology Calicut, Kerala, India, 673601.
[†]E-mail: tjsjothi@nitc.ac.in

Abstract. A wing is used in aircraft to employ lift based on the fundamental principle of aerodynamics. The cross-section of a wing gives the aerofoil shape, and thus investigation of flow characteristics of aerofoil is important to understand the lift and drag mechanisms in wing. Flaps are the components using in the wings to generate additional lift and drag when required. While aircraft take-off, flaps help in generating addition lift, and while in landing, they do aerodynamic braking. Therefore, flaps play an important role in aircraft. The present paper investigates the angle of flaps on flow characteristics in NACA 4415 aerofoil using XFoil software. The variations obtained showed notable variations in the coefficient of pressure (Cp) plots. The various aerodynamic coefficients such as lift and drag coefficients, and viscous pressure distribution were obtained corresponding to the flap deflection angle. From the results, for flap angles above 0o, the aerofoil can generate a "sudden" lift which is useful in take-off conditions. While the flap angles are made below zero degrees, the lift is observed to decrease contributing to more drag. Polar plots are made to summarize and visualize the above results graphically.

Keywords: aerodynamics, aerofoil shape, wings, flaps, flap angles, XFoil software.

1. Introduction

Aerofoil is a streamlined shape or surface or profile that provides an aerodynamic force when it interacts with the moving stream of air [1]. Generally, the high pressure or low velocity air flow section is generated beneath the aerofoil and the low pressure or high velocity air flow section is generated above the aerofoil [2]. Therefore, aerofoil plays an active and a crucial role in the aircraft motion. The aerofoils sometimes need to perform a sudden lift or drag. Thus, to achieve these requirements, the flaps are introduced in an aerofoil section of an aircraft. Flaps are the surfaces or profiles which are used to increase or decrease the lift of an aerofoil or to introduce the drag whenever required suddenly. They are usually mounted at the trailing edge of the wing [3]. The drag or lift is introduced by the flap in an aerofoil, is achieved by increasing the camber of an aerofoil. It is observed that the curvature of the wing increases as the camber of an aerofoil increases. The curvature which is achieved can be controlled by the deflection angle (α_D) produced by the flaps [4]. Thus, various combinations or cases may be studied about the lift and drag according to the requirements. Another way to achieve the curvature control is by introducing flap to the specified location from the leading edge of an aerofoil. In this report, NACA 4415 is employed with plain type of flap [4].

IOP Conf. Series: Materials Science and Engineering 1189 (2021) 012039 doi:10.1088/1757-899X/1189/1/012039

2. Methodology

In this paper, a NACA 4415 aerofoil is used [5]. It is an asymmetric aerofoil which has the lower surface asymmetric with respect to the maximum chord length drawn from the leading edge to the trailing edge of an aerofoil [6]. NACA is the abbreviation for "National Advisory Committee for Aeronautics" and 4415 is the aerofoil section designation, of which first digit "4" describes 4% maximum camber with respect to chord length, second digit "4" describes the location of camber at 40% of the chord length and last two digits "15" describes 15% thickness with respect to the chord length [5]. The 2D analysis in the viscous mode was carried out on the given NACA 4415 aerofoil with Reynolds number, Re = 4 x 10^6, and the angle of attack (α) from -10^0 to 30^0. A detailed viscous analysis on NACA 4415 was carried out successfully by using the Xfoil software. From the analysis, various polar plots like lift coefficient (C_L) variation with angle of attack (α), and C_L/C_D variation with α. All angles are measured in degrees. Furthermore, the plain flaps were introduced in the NACA 4415 aerofoil at the location 80% from the chord length measured from the leading edge and 50% of the maximum camber [4, 5]. In such a modified aerofoil, the front portion of a flap is rotated about the hinge near the trailing edge of an aerofoil [4]. The viscous analysis (Re = 4 x 10^6) on the modified NACA 4415 was carried out at $\alpha = 7^0$, 0^0 and -4^0 of an aerofoil explicitly, along with deflection angle (α_D) from -90^0 (up) to +90^0 (down). The plain flap at α = -90^0, -45^0, +45^0, +90^0 can be determined and studied using the figure 1 given below. From the analysis, various polar plots like lift coefficient (C_L) variation with angle of attack, and C_L/C_D variation with angle of attack.

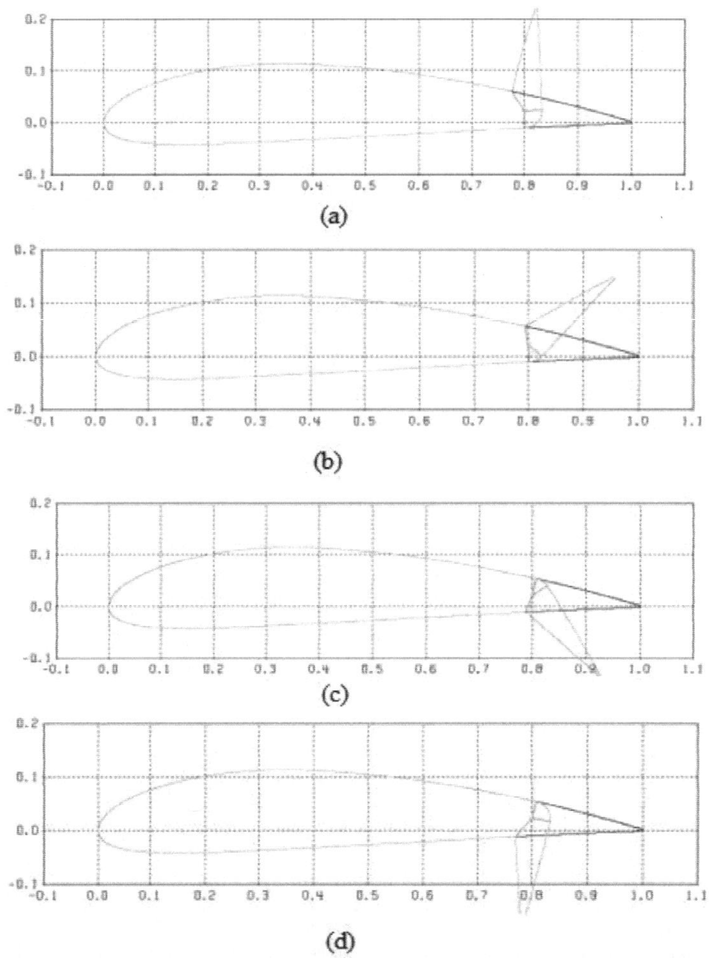

Figure 1. Aerofoil at different flap angles (a) α_D = -90^0 (b) α_D = -45^0 (c) $\alpha_D = 45^0$ (d) $\alpha_D = 90^0$.

IOP Conf. Series: Materials Science and Engineering 1189 (2021) 012039 doi:10.1088/1757-899X/1189/1/012039

3. Results and discussions

3.1. Lift and drag coefficients variation with angle of attack

The theoretical variation of the coefficient of lift (C_L) with angle of attack (α) for NACA 4415 aerofoil is calculated using the Eq. (1), and is plotted in figure 2.

$$C_L = 2\pi\alpha + C_{L0} \quad (1)$$

where, CL0 is the lift coefficient at $\alpha = 0^0$. For NACA 4415 aerofoil, $C_{L0} = 0.456$ at $\alpha = 0^0$, and $C_L = 0$ at $\alpha = 4^0$ [7, 8]. From the figure 2, we can say C_L is directly proportional to α depicting a straight line. Figure 2 can be used to summarize the theoretical approach. When viscosity is considered in the analysis, the variation of C_L with α is shown in figure 3. From the figure 3, it is observed that C_L increases when α increases till 17^0, and when $\alpha > 17^0$, the C_L decreases. This phenomenon is called as stalling and has a remarkable significance in the flight of an aircraft.

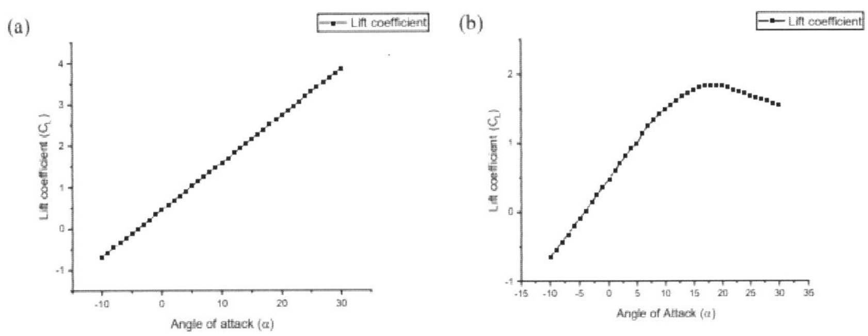

Figure 2. CL variation with α (a) theoretical (b) considering the viscosity effects.

Figure 3 shows the C_L/C_D variation for an aerofoil at different angles of attacks. It is observed that the C_L/C_D ratio increases to the maximum when α reaches 7^0. For $\alpha > 7^0$, the C_L/C_D ratio decreases. This analysis depicts the stability of an aerofoil with respect to the variation in α. Thus, the optimum value of α is between 0^0 to 7^0 above which the drag is observed to be contributing such that the C_L/C_D is seen to be decreasing.

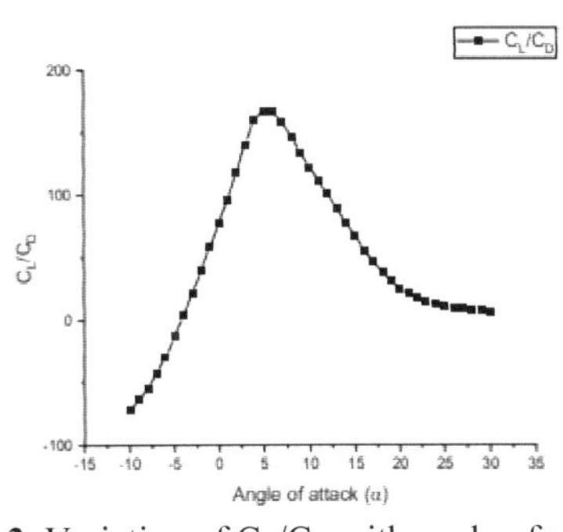

Figure 3. Variation of C_L/C_D with angle of attack, α.

3.2. Lift coefficient variation with flap deflection angle (αD)

Figure 4 shows the variation of coefficient of lift with the flap deflection angle and different angles of attack of $\alpha = -4\,^0$, 0^0, and 7^0. It is observed that when α_D is positive and is less than 70^0 (flaps down), C_L increases, and thus lift increases. This sudden lift generation is employed for short run lift generation in an aerofoil of an aircraft. Similarly, when α_D is negative and is greater than -70^0 (flaps up), C_L decreases, thus lift decreases. This condition is observed as if an aerofoil is under braking condition and this condition is thus called as "aerodynamic braking". For the considered angles of attack cases in figure 4, there is an upward shift in the values of C_L with increasing α values.

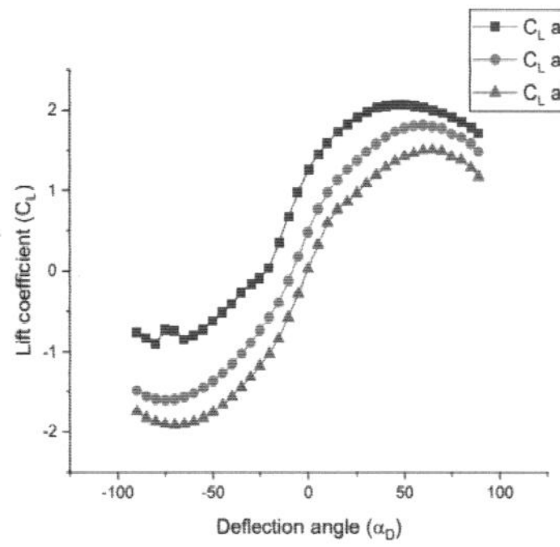

Figure 4. C_L variation with flap deflection angle at different angles of attack.

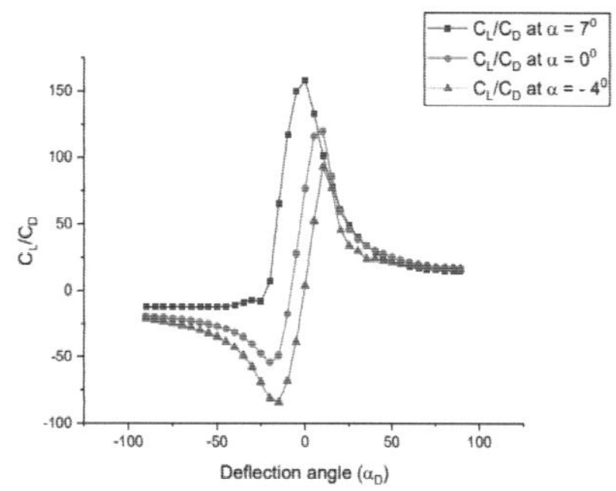

Figure 5. C_L/C_D variation with flap deflection angle at different angles of attack.

Figure 5 shows the C_L/C_D variation with flap deflection angles at the angles of attack values of $\alpha = 7^0$, 0^0, -4^0. It is observed that C_L/C_D ratio increases appreciably till $\alpha_D = +12^0$, and thereafter decreases slowly due to the contribution of drag component in the motion of the aerofoil. As α increase, this plot tends from asymmetric nature to symmetric nature.

4. Conclusions

A detailed viscous analysis on NACA 4415 was carried out successfully by using the Xfoil software. By the analysis, it was observed that, the flaps can either increase or decrease the lift according to its orientation and geometry. It was observed and concluded that the down deflection angle (positive α_D) is used to introduce the lift of an aerofoil and up deflection angle (negative α_D) is used to introduce the drag in the aerofoil. The drag introduction to the aerofoil is termed as "aerodynamic braking" and it was studied in detail using the results obtained by the analysis.

5. References

[1] Rubel R I, Uddin M K, Islam M Z, and Rokunuzzaman M (2017). Comparison of aerodynamics characteristics of NACA 0015 & NACA 4415 aerofoil blade. International Journal of Research - Granthaalayah, 5(11), 187-97.
[2] Perkins C and Hage R (1949). Airplane performance, stability and control, Chapter 2, John Wiley and Sons.
[3] Anderson J D Jr., Fundamentals of Aerodynamics, Fifth Edition, McGraw Hill Education (India) Private Limited, 2010.
[4] Gunston B, Cambridge Aerospace Dictionary, Cambridge. Cambridge University Press 2004.
[5] Jacobs E N, Ward K E and Pinkerton R M. NACA Report No. 460, The characteristics of 78 related airfoil sections from tests in the variable-density wind tunnel. NACA, 1933.
[6] Hurt H H Jr. Aerodynamics for Naval Aviators. U.S. Government Printing Office, Washington, D.C.: U.S. Navy, Aviation Training Division. (January 1965).
[7] Abbott, Ira H, von Doenhoff, Albert E (1958). Theory of wing sections. New York: Dover Publications.
[8] Clancy L J (1975), Aerodynamics, Pitman Publishing Limited, London.

IOP Conf. Series: Materials Science and Engineering 1189 (2021) 012040 doi:10.1088/1757-899X/1189/1/012040

VISION ASSIT FOR AUTONOMOUS SURGERY ROBOT

[1] Madhu Mohan R ,[2] Dr Grisha C ,[3] Kunal M S ,[4] Lokanatha Reddy V ,[5] Mahendra M and [6]Pawan N

[1] Professor, Mechanical-Engineering, A.M.C Engineering College, Karnataka, INDIA

[2] HOD, Mechanical-Engineering, A.M.C Engineering College, Karnataka, INDIA

[3,4,5,6] Student, Mechanical-Engineering, A.M.C Engineering College, Karnataka, INDIA

Abstract: The pedicle screw insertion has become the most exercised procedures in spine fusion surgery. It is performed to hold the vertebrae of the treated area in a fixed position. Each year there are about 1.3 million spinal surgeries performed in US alone. The miss-insertion rates in the lumbar spine region have been documented to be from 5% - 41%[1]. This paper provides a perspective on the development of vision assist system for an automated robot for fixation of pedicle-screw. Pedicle screw fixation devices are considered as class III medical devices. The technique of pedicle screw insertion in the spine has gone through significant development over the last two decades. Aim of this paper is to decrease the error rates by implementing an image-processing method which intends to eliminate the use of intra-operative fluoroscopy, MRI and CT.

1. Introduction

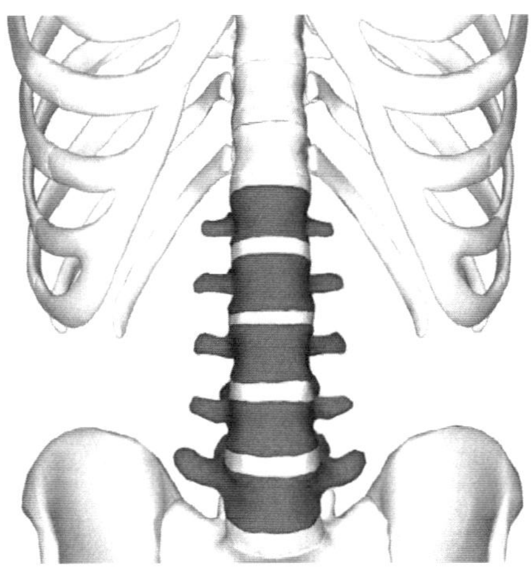

Figure 1. Lumbar region of Spine

A medical protocol that uses operative instrumental and manual techniques on a patient to treat a pathological condition such as a disease or injury is known as Surgery. Every year there are about a million spinal surgeries performed in US, 53 percentage of which is classified as spinal-fusion surgeries. Pathological condition in the spine is one of the reasons why the patient requires medical care. Hence there is increase in number of spinal surgeries performed. The surgeons or doctors diagnose and find out what the actual problem is and then if there are any major complications such as inflated intervertebral disc scoliosis, and kyphosis we resort to spinal fusion surgeries necessitating pedicle screw insertion which is a common solution to many of such complications. Most of these surgeries are performed by trained and experienced surgeons using conventional method where the surgeons operate in a blind spot. In conventional methods of operating miss-insertion of pedicle screws is quite often which is the main reason for the present error rates of about 18-20%[2]. But with the introduction of robotic technology to surgical field, the error rates can be minimized substantially and overall efficiency can be increased[3]. With this guidance system, the conventional approach of performing the surgery will be drastically improved as the whole process is automated. This system will be able to perform the operations by using a mere camera image which is then processed to find the exact point of insertion. This will be able to increase the overall efficiency and minimize the error rates and minimize or eliminate the usage of fluoroscopy, MRI and CT scans as too much of exposure to these radiations can cause cancers.

2. Objectives

Till date there have been many surgical robots for Pedicle screw insertion, which depend on intra-operative fluoroscopy, MRI, CT and CAS[4]. The use of these techniques increases the chances of bone cancers. This paper emphasizes on the vision assist system which excludes the use of such techniques and provides a cost-effective operation procedure. The main objective is to determine the exact point of insertion in the lumbar region of the spine with the help of a mere camera image. The average time taken required to insert the pedicle screw by conventional method is high which has error rates of approximately 18%, so with the help of this improved technique, the overall operating time will be substantially reduced along with increase in precision.

3. Methodology and Experimentation

The fixation of screw implants into the vertebrae is a complex procedure. Therefore, this system will be able to perform this complex procedure efficiently and without any error. The overall image processing is simplified into two steps:

1. Capturing the image of affected lumbar vertebrae: For this vision assist, a mere camera image of 720 pixels or 1080 pixels of the affected vertebrae is sufficient to obtain the point of insertion. Hence image can be captured using any device a camera or a Raspberry pi camera connected to a Raspberry Pi module can also be used to take pictures of the affected region. The operating region is illuminated by a light source which assists in capturing a better image.

Figure 2. L1 Vertebrae

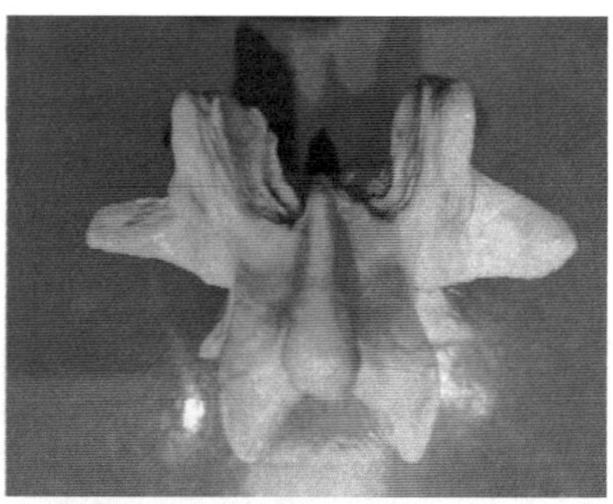

Figure 3. L4 Vertebrae

IOP Conf. Series: Materials Science and Engineering 1189 (2021) 012040 doi:10.1088/1757-899X/1189/1/012040

Table 1. Angle of insertion of lumbar vertebrae.

Vertebra	Gender	Transversal Pedicle angle	Inclination Pedicle angle
L1	F	5	6
	M	7	5
L2	F	6	5
	M	7	6
L3	F	7	6
	M	8	6
L4	F	10	7
	M	11	6
L5	F	18	8
	M	17	5

Once the image is captured by the system, it will be compared with the predefined images to identify the exact lumbar vertebrae and its angle of insertion which are also predefined, refer table 1.

2. Image processing using MATLAB: The image captured is then processed in mat-lab where it undergoes gray scaling and edge detection. Image is imported using digital camera or medical Imaging devices. The imported image undergoes grey scale conversion process, which is an M*N array whose values have been scaled to indicate intensities. A new matrix with same number of columns & rows as RGB image is created. Each RGB pixel values at location (x, y) is converted to gray scale values which is assigned to a corresponding location (x, y) in new matrix. The image then undergoes edge detection method for finding object boundaries. It works by detecting discontinuities in brightness. A function f (x, y) is represented by the digital image. In this function 'x' and 'y' are structural coordinates. The amplitude of 'f' at any structural coordinate gives the magnitude of the image at that point. Each element in an array represents an image element.

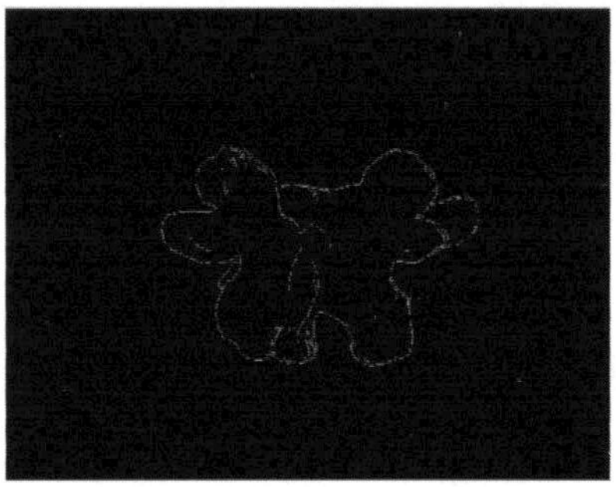

Figure 4. L1 Processed Image

Figure 5 L4 Processed Image

4. Results for point of insertion

Using the captured image, by edge detection the articular facets and the transverse facets are distinctly identified. An algorithm is then used to determine the intersection of the tangential point of transverse and articular facet which gives the point of insertion of pedicle screw as shown in figure 6 and figure 7. The outcome of the vision system coincides with the actual point of insertion as shown in figure 8 and figure 9.

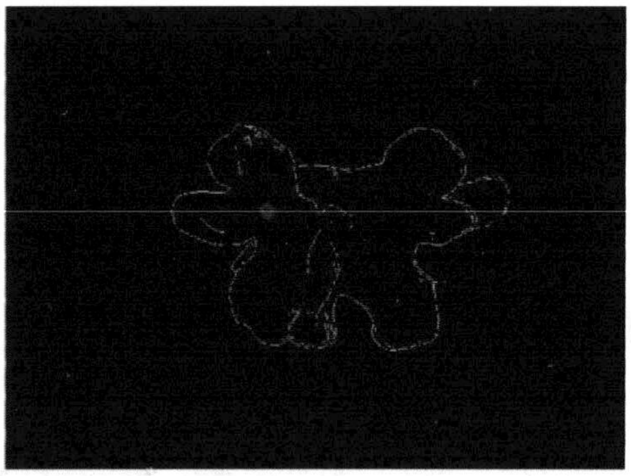

Figure 6. Point of insertion of L1 vertebrae

Figure 7. Point of insertion of L4 vertebrae

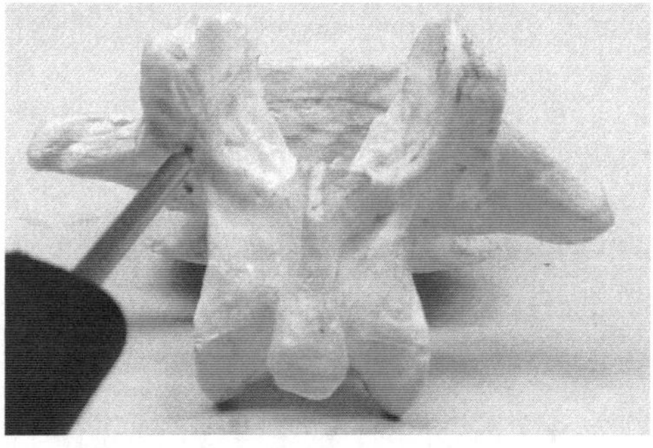

Figure 8. Actual Point of insertion

(Front View)

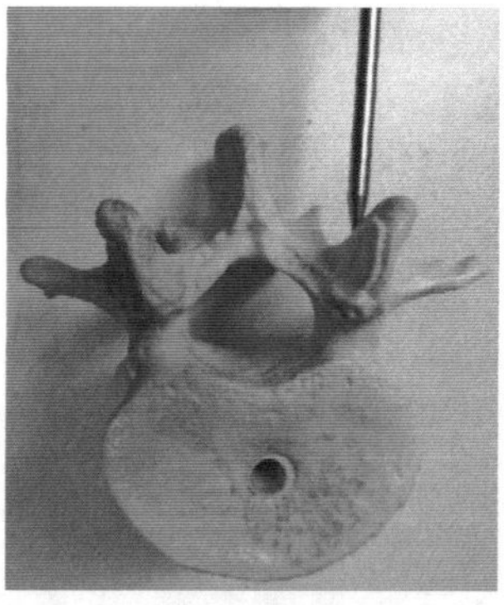

Figure 9. Actual Point of insertion

(Top View)

5. Conclusion

In previous years, the number of spinal surgeries carried out has increased due to which many surgical robots, whose operating cost is very high, came in to existence[5]. The maintenance and cost of these robots are so expensive that only few hospitals have it, on the contrary the vision assist system provides an efficient and cost effective way of operation which is affordable with improved precision and dexterity.

6. References

[1] Ringel F, Stüer C, Reinke A, Preuss A, Behr M, Auer F, et al: *Accuracy of robot-assisted placement of lumbar and sacral pedicle screws: a prospective randomized comparison to conventional freehand screw implantation. Spine* (Phila Pa 1976) **37**, 2012

[2] Pechlivanis I, Kiriyanthan G, Engelhardt M, Scholz M, Lücke S, Harders A, et al: *Percutaneous placement of pedicle screws in the lumbar spine using a bone mounted miniature robotic system: first experiences and accuracy of screw placement.* (Phila Pa 1976) **34**, 2009

[3] Benabid AL, Hoffmann D, Seigneuret E, Chabardes S: Robotics in neurosurgery: Which tools for what*? Acta Neurochir Suppl* **98**, 2006

[4] El Tecle NE, El Ahmadieh TY, Patel BM, Lall RR, Bendok BR, Smith ZA: Minimizing radiation exposure in minimally invasive spine surgery: Lessons learned from neuroendovascular surgery. *Neurosurgery Clin N* Am **25**, 2014

[5] Barzilay Y, Kaplan L, Libergall M: Robotic assisted spine surgery—a breakthrough or a surgical toy? Int *J Med Robot* **4**, 2008

[6] A method and device for efficient probing in bone region[201241010607].

[7] Bertelsen A, Melo J, Sanchez E, Borro D: *A review of surgical robots for spinal interventions. Int J Med Robot* **9**, 2013

[8] Cahill KS, Wang MY: Evaluating the accuracy of robotic assistance in spine surgery. Neurosurgery **71**, 2012

[9] El Tecle NE, El Ahmadieh TY, Patel BM, Lall RR, Bendok BR, Smith ZA: Minimizing radiation exposure in minimally invasive spine surgery: Lessons learned from neuroendovascular surgery. *Neurosurgery Clin N Am* **25**:247-260, 2014

[10] Freschi C, Ferrari V, Melfi F, Ferrari M, Mosca F, Cuschieri A: Technical review of the da Vinci surgical telemanipulator. Int *J Med Robot* **9**, 2013

[11] Gomes P: Surgical robotics: Reviewing the past, analysing the present, imagining the future. *Robot Comput Integr Manuf* **27**, 2011

[12] A ly A-H, Ginsberg HJ, Cobbold RSC (2011) On ultrasound imaging for guided screw insertion in spinal fusion surgery. *Ultrasound Med Biol* **37**.

[13] Marano A, Choi YY, Hyung WJ, Kim YM, Kim J, Noh SH: Robotic versus laparoscopic versus open gastrectomy: a metaanalysis. *J Gastric Cancer* **13:**136–148, 2013

[14] Martínez H, Obst T, Ulbrich H, Burgkart R: A novel application of direct force control to perform in-vitro biomechanical tests using robotic technology. *J Biomech* **46:**1379–1382, 2013

[15] Panjabi MM: Biomechanical evaluation of spinal fixation devices: I. A conceptual framework. *Spine* (Phila Pa 1976) **13:** 1129–1134, 1988

[16] Panjabi MM: Hybrid multidirectional test method to evaluate spinal adjacent-level effects. *Clin Biomech* (Bristol, Avon) **22:**257–265, 2007

[17] Perez-Cruet MJ, Welsh RJ, Hussain NS, Begun EM, Lin J, Park P: Use of the da Vinci minimally invasive robotic system for resection of a complicated paraspinal schwannoma with thoracic extension: case report. *Neurosurgery 71* (1 Suppl Operative):**209**, 2012

[18] Roser F, Tatagiba M, Maier G: Spinal robotics: current applications and future perspectives. *Neurosurgery 72* (Suppl 1):**12**, 2013

AUTHOR INDEX

Ahmed, Shoaib 258
Ain, Qurat Ul 59
Aman, P V 258
Amith, R 79
Ashrith, H S 147
Athreya, R Prajwal 238
Bhagyashekar, M S 11
Bharathesh, T P 101
Chaluvaraju, B. V. 251
Chandrashekar, A. 251
Chandrashekar, A 245
Davanageri, M B 108
Dinesh, P A 205
Disha, D 108
Divakar H N 65
Dixit, U Arun C 134
Doddamani, Mrityunjay 147
Durai, D Thulasi 17
Ganesh, V 6
Gangaraj 50
Girish, A R 95
Girisha C 141
Gopi, K R 169, 187
Gowda, M C Yogesh 32
Grisha, C Dr 271
Gurukiran, K 199
Halesh, M R 258
Harish 6
Harshavardhan, B 113, 134
Hussain, Shahid 164
Janamdar, I M 79
Jayavardhan, H K 193
Jeevan, T P 147
Joseph, Gonsalvis Dr. 227
Jothi, T J S 233, 267
Kadadevaramath, Rajeshwar S 214
Kaladgi, Abdul Razak 245, 251
Kalavathi, G K 205
Kamate, Vijay 221
Karthik, S 169
kollaparti, Narasimha 258
Kulkarni, Umakant P 221
Kumar, K Dilip 45
Kumar, K R Dushyanth 17, 24
Kumar, Ritesh 108
Kumar, S Mohan 221
Kumar, S Praveen 1
Kumar, T S Prasanna 32
Kunal, M S 271
Madhu, D 177
Madhu, K S 71, 128
Madhu, K.S. 122
Madhu, P. 251
Mahendra, M 271
Mallesh, G. Dr. 38
Manjunath, K. 50
Manjunatha, H.S. Dr. 38
Mohan, R Madhu 271
Nithin, H S 71
Padmanabha, Anantha 79
Pal Surjya K. 153
Pawan, N 271
Pradeep, D G 71, 128
Pradeep, D.G. 122
Pradeep, S 169, 187
Prasanna, S Ishwara 79
Raghavendra, N 6
Raj, L Bharath 32
Rajesh, A R Dr. 227
Rakesh T G 141
Rao, R Raghavendra 169, 187
Ravikumar, B N 177
Ravishankar, R 113, 134
Reddy, Sreenivasalu 245
Reddy, V Lokanatha 271
Roopashree, C 101
Rudresh, B M 177
Sabiruddin, Kazi 164
Samal, Prasanta Kumar 193, 199
Samantaray, ARUN K. 153
Sangli, Mohan 214
Sansuddi, Sagar B 24
Santhosh, G 45
Saravanan, R 101
Sathisha, H M 32
Satish V T 141
Satish, V T 38, 258
Savitha M 65
Sehgal, Rakesh 59
Shankar, B Latha 17, 24
Shankar, S. N. Ravi 153
Shantharaja, M 1, 45
Sharath, B N 71, 128
Sharath, B.N. 122
Shashidhara, Y M 95
Shivalingappa, D. 6

Shivashankar, G S 17, 24
Shivashankara, B S 169, 187
Singh, M K 59
Soundararajan, R 233
Srikanth, N S 238
Srinivas, K 11
Srinivas, S 113
Subbaiah K M 65
Suman, Saurabh 221
Suresha, B 79, 113, 134
Tejas, J 238
Umashanker, L 101
Vasundhara, M G 205
Venkatesh, C V 71, 128
Venkatesh, C.V. 122
Venson, Mascarenhas 238
Venugopal, K A Dr. 227
Vinod, G Abhishek 267
Vinutha P 65
Vishwanath, K C 245
Vishwanath, K. C 251
Wani, M F 59
Yuvaraja, B K 205

Institute of Physics
Dirac House, Temple Back
Bristol BS1 6BE UK

ISSN: 1757-8981
ISBN 978-1-7138-4649-9

CURRAN ASSOCIATES INC.
proceedings.com